DEVELOPING RESEARCH IN MATHEMATICS EDUCATION

Developing Research in Mathematics Education is the first book in the series *New Perspectives on Research in Mathematics Education*, to be produced in association with the European Society for Research in Mathematics Education (ERME). This inaugural volume sets out broad advances in research in mathematics education which have accumulated over the last 20 years through the sustained exchange of ideas and collaboration between researchers in the field.

An impressive range of contributors provide specifically European and complementary global perspectives on major areas of research in the field on topics that include:

- the content domains of arithmetic, geometry, algebra, statistics and probability;
- the mathematical processes of proving and modelling;
- teaching and learning at specific age levels from early years to university;
- teacher education, teaching and classroom practices;
- special aspects of teaching and learning mathematics such as creativity, affect, diversity, technology and history;
- theoretical perspectives and comparative approaches in mathematics education research.

This book is a fascinating compendium of state-of-the-art knowledge for all mathematics education researchers, graduate students, teacher educators and curriculum developers worldwide.

Tommy Dreyfus is Emeritus Professor of Mathematics Education at Tel Aviv University in Israel and one of the Founding Series Editors of the IMPACT Mathematics series. He is an Advisory Editor of *Educational Studies in Mathematics* and a Contributing Editor of the *Journal of Mathematical Behavior*. He has previously edited several books including one in Routledge's EARLI book series.

Michèle Artigue is Emeritus Professor of Mathematics Education at the University Paris Diderot – Paris 7 in France. She was recently awarded the Felix Klein medal for life-time achievement in mathematics education research by the International Commission on Mathematical Instruction, and is a Knight of the French National Order of the Legion of Honour.

Despina Potari is Professor of Mathematics Education at the University of Athens in Greece, and Guest Professor of Mathematics Education at Linnaeus University in Sweden. She is also Editor-in-Chief of the *Journal of Mathematics Teacher Education*.

Susanne Prediger is Professor of Mathematics Education at TU Dortmund University in Germany and vice-director of the German Centre for Mathematics Teacher Education. She has authored numerous journal articles and book chapters and is currently President of ERME, the European Society for Research in Mathematics Education.

Kenneth Ruthven is Emeritus Professor of Education at the University of Cambridge in England and Guest Professor in Mathematics Education at Karlstad University in Sweden and at the University of Agder in Norway. He is an Advisory Editor of the international journals *Educational Studies in Mathematics* and *Research in Mathematics Education* and a Fellow of the Academy of Social Sciences.

New Perspectives on Research in Mathematics Education – ERME series

ERME, the European Society for Research in Mathematics Education, is a growing society of about 900 researchers from all over Europe and beyond.

The ERME series documents the growing body of substantial research on mathematics education within the context of ERME. Volumes in the ERME series can be monographs or collections growing out of the collaboration of European researchers in mathematics education.

The volumes are written by and for European researchers, but also by and for researchers from all over the world. An international advisory board guarantees that ERME stays globally connected. A rigorous and constructive review procedure guarantees a high quality of the series.

Editors of the ERME Series

Viviane Durand-Guerrier	(France)
Konrad Krainer	(Austria)
Susanne Prediger	(Germany)
Naďa Vondrová	(Czech Republic)

International Advisory Board of the ERME Series

Marcelo Borba	(Brazil)
Fou-Lai Lin	(Taiwan)
Merrilyn Goos	(Australia and Ireland)
Barbara Jaworski	(Europe, United Kingdom)
Chris Rasmussen	(United States of America)
Anna Sierpinska	(Canada)

Developing Research in Mathematics Education

Twenty Years of Communication, Cooperation and Collaboration in Europe
Edited by Tommy Dreyfus, Michele Artigue, Despina Potari, Susanne Prediger, Kenneth Ruthven

For more information about this series, please visit: https://www.routledge.com/European-Research-in-Mathematics-Education/book-series/ERME

DEVELOPING RESEARCH IN MATHEMATICS EDUCATION

Twenty Years of Communication, Cooperation and Collaboration in Europe

Edited by Tommy Dreyfus, Michèle Artigue, Despina Potari, Susanne Prediger and Kenneth Ruthven

LONDON AND NEW YORK

First published 2018
by Routledge
2 Park Square, Milton Park, Abingdon, Oxon OX14 4RN

and by Routledge
711 Third Avenue, New York, NY 10017

Routledge is an imprint of the Taylor & Francis Group, an informa business

© 2018 selection and editorial matter, Tommy Dreyfus, Michèle
Artigue, Despina Potari, Susanne Prediger and Kenneth Ruthven;
individual chapters, the contributors

The right of Tommy Dreyfus, Michèle Artigue, Despina Potari,
Susanne Prediger and Kenneth Ruthven to be identified as the
authors of the editorial material, and of the authors for their
individual chapters, has been asserted in accordance with sections
77 and 78 of the Copyright, Designs and Patents Act 1988.

All rights reserved. No part of this book may be reprinted or
reproduced or utilised in any form or by any electronic, mechanical,
or other means, now known or hereafter invented, including
photocopying and recording, or in any information storage or
retrieval system, without permission in writing from the publishers.

Trademark notice: Product or corporate names may be trademarks
or registered trademarks, and are used only for identification and
explanation without intent to infringe.

British Library Cataloguing-in-Publication Data
A catalogue record for this book is available from the British Library

Library of Congress Cataloging-in-Publication Data
A catalog record has been requested for this book

ISBN: 978-1-138-08027-0 (hbk)
ISBN: 978-1-138-08029-4 (pbk)
ISBN: 978-1-315-11356-2 (ebk)

Typeset in Bembo
by Swales & Willis Ltd, Exeter, Devon, UK

CONTENTS

List of contributors	*x*
Series foreword	*xviii*
Preface	*xix*

The European Society for Research in Mathematics Education: introduction by its former presidents   1
Ferdinando Arzarello, Paolo Boero, Viviane Durand-Guerrier and Barbara Jaworski

1 From geometrical thinking to geometrical working competencies   8
Alain Kuzniak, Philippe R. Richard and Paraskevi Michael-Chrysanthou

2 Number sense in teaching and learning arithmetic   23
Sebastian Rezat and Lisser Rye Ejersbo

3 Algebraic thinking   32
Jeremy Hodgen, Reinhard Oldenburg and Heidi Strømskag

4 Research on probability and statistics education: trends and directions   46
Arthur Bakker, Corinne Hahn, Sibel Kazak and Dave Pratt

5 Research on university mathematics education   60
Carl Winsløw, Ghislaine Gueudet, Reinhard Hochmuth and Elena Nardi

viii Contents

6 Argumentation and proof 75
Maria Alessandra Mariotti, Viviane Durand-Guerrier and Gabriel J. Stylianides

7 Theory–practice relations in research on applications and modelling 90
Morten Blomhøj and Jonas Bergman Ärlebäck

8 Early years mathematics 106
Esther S. Levenson, Maria G. Bartolini Bussi and Ingvald Erfjord

9 Mathematical potential, creativity and talent 115
Demetra Pitta-Pantazi and Roza Leikin

10 Affect and mathematical thinking: exploring developments, trends, and future directions 128
Markku S. Hannula, Marilena Pantziara and Pietro Di Martino

11 Technology and resources in mathematics education 142
Jana Trgalová, Alison Clark-Wilson and Hans-Georg Weigand

12 Classroom practice and teachers' knowledge, beliefs and identity 162
Jeppe Skott, Reidar Mosvold and Charalampos Sakonidis

13 Mathematics teacher education and professional development 181
Alena Hošpesová, José Carrillo and Leonor Santos

14 Mathematics education and language: lessons and directions from two decades of research 196
Núria Planas, Candia Morgan and Marcus Schütte

15 Diversity in mathematics education 211
Guida de Abreu, Núria Gorgorió and Lisa Björklund Boistrup

16 Comparative studies in mathematics education 223
Eva Jablonka, Paul Andrews, David Clarke and Constantinos Xenofontos

Contents ix

17 History and mathematics education 239
Uffe Thomas Jankvist and Jan van Maanen

18 Theoretical perspectives and approaches in mathematics
education research 254
Ivy Kidron, Marianna Bosch, John Monaghan and Hanna Palmér

19 ERME as a group: questions to mould its identity? 269
Marcelo C. Borba

20 Communication, cooperation and collaboration:
ERME's magnificent experiment 276
Norma Presmeg

Index 287

CONTRIBUTORS

Paul Andrews
Stockholm University
Sweden
paul.andrews@mnd.su.se

Michèle Artigue
Université Paris-Diderot
France
michele.artigue@univ-paris-diderot.fr

Ferdinando Arzarello
Università di Torino
Italy
ferdinando.arzarello@unito.it

Arthur Bakker
Utrecht University
The Netherlands
a.bakker4@uu.nl

Maria G. Bartolini Bussi
University of Modena and Reggio Emilia
Italy
mariagiuseppina.bartolini@unimore.it

Jonas Bergman Ärlebäck
Linköping University
Sweden
jonas.bergman.arleback@liu.se

Contributors **xi**

Lisa Björklund Boistrup
Stockholm University
Sweden
lisa.bjorklund@mnd.su.se

Morten Blomhøj
Roskilde University
Denmark
blomhoej@ruc.dk

Paolo Boero
Università di Genoa
Italy
boero@dima.unige.it

Marcelo C. Borba
São Paulo State University (Unesp)
Brazil
mborba@rc.unesp.br

Marianna Bosch
Universitat Ramon Llull
Spain
mariannabosch@gmail.com

José Carrillo
Universidad de Huelva
Spain
carrillo@uhu.es

Alison Clark-Wilson
UCL Institute of Education
UK
a.clark-wilson@ucl.ac.uk

David Clarke
University of Melbourne
Australia
d.clarke@unimelb.edu.au

Guida de Abreu
Oxford Brookes University
UK
gabreu@brookes.ac.uk

xii Contributors

Pietro Di Martino
Università di Pisa
Italy
pietro.dimartino@unipi.it

Tommy Dreyfus
Tel Aviv University
Israel
TommyD@post.tau.ac.il

Viviane Durand-Guerrier
Université de Montpellier
France
viviane.durand-guerrier@umontpellier.fr

Lisser Rye Ejersbo
Aarhus University
Denmark
lre@edu.au.dk

Ingvald Erfjord
University of Agder
Norway
ingvald.erfjord@uia.no

Núria Gorgorió
Universitat Autònoma de Barcelona
Spain
nuria.gorgorio@uab.cat

Ghislaine Gueudet
University of Bretagne Occidentale
France
ghislaine.gueudet@espe-bretagne.fr

Corinne Hahn
ESCP Europe
France
hahn@escpeurope.eu

Markku S. Hannula
University of Helsinki
Finland
markku.hannula@helsinki.fi

Contributors xiii

Reinhard Hochmuth
University of Hannover
Germany
hochmuth@idmp.uni-hannover.de

Jeremy Hodgen
UCL Institute of Education
UK
jeremy.hodgen@ucl.ac.uk

Alena Hošpesová
University of South Bohemia in České Budějovice
Czech Republic
hospes@pf.jcu.cz

Eva Jablonka
Freie Universität Berlin
Germany
eva.jablonka@fu-berlin.de

Uffe Thomas Jankvist
Aarhus University
Denmark
utj@edu.au.dk

Barbara Jaworski
Loughborough University
UK
B.Jaworski@lboro.ac.uk

Sibel Kazak
Pamukkale University
Turkey
skazak@pau.edu.tr

Ivy Kidron
Jerusalem College of Technology
Israel
ivy@jct.ac.il

Konrad Krainer
University of Klagenfurt
Austria
konrad.krainer@aau.at

xiv Contributors

Alain Kuzniak
Université Paris-Diderot
France
alain.kuzniak@univ-paris-diderot.fr

Roza Leikin
University of Haifa
Israel
rozal@edu.haifa.ac.il

Esther S. Levenson
Tel Aviv University
Israel
levensone@gmail.com

Maria Alessandra Mariotti
University of Siena
Italy
mariotti21@unisi.it

Paraskevi Michael-Chrysanthou
University of Cyprus
Cyprus
pmicha@ucy.ac.cy

John Monaghan
University of Agder
Norway
john.monaghan@uia.no

Candia Morgan
UCL Institute of Education
UK
candia.morgan@ucl.ac.uk

Reidar Mosvold
University of Stavanger
Norway
reidar.mosvold@uis.no

Elena Nardi
University of East Anglia
UK
e.nardi@uea.ac.uk

Reinhard Oldenburg
University of Augsburg
Germany
reinhard.oldenburg@math.uni-augsburg.de

Hanna Palmér
Linnaeus University
Sweden
Hanna.Palmer@lnu.se

Marilena Pantziara
Cyprus Pedagogical Institute
Cyprus
marilena.p@cytanet.com.cy

Demetra Pitta-Pantazi
University of Cyprus
Cyprus
dpitta@ucy.ac.cy

Núria Planas
Universitat Autònoma de Barcelona
Spain
Nuria.Planas@uab.cat

Despina Potari
University of Athens
Greece
dpotari@math.uoa.gr

Dave Pratt
UCL Institute of Education
UK
david.pratt@ucl.ac.uk

Susanne Prediger
TU Dortmund University
Germany
prediger@math.uni-dortmund.de

Norma Presmeg
Illinois State University
USA
npresmeg@msn.com

xvi Contributors

Sebastian Rezat
University of Paderborn
Germany
srezat@math.upb.de

Philippe R. Richard
Université de Montréal
Canada
philippe.r.richard@umontreal.ca

Kenneth Ruthven
University of Cambridge
UK
kr18@cam.ac.uk

Charalampos Sakonidis
Democritus University of Thrace
Greece
xsakonid@eled.duth.gr

Leonor Santos
Universidade de Lisboa
Portugal
mlsantos@ie.ul.pt

Marcus Schütte
Technical University of Dresden
Germany
Marcus.schuette@tu-dresden.de

Jeppe Skott
Linnaeus University
Sweden
jeppe.skott@lnu.se

Heidi Strømskag
Norwegian University of Science and Technology
Norway
heidi.stromskag@ntnu.no

Gabriel J. Stylianides
University of Oxford
UK
gabriel.stylianides@education.ox.ac.uk

Jana Trgalová
Claude Bernard University Lyon 1
France
jana.trgalova@univ-lyon1.fr

Jan van Maanen
Utrecht University
The Netherlands
J.A.vanMaanen@uu.nl

Naďa Vondrová
Charles University
Czech Republic
nada.vondrova@pedf.cuni.cz

Hans-Georg Weigand
University of Würzburg
Germany
weigand@mathematik.uni-wuerzburg.de

Carl Winsløw
University of Copenhagen
Denmark
winslow@ind.ku.dk

Constantinos Xenofontos
University of Stirling
UK
constantinos.xenofontos@stir.ac.uk

SERIES FOREWORD

New perspectives on research in mathematics education – ERME series

ERME, the European Society for Research in Mathematics Education, is a growing society of about 900 researchers from all over Europe and beyond. Between 1998 and 2018, ten biannual Congresses of ERME (CERME), nine summer schools (YESS) and six ERME Topic Conferences (ETC) have taken place.

The ERME series documents the growing body of substantial research on mathematics education within the context of ERME. Volumes in the ERME series can be monographs or collections growing out of the collaboration of European researchers in mathematics education; for example, post-conference publications of selected contributions to ETCs or research in EU-funded projects.

The volumes are written by and for European researchers, but also by and for researchers from all over the world. An international advisory board guarantees that ERME stays well connected to the rest of the world and includes results of non-European research. A rigorous and constructive review procedure guarantees the high quality of the series.

The inaugural volume of the ERME Series is titled *Developing Research in Mathematics Education: Twenty Years of Communication, Cooperation and Collaboration in Europe*. We thank all involved editors, authors and reviewers for the joint work. It is a highly interesting start of the series which will – hopefully – fuel the international communication, cooperation and collaboration.

The series editors: Viviane Durand-Guerrier,
Konrad Krainer, Susanne Prediger, Naďa Vondrová

PREFACE

The aim of this book is to present, on the occasion of the 20th anniversary of the European Society for Research in Mathematics Education (ERME), the most important directions and trends of European research in mathematics education. The book reports on the main lines of development in ERME over the course of these 20 years, showing the ERME spirit in the process: a spirit of communication between different sub-areas and different countries, a spirit of cooperation between different theoretical approaches and research paradigms, a spirit of collaboration between established and developing researchers. Prior to the creation of ERME, several research traditions had already developed within Europe, each with its own identity reflecting cultural and educational particularities of its birthplace. With this book, we hope to establish shared understandings in which to ground future European research in mathematics education as well as to showcase the specific character of European research traditions for audiences inside and outside Europe.

At ERME conferences (CERMEs) participants spend most of the time in thematic working groups (TWGs). New working groups are added at each conference, typically two or three, following an open call for proposals and then selection by the International Program Committee for the conference. Equally, working groups are discontinued when they no longer attract enough interest.

Given how central the TWGs are to the scientific interaction that takes place within ERME, it was natural that the core of this book should reflect the work and achievements of these working groups. All groups that have been operational over sufficiently many CERMEs to warrant a substantial chapter have been included. The authors of each chapter have been selected from among those who have recently been active in leading the work of the relevant groups.

xx Preface

The chapters in the book have been ordered so as to begin with those focusing on specific content domains, age levels, and mathematical processes (such as proving and modelling). The following chapters then examine aspects of teaching and learning that range across contents and processes (including learning environments, affect and research about teachers). The final chapters deal with linguistic, cultural and social aspects of learning and teaching mathematics, as well as with the role of theories in mathematics education research. These 18 core chapters of the book are preceded by an introduction that provides an account of the evolution of ERME as a society, written by its former presidents.

The book concludes with two commentary chapters contributed by eminent researchers from outside Europe. Together these commentary chapters situate the research done in ERME and presented in the book in the wider context of mathematics education research worldwide. At the same time, they show how ERME might, in the future, inspire and be inspired by developments outside Europe.

Nowadays it is increasingly hard for researchers even to read all the relevant work that has been published on their topic, let alone to take comprehensive account of it in conducting and reporting their own research. There is a pressing need for greater coordination of research efforts; in particular, in the form of synthesis of frameworks and findings. Thus, this book seeks to identify cumulative achievements in research and to highlight starting points for further development within and beyond ERME. The book does not, therefore, aim to summarise 20 years of highly diverse research by hundreds of researchers but, rather, aims to display the patterns and threads that have the potential to unify these research efforts into a cumulative and coherent body of knowledge providing a view from the past into the future.

In this sense, the authors of the core chapters were asked to focus on the sustained development of ideas in the working groups over several years and ERME conferences. Also stressed were ideas that arose within a group and have become relevant for other groups, ideas that arose within ERME and initiated or influenced developments in mathematics education research beyond ERME, and ideas that originated in one geographical area, were brought into ERME and, as a consequence, were taken up more widely.

The book was initiated by the board of ERME in early 2015. When the board invited us to edit the book, we designed a process according to which initial chapter drafts would be available prior to CERME 10 (February 2017), enabling them to be discussed in the TWGs at the conference. Thus, the CERME spirit of communication and cooperation came to bear also in the process of writing the book. Chapters were then finalised, taking into account the feedback from discussions at CERME 10 as well as new developments reported at that conference. Chapter lengths were tailored to the size of the group (or, in a few cases, of groups that had recently split) and the number of years that the group had been working. We are looking forward to launching the book at CERME 11 (February 2019).

Given the nature of the book, it contains many references to CERME proceedings. In order to save space, a shortened format has been used for references to

Preface **xxi**

these proceedings in the chapters. The full references of all proceedings are listed below this preface.

In conclusion, some general patterns seem to appear across the different core chapters. In many groups, work in the earlier years of CERME was characterised by members' difficulties in understanding each other, due to the very diverse research traditions and theoretical frameworks in use within Europe. Through comparing and contrasting these, mutual understandings of the different lines of research have been reached. Even if most researchers kept on working within their respective traditions, they succeeded in finding commonalities, and understanding differences. Moreover, efforts of networking theories have been described, not only for purposes of comparing and contrasting but for coordinating or combining theoretical approaches. Indeed, in quite a number of cases where researchers have started to combine theoretical frameworks, this has led to a widening of perspectives from rather narrow research topics to more comprehensive foci that require more complex approaches.

It will, of course, be for the reader of this book to form an opinion as to what extent ERME has succeeded in showing the existence of a dynamic community which, over 20 years, has been able to create an identity for European research in mathematics education, structured around a common spirit, while remaining respectful of diversity and open to outside influences; to create original ways of collaborative research and to support the acculturation and integration of young researchers; and to contribute substantially to the advance of international research on a wide range of issues, while playing a pioneering role in some areas.

The editors: Tommy Dreyfus, Michèle Artigue,
Despina Potari, Susanne Prediger, Kenneth Ruthven

References of CERME proceedings

Note: The complete references of the CERME proceedings are listed here. All CERME proceedings are accessible via http://www.mathematik.uni-dortmund.de/~erme/index. php?slab=proceedings

CERME 1: Schwank, I. (Ed.) (1999). *European Research in Mathematics Education I: Proceedings of the First Conference of the European Society for Research in Mathematics Education* (CERME 1, August 1998). Osnabrück, Germany: Forschungsinstitut für Mathematikdidaktik and ERME.

CERME 2: Novotná, J. (Ed.) (2002). *European Research in Mathematics Education II: Proceedings of the Second Conference of the European Society for Research in Mathematics Education* (CERME 2, February 24–27, 2001). Mariánské Lázně, Czech Republic: Charles University, Faculty of Education and ERME.

CERME 3: Mariotti, M. A. (Ed.) (2004). *European Research in Mathematics Education III: Proceedings of the Third Conference of the European Society for Research in Mathematics Education* (CERME 3, February 28–March 3, 2003). Bellaria, Italy: University of Pisa and ERME.

CERME 4: Bosch, M. (Ed.) (2006). *European Research in Mathematics Education IV: Proceedings of the Fourth Congress of the European Society for Research in Mathematics Education* (CERME 4,

xxii Preface

February 17–21, 2005). Sant Feliu de Guíxols, Spain: FUNDEMI IQS – Universitat Ramon Llull and ERME.

CERME 5: Pitta-Pantazi, D., & Philippou, C. (Eds.) (2007). *European Research in Mathematics Education V: Proceedings of the Fifth Congress of the European Society for Research in Mathematics Education* (CERME 5, February 22–26, 2007). Larnaca, Cyprus: University of Cyprus and ERME.

CERME 6: Durand-Guerrier, V., Soury-Lavergne, S., & Arzarello, F. (Eds.) (2010). *Proceedings of the Sixth Congress of the European Society for Research in Mathematics Education* (CERME 6, January 28–February 1, 2009). Lyon, France: Institut National de Recherche Pédagogique and ERME.

CERME 7: Pytlak, M., Rowland, T., & Swoboda, E. (Eds.) (2011). *Proceedings of the Seventh Congress of the European Society for Research in Mathematics Education* (CERME 7, February 9–13, 2011). Rzeszów, Poland: University of Rzeszów and ERME.

CERME 8: Ubuz, B., Haser, C., & Mariotti, M. A. (Eds.) (2013). *Proceedings of the Eighth Congress of the European Society for Research in Mathematics Education* (CERME 8, February 6–10, 2013). Ankara, Turkey: Middle East Technical University and ERME.

CERME 9: Krainer, K., & Vondrová, N. (Eds.) (2015). *Proceedings of the Ninth Congress of the European Society for Research in Mathematics Education* (CERME 9, February 4–8, 2015). Prague, Czech Republic: Charles University in Prague, Faculty of Education and ERME.

CERME 10: Dooley, T., & Gueudet, G. (Eds.) (2017). *Proceedings of the Tenth Congress of the European Society for Research in Mathematics Education* (CERME 10, February 1–5, 2017). Dublin, Ireland: DCU Institute of Education and ERME.

THE EUROPEAN SOCIETY FOR RESEARCH IN MATHEMATICS EDUCATION

Introduction by its former presidents

Ferdinando Arzarello, Paolo Boero, Viviane Durand-Guerrier and Barbara Jaworski

1 Introduction

During the weekend of 2–4 May 1997, 16 representatives from European countries met in Osnabrück, Germany, to establish a new society, the European Society for Research in Mathematics Education, in short ERME, to promote Communication, Cooperation and Collaboration (the three Cs) in mathematics education research in Europe. The three Cs came, over the years, to characterize the "ERME spirit".

The full foundation of this society took place at CERME 1, the first conference of the society, in August 1998, also in Osnabrück. The first president of ERME was Jean-Philippe Drouhard (France, 1997–2001). Jean-Philippe had accepted to join us in writing this chapter, but he passed away suddenly in November 2015. We miss him. Paolo Boero (Italy), who was an active member of the group of scholars who initiated the process of establishing the new society and was a member of the first Board, became the second president of ERME (2001–2005), and Barbara Jaworski (United Kingdom), also a founding member, was the third (2005–2009). They were followed by Ferdinando Arzarello (Italy, 2009–2013) and Viviane Durand-Guerrier (France, 2013–2017). From February 2017, the ERME president is Susanne Prediger (Germany). The society has been governed from the beginning by an elected board with representation criteria from the regions of Europe.

In what follows, we first present the motivations for *establishing the new society* 20 years ago and the steps through which the original aims were achieved. In Section 3, we present a brief summary of a significant reflection that took place on *quality and inclusion* in the CERME conferences. In the last section, we report on further developments aimed at the *ongoing strengthening of the society*.

2 Establishing the new society

After CERME 1, three main challenges had to be met by the ERME Board elected in Osnabrück: (1) to take initiatives to prepare a new generation of European researchers in mathematics education according to the ERME spirit; (2) to involve a sufficiently large number of European researchers in mathematics education, in order for CERME to become the representative forum of European research in the field; and (3) to ensure stability to the new-born society through a legal status anchored in the laws of one of the European countries. At the same time, the society and its initiatives had to be opened to countries beyond Europe, in order to promote worldwide scientific exchanges in the field.

From the very beginning, the society has been linked to the organisation of the CERME conferences with a wide spectrum of themes and orientations to profit from the rich diversity in European research, as will be demonstrated in the various chapters of this book. The conferences have been structured around a number of Thematic Working Groups (TWGs), each with a designated research focus. Participants have particularly appreciated the concrete possibility, offered by the TWGs, to develop their personal research through systematic, collaborative work with other researchers engaged in the same area, and to get constructive feedback from them on their papers (before, during and after the conference). Since CERME 10 there has been an open call for new TWGs: the current call can be found on the ERME website (www.mathematik.uni-dortmund.de/~erme/). The International Programme Committee (IPC) of each CERME is elected by the ERME Board, as is the IPC chair. The choice of TWGs and their leaders is discussed in the IPC and approved by the ERME Board. The decision to include a new group is taken according to the nature, focus and relevance of the research, its potential to attract participants, and its distinction from existing groups. The choice of group leaders and co-leaders is an important lever for ensuring the quality of work in the group, for planning activities to include all those who wish to participate and for opening the society to the diversity of research in Europe.

From CERME 1, the tasks of drawing up a constitution and establishing a legal status were undertaken by the ERME Board mainly thanks to the extraordinary work performed by Board member Graham Littler (UK). On behalf of ERME, he approached the UK Charity Commission to request charitable status in the UK, and dealt with the very complicated paperwork involved. In CERME 3 a draft was presented and voted on, as a basis for the formal establishment of the society. This was proposed in the General Meeting at CERME 4 and ratified in CERME 5.

An important issue of ERME policy consists of supporting and educating new/young researchers in mathematical education. The Board elected at CERME 2 in 2001 gave the task of designing a Summer School to Paolo Boero, Barbara Jaworski and Konrad Krainer. It was to be held every two years in alternate years from the CERMEs. The main target population consisted of PhD students in mathematics education. By analogy with CERME, the Summer School was conceived as a working place for students. TWGs (led by "expert" researchers – about 60% of the whole time) offered students a unique opportunity to present the current status of their research (be it initial, or near to the conclusion), to receive constructive feedback from the "expert" and from the other participants and to establish links with other young researchers interested in the same subject.

The first Summer School, held in Klagenfurt in 2002, showed that the design of the school was suitable to meet the students' expectations. Gradually the number of applicants rose from 40 in Klagenfurt to more than 100 (resulting in 72 participants) for the last four schools, including students from other continents. In parallel with the design and the implementation of the school, the Board helped to set up an informal branch of the society, YERME (Young European Researchers in Mathematics Education), to be involved in the preparation of the school and in other initiatives of interest to young researchers. The decision was formally taken in the General Meeting in CERME 8 when the first two representatives of young researchers in the ERME Board were elected. From the institution of YERME, the summer schools took on the abbreviation of YESS – YERME Summer School. YESS 9 is being prepared as this book goes to press.

Another important issue, from the beginning of ERME, CERME and YESS, was the encouragement of mathematics education researchers from Eastern Europe to join in ERME activities. For this purpose, ERME designated funds to contribute to travel and accommodation where financial hardship was demonstrated. When, in 2009, very tragically, Graham Littler died, the ERME Board decided to name this fund the Graham Littler Fund. Since then, this fund has been topped up regularly and used to provide financial support for participants to CERME and YERME where a need has been identified.

During the years 2001–2005, when it was important to ensure the representativeness of the new society, the Board worked hard to establish relationships with several research groups and existing national societies in the field of mathematics education; also researchers from other continents were invited to join ERME initiatives. As a result, the number of participants in CERME doubled at each conference until CERMEs 5, 6 and 7, when it stabilized at about 450–500 participants. It increased anew up to about 700 participants in CERMEs 9 and 10.

3 Scientific quality and inclusion in CERME conferences

From the very beginning, the issues of quality and inclusion in CERME conferences were main concerns of the society. Quality refers to scientific standards relating to papers presented and published, and to activity in the TWGs. Inclusion refers to ways in which participants are included in activities in the groups, and in presentations and published papers. CERME's policy of encouraging presentation (after two rounds of revision) of as many papers as possible was sometimes seen to act against high scientific standards. Seeking a balance between quality and inclusion was seriously problematic.

In CERME 6, held in Lyon, France, members of the ERME Board collected data in several ways from delegates at the conference concerning issues related

4 Arzarello et al.

to quality and inclusion in CERME conferences. In response, one group leader wrote: "Being all-inclusive and academically qualitative are *a priori* incompatible." While this is an individual view, it nevertheless flags a tension between inclusivity and scientific quality. Participants acknowledged that newcomers are drawn quickly into the activity of the group. There were almost no comments that suggested that group work was not friendly and welcoming, that participants were not (overtly) encouraged to take part and join in the discussions. In scientific terms, we can see inclusion to be facilitated through the review process in which a critical review can be helpful and supportive and enable the improvement of a paper. However, to quote Gates and Jorgensen (2009, p. 164), we were aware that "the field . . . in which the participants engage recognizes and conveys power to those whose habitus is represented and privileged in the field". Thus, and according to Atweh, Boero, Jurdak, Nebres and Valero (2008, p. 445), "collaboration between educators with varying backgrounds, interests and resources may lead to domination of the voice of the more able and marginalization of the less powerful".

Such considerations led Jaworski, da Ponte and Mariotti (2011) to start characterising inclusion and quality and to relate the characterisation to the specific aims of ERME in terms of Communication, Cooperation and Collaboration, in contributing to the ERME spirit. Inclusion is characterised in affective and scientific terms. Quality is characterised scientifically through "key ideas" and their development. The key ideas need to be there for scientific quality to exist at all; they need to be engaged with for scientific quality to be recognisable in the activity of a group. We present below how it works in CERME.

Starting communication: Participants have read the accepted papers; they come together with friendliness and the sincere desire to work inclusively. There are key ideas in the accepted papers to which the review process draws attention. Activity and discussion begin to encourage communication related to the ideas where the objective is to know each other's ideas and relate them to each other.

Developing cooperation in engaging with debate: Group organisation enables a focus on the key ideas. Friendly and considerate interaction, with attention to language enables participants to start to engage with the ideas. The emphasis is on including all people attending a TWG, including those who do not present accepted papers, in order to get reactions, questions and comments. This group activity may contribute, on the one hand, to improving the accepted papers (in terms of their revision for publication in the final proceedings), and on the other hand, to improving the quality of all participants' research work.

Developing cooperation in recognising ideas: Group leaders create activity to encourage a focus on getting participants engaged with the key ideas which are recognised. The emphasis is on reaching a quality of interaction relating to scientific ideas.

Enabling collaboration: Here we see an enabling of critical inquiry into the essences of the ideas; deep engagement of a scientific quality with deep probing of ideas and corresponding critical debate. From here, collaboration can begin.

It seems clear that, for the last point to be possible, both the second and third have to be achieved. This means dealing with the organisational challenges raised in the CERME 6 survey, which are far from trivial. However, it could be that a theoretical perspective of this sort, of what is involved in achieving inclusion and quality in group work in CERME, can act as a basis for thinking about dealing with the challenges and conceptualising in practical terms what we are aiming for in CERME.

4 Carrying on strengthening the society

From 2009, the main activities of ERME, in particular CERME and YESS, continued and expanded further. We stress below some of the main ERME activities contributing to the development and visibility of the society.

During this period, the number of participants from developing countries was increasing, due to the growing effort of ERME to support people from those countries through the Graham Littler Fund. Two decisions, taken during the General Meeting in CERME 7, contributed to expanding ERME beyond the strict boundaries of Europe. The first concerned countries from North Africa. While these countries were living the exciting period of the so-called "Arab Spring", some researchers were attending CERME 7 and claimed that the possibility for them to take part in events like CERME was important also for the progress of democracy in their own countries. Thus it was decided that researchers from these countries could apply for financial support from the Graham Littler Fund. The second was the decision of the ERME Board to organize the first CERME outside the European Union, namely CERME 8 in Turkey.

Since 2011 ERME has become an affiliated society of ICMI (International Commission on Mathematical Instruction). As such, ERME has the opportunity to present a quadrennial report to the General Assembly of ICMI, so that ERME activities can be known by the whole world community working in mathematics education. More generally, an important concern was to encourage the European research community to develop relationships with mathematicians and to take part in the ICMI activities. ICME-13 in Europe in July 2016 was a wonderful opportunity to improve the international visibility of ERME. As an ICMI affiliate organisation, ERME organised two sessions, one a general presentation of ERME with a focus on teacher education research, the other on YERME.

A further action of ERME in recent years concerned the issue of rating of publication outlets. As is well known, it is very difficult to escape the influence of the ranking and grading of scientific journals for the development of researchers' careers. However, the current systems are often based on crude bibliometric analyses that have little to do with scientific quality. This represents a disadvantage for researchers in the field of mathematics education. For this reason, ERME, together with the Education Committee of the European Mathematical Society (EMS), and supported by the ICMI, decided in 2011 to appoint a joint commission, with the aim to propose a grading of research journals in mathematics education based on expert judgment. The result of their careful and joint analysis, based on consultation with 91 experts from 42 countries, was a document rating the 17 most

6 Arzarello et al.

important journals in mathematics education on a scale with four grades (Törner & Arzarello, 2012). We understand that this document has been used by some official boards charged with assessing the work and the consequent promotion of mathematics educators.

Another very important issue is the publication of the proceedings for each CERME conference. All the proceedings are available online on the ERME website and since CERME 9, thanks to Nad'a Vondrovà and Konrad Krainer, CERME Proceedings are also available on the open Archiv HAL, managed by the French CNRS (Centre national de la recherche scientifique). A detailed list appears at the end of the Preface of this book.

A further decision concerned the ERME Topic Conferences (ETC), which are conferences organised on a specific research theme or themes related to the work of ERME as presented in associated working groups at CERME conferences. Their aim is to extend the work of the group or groups in specific directions with clear value to the mathematics education research community. These initiatives take place during the year in which CERME does not take place. After careful preparatory work, in the General Meeting in Prague (2015), the rules for ETCs were approved. Three ETCs were held in 2016, and three were approved in 2017 for appearance in 2018.

5 Conclusion

A main objective of this Introduction is to present the evolution of the spirit and the activities of ERME to the reader of this book as background for the core chapters of the book (Chapters 1–18), which are based on the work of the TWGs at and between the conferences of the society. Another objective is that present and future members of the society, who have not taken part in its evolution and its activities, should know how choices have been made, and then re-thought and deepened (according to accumulated experience). In doing so, we wish to stress that the future of the society is in the hands of the members and not in the paragraphs of the Constitution.

In addition, we wish to illustrate how ideas about the society and its activities have progressively matured and how emerging needs have been taken into account. We hope that ERME's "historical" evolution will give a concrete idea of the openness of the society and of the deep reasons why CERME, YESS and the other ERME initiatives are organized in the present way.

Acknowledgements

We acknowledge all the board members, and all the IPC and LOC chairs from the very beginning, and as a whole all the TWG leaders and plenary speakers and panellists. Their names are available on the ERME website: www.mathematik. uni-dortmund.de/~erme/.

References

Atweh, B., Boero, P., Jurdak, M., Nebres, B., & Valero, P. (2008). International cooperation in mathematics education: promises and challenges. In M. Niss & E. Emborg (Eds.), *Proceedings of the 10th International Congress on Mathematical Education (ICME 10), July 4–11, 2004, Copenhagen, Denmark* (pp. 443–447). Roskilde, Denmark: Roskilde University.

Gates, P., & Jorgensen, R. (2009). Foregrounding social justice in mathematics teacher education. *Journal of Mathematics Teacher Education, 12,* 161–170.

Jaworski, B., da Ponte, J. P., & Mariotti, M. A. (2011). The CERME spirit: issues of quality and inclusion in an innovative conference style. In B. Atweh, M. Graven, W. Secada & P. Valero (Eds.), *Mapping Equity and Quality in Mathematics Education* (pp. 457–478). London: Springer.

Törner, G., & Arzarello, F. (December 2012). Grading mathematics education research journals. *EMS Newsletter, 86,* 52–54.

1

FROM GEOMETRICAL THINKING TO GEOMETRICAL WORKING COMPETENCIES

Alain Kuzniak, Philippe R. Richard and Paraskevi Michael-Chrysanthou

1 Introduction

The chapter provides a comprehensive view of the research in geometry education that has been presented during CERME conferences. To write this chapter, all the CERME conferences proceedings since the beginning have been read again in order to track the main developments in trends, theories and methodologies and identify progress in research on the teaching and learning of geometry. Undeniably, this research field in the CERME conferences benefited from the creation of the group in 2003, which allowed the development of common theoretical frameworks and research trends concerning the teaching and learning of geometry in different educational levels and systems. This facilitated discussions and collaboration between researchers coming from different countries with various experiences in the field of geometry education, with different views on what research in didactics of geometry should be.

In the first section, the evolution of topics the group dealt with is traced. The emphasis is on the close link between theoretical and empirical aspects that has always guided the researchers of the group. Then, in the second section, some methodological and theoretical tools developed within the group are introduced and we show how they constituted a common support for studies in the field. In the third and fourth sections, the main findings from the group are given by emphasising the spirit of communication and collaboration built up through the different meetings. In a concluding section, a common research agenda is considered that could be organised to achieve a deeper understanding of geometrical thinking and competencies through the whole education.

2 Linking theoretical and empirical aspects: a brief history of the thematic evolution of the group

Through this short history of the group, we will show how the researchers involved in the group displayed a constant and growing interest in using frameworks that allow the connection between theoretical and empirical aspects. Beside references to general psychological (Piaget's or Vygotski's works) or didactical theories (Brousseau's or Chevallard's theories), specific theoretical and methodological developments were provided during the meetings. The van Hiele levels, Fischbein's figural concept and Duval's registers were among the most relevant cognitive and semiotic approaches that were adopted. Regarding the epistemological and didactical approaches, a decisive importance was granted to the geometrical paradigms and the Geometrical Working Spaces, and more recently to a lesser extent, on the notion of geometrical competencies. A great number of researches focused on different aspects of geometrical understanding, such as figural apprehension, visualisation and the effort of conceptualisation in geometry using different methodologies and frameworks. Among the numerous papers presented in the Working Group on Geometry, we can distinguish some recurrent topics that we present below in parallel with the history of the group.

The first working group specifically dedicated to geometry was created in CERME 3. Since this meeting various names were given to this group including some nuances, from *thinking* of researching through *teaching* and *learning*, ending at the last edition (CERME 10) with the sole name *Geometry*. On Figure 1.1, which is drawn from the synthesis of CERME 7, themes and issues generated by the contributions are organised in a conceptual tree.

On Figure 1.2, some unifying characteristics related to geometric competencies that have been developed during CERME 8 are presented.

Further on, we will illustrate the evolution of the group with a short overview of the papers presented during these conferences, intending to show the richness of the themes developed throughout these meetings.

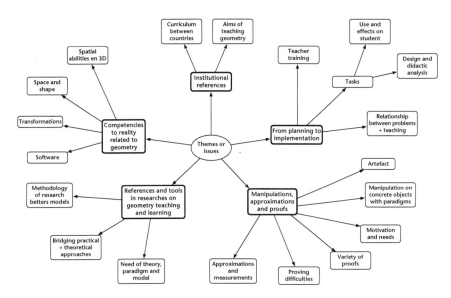

FIGURE 1.1 Main themes or issues of the group

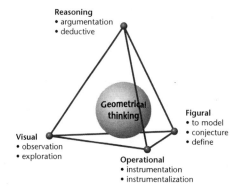

FIGURE 1.2 Geometrical competencies

2.1 Before the official creation of the group

In the first two sessions, even if the group did not officially exist, some communications were already centred on geometry with a diversity of viewpoints. Some reflexive papers (Kilpatrick, 1999) intended to provide references concerning curriculum issues and the reform of mathematics examination at a time when social and professional mobility was a major issue. Other contributions insisted on the need of metaphors and images (Parzysz, 2002) or focused on epistemological and cultural approaches about geometrical knowledge (e.g. Arzarello, Dorier, Hefendehl-Hebeke, & Turnau, 1999; Burton, 1999).

Mostly descriptive, the studies addressed the geometric content and often mentioned some general mathematical skills or cognitive and instrumental dimensions in geometric activity. They also indicated some "frictions" with other mathematical domains like algebra, and some typical interactions in the use of technologies with digital geometric software (Dreyfus, Hillel & Sierpinska, 1999).

In addition, the natural relationship of geometry to proof and proving was regularly examined, both in theoretical and philosophical essays and in reports of studies involving empirical research. They link proving, arguing, modelling and discovering processes. They also deal with the students' way for proving and their use of a logical discourse, figural signs and technology when they are faced with proof exercises in geometry. Attention was also paid to the links between the discovery process, visualisation and instrumentation processes with software (Jones, Lagrange & Lemut, 2002).

There were also studies concerned with the teacher education and professional development in geometry and the organisation of courses (Houdement & Kuzniak, 2002).

Despite the fact that several works deal with mathematical proofs, visualisation and dynamic geometry, surprisingly, other natural links with geometry are virtually missing during the first congresses. In fact, few researches focus on modelling of physical phenomena using geometrical tools.

2.2 Thematic from CERME 3 to CERME 5

During these conferences, the new geometry group starts the study of geometrical thinking using and developing the key concepts of paradigms, developmental stages and generalisation in space (Houdement & Kuzniak, 2004; Gueudet-Chartier, 2004; Houdement, 2007). At the core of many researches, visualisation is often directly connected to Duval's registers of representation (Perrin-Glorian, 2004; Kurina, 2004; Pittalis, Mousoulides & Christou, 2007). In an original study, the visualisation issue is raised for sighted and blind students (Kohanová, 2007). Other papers focus on the notion of instrumentation, both with classical drawing tools (Vighi, 2006; Bulf, 2007; Kospentaris & Spyrou, 2007) or digital tools (Rolet, 2004). Education for future teachers (Kuzniak & Rauscher, 2006) and reasoning (Ding, Fujita & Jones, 2006; Markopoulos & Potari, 2006) appear as topics of continuity, while concepts and conceptions remain key themes in many studies (e.g. Modestou, Elia, Gagatsis & Spanoudes, 2007; Marchini & Rinaldi, 2006).

2.3 Thematic from CERME 6 to CERME 7

The theoretical and methodological dimensions of research in geometry remain prominent topics, especially in further development of the notions of geometrical work and Geometrical Working Spaces. Many points were considered as a common background, as they were developed during former sessions. These points were related to the use of geometrical figures and diagrams (Deliyianni et al., 2011) and to the understanding and use of concepts and proof in geometry (Gagatsis, Michael, Deliyianni, Monoyiou & Kuzniak, 2011; Fujita, Jones, Kunimune, Kumakura & Matsumoto, 2011). For an epistemological and didactical approach, researchers used the geometrical paradigms and geometrical work spaces. Attention is also paid to 3D geometry forms of representation through the possible use of digital tools (Mithalal, 2010; Hattermann, 2010; Steinwandel & Ludwig, 2011). In addition to the usual geometrical topics, special attention is paid to general or cross-cutting aspects, such as educational goals and curriculum in geometry (Girnat, 2011; Kuzniak, 2011), communication and language (Bulf, Mathé & Mithalal, 2011), the teaching, the thinking and the learning processes in geometry. Moreover, Mackrell (2011) questions the interrelations between numbers, algebra and geometry, especially in digital environments.

2.4 Restructuring in CERME 8 to CERME 10

More recently, the working group sought to revisit and extend the issue of geometrical thinking and geometrical work by reformulating it in terms of four geometrical competencies (reasoning, figural, operational and visual) organised in a tetrahedron (see Figure 1.2). Each geometric competence constitutes a pole of geometrical thought and it is the study of the link between these competences that makes it possible to better understand the global functioning of geometrical thought (Maschietto, Mithalal, Richard & Swoboda, 2013). Further on, at CERME 9, there were more specific contributions about the way geometry is, or should be, taught and the four competencies were used as a general way of describing the geometrical activity and for creating links between different points of view (theoretical and empirical). In this group the discussions were related to geometry teaching and learning (Douaire & Emprin, 2015; Kuzniak & Nechache, 2015) and issues such as teaching practices and task design (Mithalal, 2015; Pytlak, 2015). Furthermore, cultural and educational contexts modifying the geometry curricula were also discussed, introducing a new issue about the role of language and social interactions in the teaching and learning of geometry. In CERME 10, the four competencies were used to describe geometrical thinking: reasoning, figural, operational and figural. The group took these dimensions as a background that was very helpful to understand each other and to compare our approaches to the issue of what is at stake in the teaching and learning of geometry.

Three main issues were addressed during the working group: The role of material activity in the construction of mathematical concepts, including using instruments, manipulation, investigation, modelling . . .; Visualisation and spatial skills; Language, proof and argumentation. In comparison to the previous CERME, this time psychological points of view, among others, were represented. This raised new questions, often with very different theoretical and methodological backgrounds sometimes far from mathematics.

3 Studying the teaching and learning of geometry through a common lens

Since the creation of the group, the development of shared theoretical frameworks was central to ground collaboration between participants. This point is particularly evident in CERME 3, where the main theoretical concerns of the group were summarised by Dorier, Gutiérrez and Strässer (2004). Their classification is used to highlight some theoretical approaches which have been reinforced by their presentation and discussion during CERME conferences.

3.1 Geometrical paradigms

The history of geometry shows two contradictory trends. First, geometry is used as a tool to deal with situations in real life but, on the other hand, geometry for more than two thousand years was considered the prototype of logical, mathematical thinking and writing after the publication of Euclid's "Elements". These contradictory perspectives are taken into account in geometry education by Houdement and Kuzniak (2004) with geometrical paradigms. Three paradigms are distinguished. "Geometry I: natural geometry" is intimately related to reality; experiments and deduction grounded on material objects. "Geometry II: natural/axiomatic geometry" is based on hypothetical deductive laws as the source of validation with a set of axioms as close as possible to intuition and may be incomplete. "Geometry III: formal axiomatic geometry" is formal, with axioms that are no more based on the sensory reality.

These various paradigms are not organised in a hierarchy making one preferable to another, but their work horizons are different and the choice of a path toward the solution is determined by the purpose of the problem and the researcher's viewpoint.

Useful to provide a method to classify geometrical thinking, geometrical paradigms have also been helpful to interpret tasks eventually given to students and future teachers and can be used to classify the students' productions, offering an orientation for the teacher of geometry. In contrast to van Hiele's theory, this approach is not dependent on the general thinking and reasoning development, but relates more to an epistemological viewpoint.

3.2 Development stages

The so-called "van Hiele levels" of geometrical reasoning are among the most used theoretical frameworks for organising the teaching and learning of geometry according to development stages. The description of the "van Hiele levels" already gives hints to fundamental links between these levels and the model suggested by Houdement and Kuzniak (2004). There was some discussion whether geometrical knowledge progresses through sequences of stages. Some papers presented (Braconne-Michoux, 2011) appear as contradictory to the traditional "van Hiele levels". This is especially true if the levels are linked to clearly identified and fixed ages or if individual persons are thought to necessarily follow the order of the respective levels. Positioning itself on ideas of stages of development and/or learning is obviously important for constructing and discussing geometric tasks or research projects. The stage levels can help to find and further develop appropriate tasks, and they are obviously helpful for exploratory activities.

3.3 Registers of representation

Semiotic consideration has always been important in geometry and the distinction of figure, in the sense of the most general object of geometry (either 2D or 3D) versus its material representation, is classic in the field. Three main groups of semiotic representation can be distinguished: material representation (in paper, cardboard, wood, plaster, etc.), a drawing (made either with paper and pencil, or on a computer screen), and a discursive representation (a description with words using a mixture of natural and formal languages). Each register bears its own internal functioning, with rules more or less explicit. Moreover, students have to move from one register to another, sometimes explicitly, sometimes implicitly, sometimes back and forth. Questions about registers of semiotic representation and cognitive processes have been studied in depth by Duval (1993). He defines "semiotic representations as productions made by use of signs belonging to a system of representation which has its own constraints of meaning and functioning". Semiotic representations are necessary to mathematical activity, because its objects cannot be directly perceived and must, therefore, be represented.

3.4 Instrumentation with artefacts and digital tools

There is a broad consensus in the community of mathematics teachers and educators that learning geometry is much more effective if concepts, properties, relationships, etc. are presented to students materialised by means of instruments modelling their characteristics and properties. Furthermore, the use of didactic instruments is very convenient, if not necessary, in primary and lower secondary grades. There is a huge pile of literature reporting the continuous efforts devoted by mathematics educators since long ago to explore the teaching and learning of geometry with the help of manipulative, computers, and other tools. Unfortunately, many researchers interested in this topic have generally preferred to participate in the CERME group on technology and thus have deprived, in some part, the group of discussions on this subject.

3.5 On Geometrical Working Spaces

In relation to semiotic, instrumental and discursive proof, the idea of global thinking on geometric work was first introduced in CERME 4 (Kuzniak & Rauscher, 2006) and was based on the model of Geometrical Working Spaces (GWS) presented during a plenary lecture at CERME 8 (Kuzniak, 2013). The basic idea behind this approach is that some real geometrical work appears only when a student's activity is both coherent and complex to develop a reasoning using intuition, experimentation and deduction. The GWS is conceived as a dynamic abstract place that is organised to foster the work of people solving geometry problems. The model articulates two main concerns, the one of epistemological nature, in a close relationship with mathematical content of the studied area, and the other of cognitive nature, related to the thinking of the person solving mathematical tasks. This complex organisation is generally summarised in the diagrams in Figure 1.3 (for details, see Kuzniak, Tanguay & Elia, 2016).

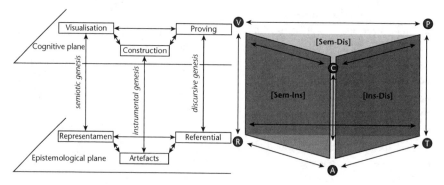

FIGURE 1.3 The Geometrical Working Space diagrams

For describing the work in its epistemological dimension, there are three components in interaction, organised according to purely mathematical criteria: a set of concrete and tangible objects, the term sign or representamen is used to summarise this component; a set of artefacts such as drawing instruments or software; and a theoretical system of reference based on definitions, properties and theorems. In close relation to the components of the epistemological level, three cognitive components are introduced as follows: visualisation related to deciphering and interpreting signs; construction depending on the used artefacts and the associated techniques; and proving conveyed through processes producing validations, and based on the theoretical frame of reference.

The process of bridging the epistemological plane and the cognitive plane can be identified through the lens of GWSs as genesis related to a specific dimension in the model: semiotic, instrumental and discursive geneses. In order to understand this complex process of interrelationships, the three vertical planes of the diagram are useful and can be identified by the genesis which they implement [Sem-Ins], [Ins-Dis] and [Sem-Dis] (Figure 1.3).

To the coordination of the geneses in the GWS model, there are three internal fibrations that focus on the role of tools, controls and representations in the mathematical work which allow us to describe the complexity of managing the relationships between all the components involved in geometrical activity. Richard, Oller and Meavilla (2016) have introduced three internal fibrations that focus on the role of tools, controls and representations in the mathematical work. This model was briefly discussed in CERME 10.

3.6 Within and beyond the model of GWS

The GWS model results largely from the work done in the group of geometry where its use led to understanding that the model can easily be adapted to relate with other theoretical approaches. This means that, depending on the issues and problems dealt with, it can be complementarily associated with other theories: (a) to increase its contextual perspective, such as the Mathematics Teachers' Specialised Knowledge (MTSK), the Action, Process, Object and Schema (APOS) theory, the Activity Theory (AT) or the Theory of Didactical Situations in Mathematics (TDSM); (b) to refine it according to the needs of the studies, such as Peirce's semiotics (1931), Rabardel's cognitive ergonomics (1995) or Richard and Sierpinska's (2004) functional-structural approach.

Using the GWS model makes it possible to question in a didactic and scientific way – mostly non-ideological – the teaching and learning of geometry, straddling the mathematical education and the didactics of the field. For example, we know that curricula change frequently or that the meanings of the words or concepts of reference vary from one region to another, which is a problem from a research point of view.

4 Building and developing a spirit of communication and collaboration

As previously underlined, a strong point of the group is participants' emphasis on relating theoretical and empirical aspects of research in geometry education. The continuity of this trend was evident through the meetings of the group so far, as the participants were mainly discussing results of empirical or developmental research studies and theoretical reports about the teaching and learning of geometry. The need for a common framework related to geometry education appeared necessary in the working group in order to stimulate the discussions among members and to allow the capitalisation of knowledge in the domain.

Due to collaborations initiated during CERME meetings with colleagues from France, Cyprus, Spain, Canada, Mexico and Chile, it has been made possible to develop a joint theoretical framework. The framework should be dedicated to study the teaching and learning of geometry, space and shape on the whole educational system and should be neutral in the sense that it can be used to facilitate exchanges in different countries and institutions. According to this need, the framework of geometrical paradigms, as explained above, was introduced in the CERME 3 conference. In the pursuit of results at the CERME 3, the geometry working group of CERME 4 and 5 continued by looking into geometrical paradigms.

In particular, in CERME 4, Kuzniak and Rauscher (2006) analysed pre-service schoolteachers' geometrical approaches, based on the notion of geometrical paradigms and levels of argumentation. They found that students' levels of understanding and memorisation of the bases of the elementary geometry differ greatly and that they keep the practical use of geometry. Although their study was conducted with a particular population, their results can be useful for evaluating the long-term effects of education in geometry.

Moving forward, in CERME 5, during discussions about the possible uses of geometrical paradigms, new participants of the group initiated a discussion about the real benefit of this approach. Perspectives in these directions were given by Houdement's (2007) and Bulf's (2007) papers. In fact, Houdement highlighted the uses of this approach for comparing curricula in different countries and for reflecting on the necessity to teach Geometry II and the proper way to introduce it. In Bulf's effort to examine the link between geometrical knowledge and the reality in relation to the concept of symmetry, this approach was useful for tracing a double play between the Geometry I and Geometry II on one side and Reality/Theory on the other side. Furthermore, Kospentaris and Spyrou (2007) used the approach of paradigms in teachers' education, as done in CERME 4 by Kuzniak and Rauscher (2006). Their results were in line with previous results presented at CERMEs 3 and 4 about pre-service teachers' geometrical thinking. They actually found that visual strategies or measurement using tools are used by students at the end of secondary school, interpreting it not by a developmental approach, but based on the geometrical paradigms. The discussion of the aforementioned papers gave a future perspective for the group in order to make precise the sufficiency of the so far existing theoretical tools for determining the nature and the construction of the GWS used by students and teachers. In fact, no paper was traced in the following CERMEs that challenged these new theoretical tools.

Following up the spirit of the previous years, the participants in CERME 6, who came from both Europe and America, have extended and enriched the results obtained so far. Until then a common background was built and known by experienced participants, thus the participants worked within the continuity of the former sessions of CERME and their discussions were effectively facilitated by this common culture. In fact, the participants came to two main conclusions regarding the use of theory in research: (1) theory can serve as a starting point for initiating a research study and (2) theory can act as a lens to look into the data. An example of such research is that of Kuzniak and Vivier (2010) who examined the Greek Geometrical Work at secondary level from the French viewpoint, using a theoretical frame based on paradigms and GWSs. Also, Panaoura and Gagatsis (2010) compared the geometrical reasoning of primary and secondary school students, based on the way students confronted and solved specific geometrical tasks, finding difficulties and phenomena related to the transition from Natural Geometry to Natural Axiomatic Geometry. Therefore, a perspective for future research on geometry theories and their articulation for the group was the use of geometrical paradigms as a tool for analysing existing curricula and students' behaviour.

The creation of a common spirit of communication has also built ideas of collaboration between participants through the discussions of the group. This was evident regarding the focus of research in specific educational levels. Actually, at the first discussions of the working group the attention was given to primary education, as many of the papers were about young students' geometrical concepts (Marchetti, Medici, Vighi & Zaccomer, 2006; Marchini & Rinaldi, 2006) and the role of specific tools for the teaching and learning of geometry at that age level (Vighi, 2006). However, in subsequent meetings of the group (in CERME 5), collaborations were envisaged about the transition from a lower to a higher educational level and also the adaptation of a common framework to work out such kinds of studies as paradigms, GWS, spatial abilities and conceptions about the figure.

This was succeeded in the next meeting of the group in CERME 6 as, among the research presented in the group, the dimension of the students' transition from primary to secondary school was also taken into account. For example, Deliyianni, Elia, Gagatsis, Monoyiou, and Panaoura (2010) investigated the role of various aspects of figural apprehension in geometrical reasoning in relation to the students' transition from primary to secondary education, revealing differences between the two groups of students' performance and strategies in solving geometrical tasks. In a similar sense, Panaoura and Gagatsis (2010) compared primary and secondary school students' solutions of geometrical tasks and stressed the need for helping students progressively move from the geometry of observation to the geometry of deduction as they transit to a higher educational level.

Finally, in an effort to build a spirit of communication and collaboration in the group, collaborations between experienced and new researchers were

18   Kuzniak, Richard & Michael-Chrysanthou

accomplished (e.g. the work of Deliyianni et al. (2011) and Gagatsis et al. (2011)). These common works facilitated not only the communication between old and new research, but also the collaboration between researchers from different countries (e.g. France and Cyprus).

5 Toward a common research agenda?

Even if exchange within the group is still very rich and exciting, the geometry group seems to have gradually forsaken some of its initial ambitions, because of the existence of various groups specifically interested in technology use, proof, teacher education, semiotic aspects, etc. It has been partially disembodied and deprived of what has always been the strength of geometry: its transdisciplinary contribution to human thinking.

Another challenge faced by the group comes from its difficulty to capitalise its results and findings because of two kinds of volatility. The first is natural and comes from the renewal of participants who are younger and younger, and sometimes beginners in the field. Experienced researchers were attracted by other groups, developing topics closer to their own researches. On the other hand, this movement is also positive in the sense that new participants might give new ideas and perspectives to these groups.

Another reason for the difficulty to capitalise in the field is the constant curriculum changes in geometry. This can be illustrated by the erratic presence of geometric transformations. Moreover, and fundamentally, geometrical activity seems more and more oriented to other mathematical fields through modelling activity based on geometrical support, such as physics, geography, etc.

To overcome some of these problems, we suggest that the viewpoint on geometrical work could help to shape a common research agenda aiming at understanding better the competencies involved in geometrical work through and beyond the whole education. It requires coordination between cognitive, epistemological and sensible approaches, structured around three complementary dimensions which relate to visual, experimental and reasoning competencies.

5.1 On semiotic work and visual competency

Geometry is traditionally viewed as a work on geometrical configurations which are both tangible signs and abstract mathematical objects. This difference is clearly identified under the classical opposition *drawing versus figure*, which focuses on strong interactions and differences between semiotic and discursive dimensions. The semiotic dimension, especially worked through the visualisation process, is at the centre of Duval's research, which developed very powerful tools, such as registers of semiotic representation, to explore the question. In his view, a real understanding of mathematical objects requires interplay between different registers which are the sole tangible and visible representations of the mathematical objects. Within a multi-modal approach on semiotic resources, it is possible to study the entanglement of mathematical objects and their various semiotic representations. One of the main issues will be to see how these links are formed and reformed and can hide the very nature of geometrical objects to students.

5.2 On instrumental work and experimental competency

Geometry could not exist without instrumented activities or drawing tools, and the study of their different uses allows identification of two types of geometry which are well described by Geometry I and Geometry II paradigms. Some constructions are possible or impossible depending on tools or milieus. Thus, the trisection of the angle is impossible with ruler and compass, but quite possible with origami or mechanical instruments. The same is true for duplicating the cube. In general, it is possible to understand how, and justify why, constructions with drawing tools are effective, by questioning the structure of the mathematical objects, especially the nature of numbers, involved in these constructions. From precise drawings based on mathematically wrong constructions (like the Dürer pentagon) to mathematically correct but imprecise, in the sense of their measure, constructions (like Euclid's pentagon), it is possible to see all the epistemic conflicts opposing constructions based on an approximation to constructions based on purely deductive arguments. The tension between precise and correct constructions has been renewed with the appearance of dynamic geometry software (DGS). As Strässer (2002) suggested, we need to think more about the nature of geometry embedded in tools, and reconsider the traditional opposition between practical and theoretical aspects of geometry.

5.3 On the work around proof and reasoning competency

Since antiquity, demonstration work in its demonstrative form has always been emphasised in geometry, considered as a kind of ideal of rational thought grasped in its most intuitive and visible form. But in education, this idealised form of advanced mathematical work is not so obvious to discover and to implement, because it is hidden behind the play between practical and formal geometric paradigms. Moreover, thanks to its graphical precision, the software finally convinces users of the validity of the results. Proofs are no longer only formal and the modes of argumentation are enriched by experiments that give new meaning to the classical epistemological distinctions between the iconic and non-iconic reasoning and relate to discursive-graphic reasoning (Richard, 2004). Increasingly, human reasoning is supported on representation and processing delegated to a machine (Richard et al., 2016). It remains to understand how to animate the figures (instrumental aspect) by coordinating better the semiotic aspects associated with figures and the discourse of proof in natural language.

5.4 On modelling competency in geometrical work

Being as old as the first forms represented on walls of prehistoric caves, the very constitution of the geometric model is certainly an incarnation of what is the modelling of space and forms. Unfortunately, modelling is not widely practised in compulsory education, and problem solving in geometry classes too often relies on classical and not open tasks.

In several European curricula, emphasis is now placed on the notion of measurement in order to understand certain geometric properties. The passage from synthetic geometry to arithmetic and algebra is thus favoured. However, the geometric model remains necessary to develop the visual discovery of new properties, but also of proofs without recourse to measurement and algebra. Geometry is always one of the ingredients of the discovery of the beauty of mathematics and is also deeply linked to geography, physics or the arts and thus to the understanding of the world.

References

Arzarello, F., Dorier, J.-L., Hefendehl-Hebeker, L., & Turnau, S. (1999). Mathematics as a cultural product. *CERME1* (pp. 70–77).

Braconne-Michoux, A. (2011). Relations between geometrical paradigms and van Hiele levels. *CERME7* (pp. 618–627).

Bulf, C. (2007). The use of everyday objects and situations in mathematics teaching: The symmetry case in French geometry teaching. *CERME5* (pp. 962–971).

Bulf, C., Mathé, A.-C., & Mithalal, J. (2011). Language in the geometry classroom. *CERME7* (pp. 649–659).

Burton, L. (1999). Mathematics and their epistemologies: And the learning of mathematics. *CERME1* (pp. 87–102).

Deliyianni, E., Elia, I., Gagatsis, A., Monoyiou, A., & Panaoura, A. (2010). A theoretical model of students' geometrical figure understanding. *CERME6* (pp. 696705).

Deliyianni, E., Gagatsis, A., Monoyiou, A., Michael, P., Kalogirou, P., & Kuzniak, A. (2011). Towards a comprehensive theoretical model of students' geometrical figure understanding and its relation with proof. *CERME7* (pp. 598–607).

Ding, L., Fujita, T., & Jones, K. (2006). Developing geometrical reasoning in the classroom: Learning from highly experienced teachers from China and Japan. *CERME4* (pp. 727–737).

Dorier, J. L., Gutiérrez, A., & Strässer, R. (2004). Introduction to the thematic working group on geometry. *CERME3*: www.mathematik.uni-dortmund.de/~erme/CERME3/Groups/TG7/TG7_list.php.

Douaire, J., & Emprin, F. (2015). Teaching geometry to students (from five to eight years old). *CERME9* (pp. 529–535).

Dreyfus, T., Hillel, J., & Sierpinska, A. (1999). Cabri based linear algebra: Transformations. *CERME1* (pp. 209–221).

Duval, R. (1993). Registres de représentation sémiotique et fonctionnement de la pensée. *Annales de didactique et de sciences cognitive, 5*, 37–65.

Fujita, T., Jones, K., Kunimune, S., Kumakura, H., & Matsumoto, S. (2011). Proofs and refutations in lower secondary school geometry. *CERME7* (pp. 660–669).

Gagatsis, A., Michael, P., Deliyianni, E., Monoyiou, A., & Kuzniak, A. (2011). Secondary students' behaviour in proof tasks: Understanding and the influence of the geometrical figure. *CERME7* (pp. 608–617).

Girnat, B. (2011). Geometry as propaedeutic to model building: A reflection on secondary school teachers' beliefs. *CERME7* (pp. 628–637).

Gueudet-Chartier, G. (2003). Geometric thinking in a n-space. *CERME3*: www.mathematik. uni-dortmund.de/~erme/CERME3/Groups/TG7/TG7_list.php.

Hattermann, M. (2010). The drag-mode in three dimensional dynamic geometry environments: Two studies. *CERME6* (pp. 786–795).

Houdement, C. (2007). Geometrical working space: A tool for comparison. *CERME5* (pp. 954–972).

Houdement, C., & Kuzniak, A. (2002). Pretty (good) didactical provocation as a tool for teacher's training in geometry. *CERME2* (pp. 292–303).

Houdement, C., & Kuzniak, A. (2004). Elementary geometry split into different geometrical paradigms. *CERME3*: www.mathematik.uni-dortmund.de/~erme/CERME3/Groups/TG7/TG7_list.php.

Jones, K., Lagrange, J.-B., & Lemut, E. (2002). Introduction to WG2: Tools and technologies in mathematical didactics. *CERME2* (pp. 125–127).

Kilpatrick, J. (1999). Ich bin europäisch. *CERME1* (pp. 49–68).

Kohanová, I. (2007). Comparison of observation of new space and its objects by sighted and non-sighted pupils. *CERME5* (pp. 982–991).

Kospentaris, G., & Spyrou, P. (2007). Assessing the attainment of analytic-descriptive geometrical thinking with new tools. *CERME5* (pp. 992–1011).

Kurina, F. (2004). Geometry: The resource of opportunities. *CERME3*: www.mathematik. uni-dortmund.de/~erme/CERME3/Groups/TG7/TG7_list.php.

Kuzniak, A. (2011). Geometric work at the end of compulsory education. *CERME7* (pp. 638–648).

Kuzniak, A. (2013). Teaching and learning geometry and beyond: Plenary conference. *CERME8* (pp. 33–49).

Kuzniak, A., & Nechache, A. (2015). Using the geometric working spaces to plan a coherent teaching of geometry. *CERME9* (pp. 543–549).

Kuzniak, A., & Rauscher, J. C. (2006). On the geometrical thinking of pre-service school teachers. *CERME4* (pp. 738–747).

Kuzniak, A., & Vivier, L. (2010). A French look on the Greek Geometrical Working Space at secondary school level. *CERME6* (pp. 686–695).

Kuzniak, A., Tanguay, D., & Elia, I. (2016). Mathematical Working Spaces in schooling: an introduction. *ZDM-Mathematics Education, 48*(6), 721–737.

Mackrell, K. (2011). Integrating number, algebra, and geometry with interactive geometry software. *CERME7* (pp. 691–700).

Marchetti, P., Medici, D., Vighi, P., & Zaccomer, E. (2006). Comparing perimeters and area: Children's pre-conceptions and spontaneous procedures. *CERME4* (pp. 766–776).

Marchini, C., & Rinaldi, M. G. (2006). Geometrical pre-conceptions of 8 year old pupils. *CERME4* (pp. 748–755).

Markopoulos, C., & Potari, D. (2006). Using dynamic transformations of solids to promote children's geometrical reasoning. *CERME4* (pp. 756–765).

Maschietto, M., Mithalal, J., Richard, P., & Swoboda, E. (2013). Introduction to the papers and posters of WG4: Geometrical thinking. *CERME8* (pp. 578–584).

Mithalal, J. (2010). 3D geometry and learning of mathematical reasoning. *CERME6* (pp. 796–805).

Mithalal, J. (2015). Combining epistemological and cognitive approaches of geometry with cK¢. *CERME9* (pp. 557–563).

Modestou, M., Elia, I., Gagatsis, A., & Spanoudes, G. (2007). Problem solving in geometry: The case of the illusion of proportionality. *CERME5* (pp. 1052–1061).

Panaoura, G., & Gagatsis, A. (2010). The geometrical reasoning of primary and secondary school students. *CERME6* (pp. 746–755).

Parzysz, B. (2002). Introduction to WG7: Working together on metaphors and images. *CERME2* (pp. 531–532).

Peirce, C. S. (1931). *Collected Papers, vols. 1–6.* Cambridge, MA: Harvard University Press.

Perrin-Glorian, M.-J. (2004). Studying geometric figures at primary schools: From surfaces to points. *CERME3*: www.mathematik.uni-dortmund.de/~erme/CERME3/Groups/TG7/TG7_list.php.

Pittalis, M., Mousoulides, N., & Christou, C. (2007). Spatial ability as a predictor of students' performance in geometry. *CERME5* (pp. 1072–1081).

Pytlak, M. (2015). Learning geometry through paper-based experiences. *CERME9* (pp. 571–577).

Rabardel, P. (1995). *Les hommes et les technologies: Une approche cognitive des instruments contemporains.* Paris: Armand Colin.

Richard, P. R. (2004). *Raisonnement et stratégies de preuve dans l'enseignement des mathématiques.* Berne: Peter Lang.

Richard, P. R., & Sierpinska, A. (2004). Étude fonctionnelle-structurelle de deux extraits de manuels anciens de géométrie. *Revue des Sciences de l'Education, 30*(2), 379–409.

Richard, P. R., Oller, A. M., & Meavilla, V. (2016). The concept of proof in the light of mathematical work. *ZDM-Mathematics Education, 48*(6), 843–859.

Rolet, C. (2004). Teaching and learning plane geometry in primary school: Acquisition of a first geometrical thinking. *CERME3*: www.mathematik.uni-dortmund.de/~erme/CERME3/Groups/TG7/TG7_list.php.

Steinwandel, J., & Ludwig, M. (2011). Identifying the structure of regular and semiregular solids: A comparative study between different forms of representation. *CERME7* (pp. 670–680).

Strässer, R. (2002). Cabri-géomètre: Does Dynamic Geometry Software (DGS) change geometry and its teaching and learning? *International Journal of Computers for Mathematical Learning, 6*(3), 319–333.

Vighi, P. (2006). Measurement on the squared paper. *CERME4* (pp. 777–785).

2

NUMBER SENSE IN TEACHING AND LEARNING ARITHMETIC

Sebastian Rezat and Lisser Rye Ejersbo

1 Introduction

Starting in 2011, a separate Thematic Working Group (TWG) has been devoted to research on teaching and learning arithmetic and number systems at CERME. While the topics of the group encompassed grades 1 to 12, the majority of contributions related to the elementary and lower secondary grades. The TWG covers a range of issues related to the teaching and learning of number including number relations such as equality and proportion as well as related argumentation and reasoning. A dominant theme has been the conceptual understanding of natural and rational numbers and related representations mostly combined with the theme of flexibility and adaptiveness in mental calculation. Research in this field often refers to the concept of 'number sense', which seems to be both the starting point and the goal of the development of a conceptual understanding of numbers.

Already at CERME 7 (2011), where the TWG 'Arithmetic and Number Systems' was introduced, there were several contributions referring to the concept 'number sense'. The number of such contributions has increased and at CERMEs 8 (2013) and 9 (2015) more than half of the papers in this TWG referred to number sense. At CERME 10 (2017), a plenary talk by Lieven Verschaffel was devoted to 'Young children's early mathematical competencies', in which he also related to major aspects of number sense.

In fact, the concept number sense has been a theme at CERME conferences from the beginning, but through other TWGs. However, the concept appears to be vague in the papers. A closer look reveals different meanings of the concept 'number sense' and sometimes the very meaning is not even explicated. In many papers the development of number sense is mentioned as a goal without clarifying what it means to develop or possess number sense. In some papers, it is implicitly defined by the research that is supposed to contribute to the development of number sense.

24 Rezat & Ejersbo

In this chapter, we attempt to unveil the different notions of number sense that are used at CERME and complement these by definitions from other perspectives, especially from cognitive- and neuro-psychology. We will address three main questions:

1. What are the different understandings of number sense within CERME and how do they relate to one another?
2. What is suggested in order to foster the development of number sense?
3. What is the scope of number sense in terms of number domains?

Each question will be discussed in a separate section. Our main aim is to clarify different perspectives on abilities related to numbers that are subsumed under the term 'number sense' and to differentiate between these different abilities.

In order to identify papers related to number sense we searched for the term 'number sense' in CERME proceedings 1–9. This search revealed altogether 39 papers that contained the term number sense. We analysed these 39 papers according to the questions 1–3 above, and summarise our findings in this chapter.

2 Different understandings of number sense and related concepts within CERME

A number of papers that refer to number sense do not explicate their understanding of the very term, but use it as an undefined term. In fact, many papers (e.g. Carvalho & da Ponte, 2013; Chrysostomou, Tsingi, Cleanthous & Pitta-Pantazi, 2011; Hedrén, 1999; Meissner, 2011; Morais & Serrazina, 2013) that characterise their understanding of number sense explicitly, or implicitly relate to the definition provided by McIntosh, Reys and Reys (1992, p. 3) who define number sense as

> a person's general understanding of number and operations along with the ability and inclination to use this understanding in flexible ways to make mathematical judgements and to develop useful strategies for handling numbers and operations. It reflects an inclination and an ability to use numbers and quantitative methods as a means of communicating, processing and interpreting information. It results in an expectation that numbers are useful and that mathematics has a certain regularity.

This definition lays the ground for the earlier mentioned mutual relation of number sense, flexibility and strategy choice, which we can observe in the papers of the TWG 'Arithmetic and Number Systems'. Furthermore, number sense is characterised as a person's understanding combined with an ability and inclination to make use of this understanding. In that it shows similarities to the definition of competencies according to Weinert (2001). Although it is not said explicitly in McIntosh et al.'s definition, the close relation to competencies is likely to evoke the idea that number sense can be developed through teaching. Therefore, we call the understanding of number sense according to McIntosh et al. 'didactical understanding of number sense', as opposed to an innate sense.

Stanislav Dehaene is one of the most prominent representatives of the understanding of number sense as an innate sense, which he describes as a rudimentary number sense that is wired into the brain. As opposed to McIntosh et al. (1992), Dehaene (1997) explains his definition of number sense from a biological view. In his research he found an area in the brain, the Intra Parietal Sulcus (IPS) in the parietal lobe, which enables us to sense exactly the size of a number, to compare numbers, and to determine the larger amount between two sets. This area he calls number sense. We are born with this area/module and so are other biological species. Number sense in the understanding of Dehaene (1997) is an evolutionarily derived sense, and it is a part of our natural intuition. The innate conception of number sense is closely related to the Approximate Number System (ANS). The ANS is a part of the cognitive system that supports estimation of the magnitude of a group without relying on language or symbols. It means that even children of five to six years can differentiate between magnitudes up to 100. This was shown in experiments, in which an adult presents different situations to a child. In these situations two to three children give and take different amounts of candy from each other. Instead of giving a precise number, the child should only mention the name of the one who has most. In that situation, the children can both add and subtract big numbers without counting, only using estimation to give an answer (Gilmore, McCarthy & Spelke, 2007; Ejersbo, 2016). Using that research as a basis for didactical research causes an ontology, in which we believe that children are able to differentiate between different magnitudes and use this skill in the development of arithmetic understanding.

In fact, Turvill's (2015) paper is one of the very few examples that provide a detailed description of the two different systems for processing numerosity, offered from cognitive psychology. The two systems are subitising and ANS, which both are taken to be innate and are contained in the cognitive concept of number sense. She refers to Xu and Spelke (2000), who suggest that number sense develops spontaneously and have demonstrated how children as young as six months can discriminate between sets of two and three, which is subitising. Furthermore, she also refers to Halberda and Feigenson (2008), who demonstrate how number sense is developed throughout infancy and can be used in schools, while Feigenson, Libertus and Halberda (2013) show how regular use of number sense as an ANS influences individual mathematics abilities because it is lifelong. Turvill even raises the question: How can we use results from neuroscience?

Some papers at CERME acknowledge that there are different understandings of number sense and mention both the innate and the didactical understanding of number sense. However, in most of the cases this is done in order to contextualise the study within one of the understandings, which is almost exclusively the didactical understanding.

Other papers introduce new concepts, which are closely related to number sense – either in the didactical or innate sense.

26 Rezat & Ejersbo

In their article, Rechtsteiner-Merz and Rathgeb-Schnierer (2015) describe how to develop flexible mental calculation and relate it to problem characteristics, number patterns and numerical relationships. In line with that, they introduce the concept 'Zahlenblick'. Referring to Schütte (2004), they define Zahlenblick as "the competence to recognise problem characteristics, number patterns and numerical relationships immediately, and to use them for solving problems" (Rechtsteiner-Merz & Rathgeb-Schnierer, 2015, p. 355). According to the authors, Zahlenblick refers to both number sense and structure sense. Likewise, Zahlenblick is also a product of development. The authors explain that "Comparing number sense, structure sense and Zahlenblick it is obvious that the meaning of number and structure sense as acquired skills that can be developed by special activities is quite similar to our notion of Zahlenblick." Zahlenblick is to be developed through a programme that focuses on developing flexible mental calculation. The programme can be understood as an essential principle of teaching arithmetic and is, in that way, a suggestion of a theoretical framework. Even though they do not mention McIntosh et al. (1992, p. 3), it is still this understanding they work into. However, the commonalities, differences and relations of the notions of number sense, structure sense and Zahlenblick need some further clarification. It seems as if Zahlenblick incorporates a relation between problem characteristics and the understanding of numbers, which is conceptualised mainly as a recognition of patterns.

Sayers and Andrews (2015) differentiate preverbal, foundational and applied/relational number sense. According to the authors, "preverbal number sense reflects those number insights innate to all humans and comprises an understanding of small quantities that allows for comparison" (p. 361), whereas applied/relational number sense "concerns those number competencies related to arithmetical flexibility that prepare learner for an adult world" (p. 361). While preverbal refers to the innate number sense and applied refers to the didactical understanding, the authors derive a third notion of number sense from a systematic literature review, which they call "foundational number sense". According to the authors, foundational number sense "comprises those understandings that precede applied, typically arise during the first year of school and require instruction" (p. 361). They specify foundational number sense based on eight components: number recognition, systematic counting, awareness of the relationship between number and quantity, quantity discrimination, an understanding of different representations of number, estimation, simple arithmetic competence, and awareness of number patterns (Sayers & Andrews, 2015, p. 362). Sayers' and Andrews' (2015) definition of foundational number sense relates to applied/relational number sense using the notion of 'preceding'. Although the authors do not say explicitly that preverbal also precedes foundational number sense, their definition of the foundational number sense conveys a notion of three different stages of number sense. In this sense, they argue that the acquisition of the latter is unlikely without the achievement of the former. Thus, foundational number sense seems to be introduced as a link between the innate and the didactical number sense. However, beyond the justification provided by a systematic literature review, it is not yet clear how foundational number sense actually provides the link, i.e. how foundational number sense builds on preverbal number sense in order to develop applied number sense.

Aunio, Aubrey, Godfrey, Liu and Yuejuan (2007) take a different approach to number sense. The authors have a very general understanding of number sense. For them "number sense entails operating with quantities and number-word systems" (p. 1577). In their study their reference for measuring number sense is the Utrecht Early Numeracy Test (UENT). They investigate the influence of age, gender and culture on number sense (as measured by UENT). Their findings with children of about five years from three different countries indicate that number sense develops differently in the three countries. "UENT measures two different aspects of the number-sense construct with the relational skills being less influenced by teaching and language difference and, hence culture than counting skills" (p. 1583). They conclude that there are two different structures of numerical skills: general numerical skills and specific numerical skills with the latter being affected by teaching.

Although the authors do not relate their results to the innate or didactical understanding of number sense, it seems reasonable that the general, relational skills are related to the innate number sense, while the specific, counting skills relate to the didactical number sense. Thus, their results challenge the view that the different understandings of number sense can be regarded as stages. The two notions rather seem to be two aspects of number sense that are intertwined and work together. However, the question of how the innate and didactical number sense interrelate and how the development of the didactical number sense can be supported by building on the innate number sense do not seem to have been tackled so far. While CERME papers introduce new notions related to number sense, they do not seem to provide a link between the innate and the didactical understanding of number sense.

3 Fostering number sense

In line with the didactical understanding of number sense some papers make suggestions how the development of number sense might be fostered. Ferreira and Serrazina (2011) summarise some of the main lines of argumentation in this context. Referring to Yang (2003, p. 132) they argue that "students' number sense can be effectively developed through establishing a classroom environment that encourages communication, exploration, discussion, thinking and reasoning" (p. 308).

The majority of papers tackling number sense closely associate number sense and mental calculation. However, the role of number sense related to mental calculation does not seem to be completely clear. On the one hand, it is argued that number sense is exhibited by manipulating numbers mentally and by applying efficient mental calculation strategies. In this line of argumentation, flexible and adaptive mental calculation is regarded as a competence and number sense is regarded as one of the main influential factors on it (e.g. Rezat, 2011). On the other hand, mental calculation is regarded as a way to develop number sense.

In fact, studies sharing this line of argumentation investigate factors in the teaching and learning of mental calculation that have an effect on the development of number sense. For example, Ferreira and Serrazina (2011) refer to studies that contend that the development of number sense is fostered if students are encouraged to formulate their own mental computation strategies.

As mentioned before, a different approach is taken by Rechtsteiner-Merz and Rathgeb-Schnierer (2015). In order to develop *Zahlenblick* they do not only focus on mental calculation, but generally suggest to "provide activities, which highlight problem characteristics, patterns and numerical relationships" (p. 355).

4 The scope of number sense in terms of number domains

Whereas the notion of number sense typically applies to the domain of natural numbers, some papers within TWG 'Arithmetic and Number Systems' also refer to number sense related to other number domains. This is also connected to research investigating students' mental computation strategies in these number domains. These authors' line of argumentation is that mental computation in other number domains will foster students' number sense. Therefore, the authors argue for promoting mental computation also in other number domains. The underlying general assumption is that number sense can be developed by mental calculation, i.e. is not innate. Therefore, the line of argumentation already indicates that the authors refer to a didactical notion of number sense.

The findings of Carvalho and da Ponte (2013) reveal that students develop particular mental calculation strategies in the domain of fractions. In this context, it has to be acknowledged that working in new number domains also poses semiotic challenges on students, because they have to understand a new number representation. For example, if we look at the fraction $\frac{13}{17}$ it seems to consist of two natural numbers, but they are put on top of each other with a line in between. Furthermore, the value of the fraction, which consists of two natural numbers much bigger than 1, is actually less than 1. This is a semiotic challenge, which might interfere with conceptual challenges. Therefore, it is difficult to distinguish strategies that indicate a different number sense in the domain of fractions from strategies that arise due to the particular challenges of the new number domain.

In contrast, Rezat (2011) concludes in his investigation of students' mental calculation strategies with negative numbers that "all the problems were solved using familiar number-transformation strategies from the set of natural numbers. No strategies specifically related to rational numbers were observed" (p. 403). These opposite findings lead to the question of whether there is a general notion of number sense, which relates to all number domains, or if there is particular number sense for every number domain. While Carvalho's findings support the hypothesis that a particular number sense related to fractions is helpful for mental calculation in this number domain, Rezat's (2011) findings might be interpreted in the way that mental calculation in number domains other than the natural numbers, draws on the number sense (of natural numbers) wherever possible.

According to Carvalho & da Ponte (2013, p. 284) "concerning fractions, number sense is related with the ability to use relational thinking [. . .] to compare and work with this rational number representation and to understand the quantities involved". With their definition, they add relational thinking as an important feature of number sense in fractions, and stress the number representation and an understanding of rational quantities. None of these three features occur in the definition by McIntosh et al. (1992).

The seminal definition of number sense provided by McIntosh et al. (1992) does not make a number domain explicit, while the innate number sense only applies to natural numbers. Consequently, the meaning of number sense in other number domains is not clear and needs to be specified in future research.

5 Discussion

The synopsis of research referring to number sense in CERME shows that number sense is a multi-faceted concept. Therefore, it is important to be clear about the actual notion of number sense whenever the term is used.

Broadly speaking, two different notions of number sense can be differentiated: (1) a notion of number sense from the biological/cognitive area and (2) a different notion from the educational/didactical perspective. While the first notion regards number sense as an innate sense, the second notion sees number sense as an ability, which humans develop through learning and instruction. The majority of CERME papers refer to the latter perspective. The understandings and skills described in the didactical view of number sense are developed through teaching.

For us, the main question that arises from the multi-faceted concept of number sense is how to bridge the gap between the knowledge of number sense gained from the biological/cognitive area and the educational/didactical perspective, respectively. Why do we believe that this is important? Research in cognitive psychology and mathematics education has developed important insights into the development of a conceptual understanding of numbers. While one area focuses on innate prerequisites of the development of such a conceptual understanding, the other area is concerned with mathematical activities that foster this development. We believe that both fields can profit from each other. If, on the one hand, mathematical activities do not build on innate predispositions they might be idle. On the other hand, the aim of developing activities and learning arrangements in order to foster conceptual understanding of numbers might provide a direction for cognitive research.

Bridging the gap demands that the two different research communities, who use the concept 'number sense' differently, can communicate with each other and understand how they can supply each other and use each other's research results, instead of being extremely critical or, even worse, ignoring each other. Bridging the gap could be worthwhile because knowledge from educational neuroscience can complement the education with new knowledge that might change our beliefs and habits when planning teaching. Dehaene's research on number sense has inspired teachers to prepare and run several teaching episodes that make use of his concepts. On the one hand, Halberda, Mazzocco and Feigenson (2008), showed in a longitudinal study how ANS can be used for estimating answers for mathematical questions. However, the conceptual components of computational estimation, as Lübke (2015) summarised them, do not relate to these results from cognitive psychology and make use of them. On the other hand, the teaching and learning of arithmetic, even in other number domains, might build upon the innate abilities in order to develop didactical number sense.

The gap between the biological/cognitive research in mathematics and the didactical research in mathematics can be overcome if we see it as an addition instead of comparison and it goes both ways. First attempts have been made to foster the communication between the two communities. ZDM Mathematics Education has published two issues – the first in 2010 and the second in 2016 – focusing on cognitive neuroscience and mathematics learning. These ZDM issues and this overview of CERME papers related to the special theme of number sense show that we are only at the beginning of the two communities taking notice of each other. If we continue this, it is the students who will be the winners.

References

Aunio, P., Aubrey, C., Godfrey, R., Liu, Y., & Yuejuan, P. (2007). Young children's number sense in China, England and Finland. *CERME5* (pp. 1577–1586).

Carvalho, R., & da Ponte, J. P. (2013). Student's mental computation strategies with fractions. *CERME8* (pp. 283–292).

Chrysostomou, M., Tsingi, C., Cleanthous, E., & Pitta-Pantazi, D. (2011). Cognitive styles and their relation to number sense and algebraic reasoning. *CERME7* (pp. 287–296).

Dehaene, S. (1997). *The number sense: How the mind creates mathematics.* London: Penguin Press.

Ejersbo, L. R. (2016). Number sense as a bridge to number understanding. In L. Lindenskov (Ed.), *Special needs in mathematics education* (pp. 189–202). Aarhus: Aarhus University.

Feigenson, L., Libertus, M. E., & Halberda, J. (2013). Links between the intuitive sense of number and formal mathematics ability. *Child Development Perspectives, 7*(2), 74–79. doi: 10.1111/cdep.12019.

Ferreira, E., & Serrazina, L. (2011). Strategies and procedures: What relationship with the development of number sense of students? *CERME7* (pp. 307–315).

Gilmore, C. K., McCarthy, S. E., & Spelke, E. S. (2007). Symbolic arithmetic knowledge without instruction. *Nature, 447,* 589–592. doi: 10.1038/nature05850.

Halberda, J., & Feigenson, L. (2008). Developmental change in the acuity of the "Number Sense": The Approximate Number System in 3-, 4-, 5-, and 6-year-olds and adults. *Developmental Psychology, 44*(5), 1457–1465. doi: 10.1037/a0012682.

Halberda, J., Mazzocco, M., & Feigenson, L. (2008). Individual differences in non-verbal number acuity correlate with maths achievements. *Nature, 455,* 665–669.

Hedrén, R. (1999). The teaching of traditional standard algorithms for the four arithmetic operations versus the use of pupils' own methods. *CERME1* (pp. 233–244).

Lübke, S. (2015). Investigating fourth graders' conceptual understanding of computational estimation using indirect estimation questions. *CERME9* (pp. 302–308).

McIntosh, A., Reys, B. J., & Reys, R. E. (1992). A proposed framework for examining basic number sense. *For the Learning of Mathematics, 12*(3), 2–8.

Meissner, H. (2011). Teaching arithmetic for the needs of the society. *CERME7* (pp. 346–356).

Morais, C., & Serrazina, L. (2013). Mental computation strategies in subtraction problem solving. *CERME8* (pp. 333–343).

Rechtsteiner-Merz, C., & Rathgeb-Schnierer, E. (2015). Flexible mental calculation and "Zahlenblickschulung". *CERME9* (pp. 354–361).

Rezat, S. (2011). Mental calculation strategies for addition and subtraction in the set of rational numbers. *CERME7* (pp. 396–406).

Sayers, J., & Andrews, P. (2015). Foundational number sense: Summarising the development of an analytical framework. *CERME9* (pp. 361–368).

Schütte, S. (2004). Rechenwegnotation und Zahlenblick als Vehikel des Aufbaus flexibler Rechenkompetenzen. *Journal für Mathematik-Didaktik, 25*(2), 130–148.

Turvill, R. (2015). Number sense as sorting machine in primary mathematics education. *CERME9* (pp. 1658–1664).

Weinert, F. E. (2001). Concept of competence: A conceptual clarification. In D. S. Rychen & L. H. Salganik (Eds.), *Defining and selecting key competencies* (pp. 45–65). Seattle, WA: Hogrefe & Huber.

Xu, F., & Spelke, E. S. (2000). Large number discrimination in 6-month-old infants. *Cognition, 74*, B1–B11.

Yang, D. C. (2003). Teaching and learning number sense: An intervention study of fifth grade students in Taiwan. *International Journal of Science and Mathematics Education, 1*(1), 115–134.

3

ALGEBRAIC THINKING

Jeremy Hodgen, Reinhard Oldenburg and Heidi Strømskag

1 Introduction

Algebra is one of the most extensively researched areas in mathematics education. Over the past 40 years, many researchers have addressed the problems associated with the learning and teaching of algebra in school and beyond. Hence, as Radford noted in his plenary lecture at CERME 6, this raises the question of "whether or not there is really something new to say about algebraic thinking" (p. XXXIV). Having reviewed the work on algebraic thinking at CERME, our answer to Radford's question is most emphatically "yes". While this body of work at CERME has extended our understandings of algebraic thinking, it also demonstrates how we have yet to reach a consensus on some of the fundamental questions associated with the teaching and learning of algebra. Like the discipline as a whole, the Algebraic Thinking Working Group has a long history. The group has featured at all CERME conferences except CERME 2. A total of 146 papers have been presented with authors representing 29 countries across the world. In this chapter, we cannot refer to all these papers individually. Rather, we highlight the main themes that have been discussed, highlighting what are, in our view, the key papers that contribute to these themes.

We begin with a clarification of what algebraic thinking is. On this basis, various topics of algebra are described before the particular issues of their teaching and learning are discussed. We conclude with an evaluation and critique of CERME algebraic thinking research as a whole. Finally, we consider potential future avenues of work.

2 The nature of algebraic thinking

2.1 Definitions of algebraic thinking

Drawing on Kaput (2008), we try to provide concise definitions of algebra and algebraic thinking: Whereas algebra is a cultural artefact – a body of knowledge

Algebraic thinking **33**

embedded in educational systems across the world, algebraic thinking is a human activity – an activity from which algebra emerges. Since CERME 3, the title of the group is Algebraic Thinking. This title reflects that the research reported in the group is into students' ways of doing, thinking, and talking about algebra, and further, into teachers' ways of dealing with algebra in terms of instructional design and implementation. According to Kaput (2008), school algebra has two core aspects: algebra as generalisation and expression of generalisations (see Section 2.4) in increasingly systematic, conventional symbol systems; and algebra as syntactically guided action on symbols within conventional symbol systems. He claims, further, that these aspects are embodied in three strands of school algebra: algebra as the study of structures and systems abstracted from computations and relations; algebra as the study of functions, relations, and joint variation; and, algebra as the application of a cluster of modelling languages (both inside and outside of mathematics).

Another model of school algebra is proposed by Kieran (2004), where she describes three interrelated principal activities of algebra: generational activity; transformational activity; and global/meta-level activity. The generational activities involve the creation of algebraic expressions and equations such as (1) equations that represent quantitative problem situations; (2) expressions of generality arising from shape patterns or numerical sequences; and (3) expressions of the rules that determine numerical relationships. The transformational activities involve syntactically guided manipulation of formalisms including: collecting like terms; factoring; expanding brackets; simplifying expressions; exponentiation with polynomials; and solving equations. The global/meta-level activities involve activities for which algebra is used as a tool, and include: problem solving; modelling and predicting; studying structure and change; analysing relationships; and generalising and proving.

In comparison, generational activity in Kieran's model parallels (but is not equivalent to) Kaput's first core aspect; transformational activity parallels Kaput's second core aspect; and, global/meta-level activity contains Kaput's second and third strands of algebra.

2.2 Theoretical papers and research frameworks

Debates about theory have been a constant within the Working Group. However, there are relatively few purely theoretical papers. In this section, we discuss three such papers (Bergsten, 1999; Dörfler, 2007; Godino et al., 2015). A more common approach is to make a theoretical contribution that is rooted in an analysis of empirical data (Schwartz, Herschkowitz & Dreyfus, 2002; Rinvold & Lorange, 2010).

On semiotics, Bergsten (1999) discusses figurative aspects of algebraic symbolism in light of Lakoff and Johnson's theory of image schemata in order to better understand the development of Arcavi's symbol sense. Bergsten's hypothesis is that there is a dynamical interplay between form and content, facilitated by the use of image schemata (e.g. the notion of equation can be seen as formalisation of properties of the balance schema). Further, on semiotics, Dörfler (2007) gives a narrative account of a hypothetical learning process concerned with matrices.

Drawing on Peirce and his own publications, Dörfler makes the hypothesis that the matrix – in the form of a diagram (iconic sign) – is not just a means, but rather the very object of mathematical activity (referred to as diagrammatic reasoning). A third contribution on semiotics is that of Radford (2010). Based on recent conceptions of thinking as it is conceptualised by anthropology, semiotics and neurosciences, he suggests that thinking is a complex form of reflection mediated by the senses, the body, signs and artefacts. Exemplified by the context of pattern generalisation, Radford suggests a classification of three forms of algebraic thinking: factual; contextual; and symbolic. The classroom data he presents provides a glimpse of the ontogenetic journey of students on their route to algebraic thinking. It stresses some of the challenges they have to overcome when passing from factual to contextual to symbolic thinking, in particular the changes that have to be accomplished in modes of signification.

Godino et al. (2015) discuss their developing model of algebraic thinking. They extend their previous three-level model of proto-algebraic reasoning in primary education by including three more advanced levels of algebraic reasoning in secondary education. The six-level model of algebraic reasoning is based on an onto-semiotic approach to mathematical knowledge and instruction, where the advanced levels involve use of parameters to represent families of functions, and the study of algebraic structures themselves. By describing and exemplifying (theoretically) the six algebraisation levels, Godino and colleagues point at the potential impact of the model on teacher education.

Abstraction is naturally a recurrent theme of algebraic thinking. Whereas many researchers take a cognitive stance to abstraction (rooted in Piaget), Schwartz et al. (2002) take a context-dependent stance to abstraction. They analyse an interview with a pair of Grade 7 students carrying out an algebraic proof, using their previously proposed model for the genesis of abstraction. The model is operational in the sense that its components are three observable epistemic actions: recognising; building with; and constructing. Another contribution on abstraction is made by Rinvold and Lorange (2010), where they propose a cognitive allegory theory in analogy with metaphor theory. Their hypothesis is that narrative text problems that can represent or create something else that is more abstract, have the potential to become prototypical text problems (allegories) for mathematical models. Rinvold and Lorange's argument is based on empirical data from teacher education, where a narrative text problem (corresponding to a linear congruence equation) is used by three student teachers as a prototype for a subsequent task that corresponds to the same mathematical topic.

Of course, theory has a crucial role in framing research, and the examination of how theoretical and research frameworks can be employed to investigate algebraic thinking has featured at every CERME conference. It is striking how the number and range of theories drawn on in papers has increased over time and, at CERME 9 in 2015, a total of 25 different theoretical frameworks were used. This presents some challenges in terms of communicating between and across these frameworks (and this relates to Chapter 18 of this book: "Theoretical perpectives and approaches in mathematics education research"). The theoretical frameworks used in research on algebraic thinking can be categorised in three groups that have different scales. (1) Conceptual frameworks are skeletal structures of justification, rather than structures of explanation based on a formal theory. Some of these are: models for conceptualising algebra and algebraic thinking; frameworks of variables and equation solving; frameworks of teaching of linear algebra; frameworks of functions and functional thinking. (2) General theories of teaching and learning are frameworks where algebra is the focal topic "imported" into the framework by the researcher using it. Some of these are: semiotic theory; genetic epistemology; theory of sense and reference; theory of mediating tools; cognitive theory of instrument use. (3) Holistic theories are frameworks that encompass a methodo-logy for instructional design. These include: the theory of didactical situation in mathematics; the Anthropological Theory of the Didactic; and variation theory. This plethora of theories raises two important questions. First, what research problems in the teaching and learning of algebra are related to which theoretical frameworks? It seems to us that many theoretical approaches could benefit from better articulation in terms of its description, explanation, prediction and scope – that is, what is it a theory of and for? Second, to what extent are these different theoretical approaches complementary or contradictory? Clearly, there is potential here to draw on the research outlined in Chapter 18 of this book.

2.3 Insights from the historical studies of algebra and mathematics

From an epistemological point of view, algebra is a complex subject. It is domi-nated by abstract concepts that relate to more concrete entities in a subtle way that has evolved in the history of mathematics over many centuries. Thus, it is natural that several contributions clarified foundational issues by incorporating insights from history and philosophy, as noted by Drouhard, Panizza, Puig, and Radford (2006).

The genetic development of algebra as seen in the history of algebra gives valu-able insights that are interesting not just for their own sake, but also because they can influence the development of modern teaching approaches. This has been exemplified by Chiappini (2011) who used Peacock's (1940) distinction between symbolic and arithmetical algebra to guide activities in a modern microworld and explain students' learning needs in coping with negative numbers. In Peacock's (as referred by Chiappini) arithmetical algebra, the meaning of expressions involv-ing letters is completely determined by the laws of arithmetic. An expression like $a - b$ over the domain of natural numbers is thus only sensible if $a > b$, while in symbolic algebra it has sense by transferring the characteristics of the operation to new objects. The ALNUSET microworld (essentially an interactive number line) allows students to discover the meaning of expressions like $a - b$ by extending the domain while keeping operational characteristics – and thus passing from arith-metical to symbolic algebra.

36 Hodgen, Oldenburg & Strømskag

Bagni (2006) has studied the development of equations and in-equations and inequality. He concludes that in order to avoid breaks between sense and denotation of algebraic expressions, an integrated introduction of equations and inequalities from a functional point of view is adequate and should focus on the concept of boundary points as they are solutions of the corresponding equation.

Both papers (as well as others not mentioned) support the didactical version of Haeckel's law, namely that the historical and the individual genesis of meaning have parallels.

2.4 Generalisation

Generalisation is a topic that is deeply integrated into the nature of algebra as has been shown in Section 2.1. Virtually all contributions touch on it to some extent. However, some papers have dealt with generalisation explicitly. Chua and Hoyles (2011) investigate the generalisations used by different groups of Singaporean students working with number patterns. They found no differences for linear relationships, where both normal programme (average attaining) and express programme (higher attaining) students tended to use a numerical approach. However, for patterns based on quadratic relationships, a greater proportion of express programme students favoured a constructive diagrammatic strategy, while the normal programme students tended to use a numerical method. Similarly, Cañadas, Castro and Castro (2011) examined how the presentation of generalisation tasks affects the approach used by Spanish Grade 9 and 10 students. They discuss three tasks presented in different ways, diagrammatically, verbally and numerically. They found that students had a very strong tendency to use a numerical approach and that students were more likely to use a formal algebraic approach where a problem was presented numerically. Bolea, Bosch and Gascón (2004) also discuss the dominance of numerical approaches, which they suggest is strongly related to teachers' understandings of school algebra. They argue for the introduction of an algebraic modelling approach to counter this.

2.5 Early algebra

The issue of early algebra and the relationship (or transition) between arithmetic and algebra has been a recurrent – and hotly debated – theme, which touches on the nature of algebraic thinking. Some systems, e.g. in Portugal, have introduced an explicit strand of early algebra within their curriculum and a number of papers have examined the implementation of this in primary classrooms. Mestre and Oliveira (2013) describe and analyse a teaching experiment in which Grade 4 students are introduced to the use of informal symbols as quasi-variables. They find that such an approach has benefits for the development of algebraic thinking, particularly in moving from equations involving specific unknowns to equations expressing generalisations about arithmetic. In contrast, however, Gerhard (2013) argues that a generalised number approach could hinder the development of algebraic thinking. She highlights in particular an over-reliance on repeated addition

Algebraic thinking **37**

as a model of multiplication and suggests that Davydov's more geometrically based approach has the potential to overcome this gap.

Others have examined the role of discourse in early algebraic thinking. Caspi and Sfard (2011) investigate the discourse of Israeli Grade 7 students as they move from informal meta-arithmetic toward formal algebra. By examining a historical example, they show how students' discourse, while informal and ambiguous, contains some algebra-like features, not normally found in everyday discourse. Dooley (2011) examines a group of primary students in Ireland aged nine to 11 years. She uses the epistemic actions of recognising, building with and constructing to analyse and describe the development of algebraic reasoning among the students. She argues that in some case the use of "vague" language facilitated this development.

Pittalis, Pitta-Pantazi and Christou (2015) take a different approach by examining the development of number sense among Grade 1 students. Based on their analysis of test data over the course of 12 months, they suggest that algebraic arithmetic has a positive effect on the development of the other number sense components, particularly conventional arithmetic.

The promise of early algebra can only be fulfilled if it improves algebraic competence in the long run. Isler et al. (2017) show that students who had taken part in an early algebra programme in Grade 3 show significantly better abilities in representing functional relationships when they are in Grade 6. This applies for the ability to express the relationship verbally as well as symbolically. Interestingly, both age groups were more successful with symbolic representations than with verbal representations.

3 Topics within algebra

Within the wide range of algebraic topics or more general topics that can be handled by algebraic methods, research has focused mainly on some central issues that are taught in many curricula, such as functions and linear equations.

An early example is the work of Bazzini (1999) where the difficulty in symbolising relations between variables is addressed by using questionnaires and interviews. It turned out that the translation from a situation into the symbolic language is a key issue, i.e. often the change from one semiotic register to another (natural language and symbolic language) does not occur appropriately.

Quadratic equations have been investigated by Didiş, Baş and Erbaş (2011). They found that students tend to solve quadratic equations as quickly as possible without paying much attention to their structures, and thus ignore special cases. Similarly, in a study probing the features of quadratic equations perceived by Grade 9 and 10 students in Germany, Block (2015) used the novice-expert paradigm to identify flexible algebraic action. He found that students tend to focus on just one feature and thus their ability to act flexibly is limited. He, therefore, proposes that students should spend time categorising quadratic equations in different ways.

Basic models of logarithmic functions are discussed in Weber (2017). He identifies four such models, namely multiplicative measure (how often you can divide by the base), digit counting, decrease of hierarchy level and inverse to exponentiation. While most textbooks give preference to the last model, Weber suggests that the first gives a better operational start.

For general polynomial functions, Douady (1999) worked with qualitative reasoning (topological arguments, e.g. does the sign change around a zero?) and found that this fosters students' understanding. She argues that the study of zeros and their multiplicity should be augmented by (at least implicit) arguments about continuity.

In recent years, several papers have addressed issues relating to equivalence and students' understanding of the equal sign, dealing with the well-known distinction between operational and relational meanings of the equal sign. Alexandrou-Leonidou and Philippou (2011) report on a teaching experiment with Cypriot primary students that enabled students to develop this dual meaning of the equal sign and showed that this in turn had a significant effect on students' ability to solve equations. Zwetzschler and Prediger (2013) argue that, while previous research has examined the learning of equivalence in *transformational* situations, little attention has been devoted to the equivalence of expressions in *generational* activities, where algebraic expressions are understood as "pattern generalizers of arithmetical or geometrical pattern" (p. 559). They highlight student understanding of the connection between geometric shapes and algebraic expressions as a conceptual barrier. Jones (2010) reports on a computer-based task designed to enable primary students to develop a relational understanding of the equal sign. This work suggests that students experience difficulties in coordinating two aspects of the relational meaning: sameness and exchanging.

4 Teaching and learning algebraic thinking

4.1 Students' difficulties, misconceptions and partial understandings

As is evident from the preceding discussion, the issue of student difficulties has been a central theme. Several authors have analysed difficulties in new areas such as Postelnicu's (2013) work on linear functions. Bazzini (1999) highlights the persistence of such errors even when students receive what is considered good teaching.

Building on critiques of the cognitivist literature on misconceptions, discussions at early CERME conferences have shifted attention away from categorising errors and misconceptions toward analysing students' algebraic activity from non-cognitive perspectives. Drawing particularly on socio-cultural/historical, anthropological and semiotic theories (e.g. Radford (2010), see discussion above), discussion has focused on how context can enable students to understand algebraic symbolism (Drouhard et al., 2006, p. 638).

In recent conferences, there has been a resurgence of more cognitive perspectives, but with a focus on how these difficulties can be overcome. Several papers, for example, have replicated aspects of Küchemann's (1981) work relating to generalised number and the meaning of letters. Broadly these papers indicate that these

Algebraic thinking **39**

findings still hold, subject to some minor variation due to curricular or cultural factors (e.g. Hadjidemetriou, Pampaka, Petridou, Williams & Wo, 2007). Hodgen, Küchemann, Brown and Coe (2010) suggest that the earlier teaching of algebra in England does not appear to have produced better understanding.

Others have examined the interrelationship of syntactical and semantic understanding in order to better understand how to enable the learning of algebraic symbolisation. Malara and Iaderosa (1999) argue that there is a conflict between additive and multiplicative notation that creates difficulties as students move from arithmetic to algebra, and suggest that this difficulty might be overcome by promoting semantic and metacognitive activity. However, Oldenburg, Hodgen and Küchemann (2013) show that the distinction between syntactic and semantic aspects of algebra is not straightforward to make empirically.

4.2 Teaching experiments and design research studies

Design-based research is where a *tool* (a product or process) is designed, developed and refined through cycles of enactment, observation, analysis and redesign, with systematic feedback from those involved. The goal is transformative; new teaching and learning opportunities are created and studied in terms of their impact on teachers, students and other actors. Examples of *principled* design-based research are curriculum development and didactical engineering (e.g. Brousseau, 1997). The limited length of CERME papers is a challenge for design studies, so small-scale teaching experiments are more frequently reported.

The focus of Douady (1999) is the elements of a didactical engineering related to teaching of polynomial functions in secondary school, where the didactical hypothesis is that students need to conceive polynomial functions both from an algebraic and a topological perspective. This involves the premise that the study of zeros and their multiplicity must be performed in relation to properties of continuity and differentiability, at least implicitly. An important principle of the design, which is rooted in the Theory of Didactical Situations (TDS), is illustrated by the way the class is organised and how the knowledge unfolds: Conjectures proposed by the teacher are examined by the students, where the decision about the validity of the conjectures is instrumental in students' development of the target knowledge. Douady describes how the engineering has been inserted in a work decided at national level. Strømskag (2015) uses the TDS methodological principle – that the knowledge at stake is integrated in a *situation* as the optimal solution to a problem – in a teaching experiment within a teacher education programme. The experiment is concerned with a general numerical statement about odd numbers and square numbers, and she describes how the design of the milieu is related to the nature of the knowledge aimed at (an equivalence statement).

Principles of task design are the topic of Ainley, Bills and Wilson's (2004) paper. They propose *purpose* and *utility* as design principles of a sequence of tasks for the teaching and learning of algebra in the first years of secondary school, based on the use of spreadsheets. Kieran's (2004) triadic model of activities of school

40 Hodgen, Oldenburg & Strømskag

algebra is used as a framework for the task design, where Ainley and colleagues have attempted to achieve a balance between generational-, transformational-, and global/meta-level activities.

Drawing on a design science approach, Gerhard (2011) uses task-centred interviews with secondary students to exemplify the use of an analytic tool – an interdependence model that examines how new algebraic knowledge interacts with prior arithmetic knowledge. In the interview tasks, arithmetic rather than algebraic word problems were chosen, where a modification was done in terms of substituting numbers by letters. Gerhard argues that it is important to distinguish between the transition from arithmetical to algebraic thinking, and the transition from numbers to variables. Framed within Bartolini Bussi and Mariotti's (2008) theory of semiotic mediation, Maffei and Mariotti (2011) use Aplusix CAS to examine the interplay between different representations of algebraic expressions. They emphasise how the semiotic potential of the different artefacts does not emerge spontaneously; hence, a specific didactic organisation in terms of task design and teacher's actions is necessary.

Design-based research is theory driven. It is important to describe the principles of the underlying theory and explain how these are related to the design itself. For this to happen, it is necessary to be explicit about the principles of the design. In the studies referred to above, this is done to varying degrees. The relationship between theory and design principles (including tasks) is important to include at future CERME conferences.

4.3 Technology

The last two decades have seen a significant increase in the availability and sophistication of digital technology, and discussions of the potential impact of these new tools on the teaching of algebra have been going on over the whole period. It is remarkable that even early CERME papers show a great awareness of the challenge to turn computers into useful instruments that support the learning process.

Spreadsheets have been a class of tools that have been investigated by many researchers, mainly to support functional approaches to algebra and to bridge the gap between arithmetic and algebra (e.g. Ainley et al., 2004). In contrast, Hewitt (2011) reports on the use of a bespoke package, Grid Algebra, based on an underlying multiplication grid, which has been designed to help students create and discuss expressions with numbers and symbols. He shows how this can be used to enable primary students to engage with relatively complex manipulation and thus begin to "express generality".

Computer algebra systems (CAS) have been a big issue for some time, although interest in this technology seems to be decreasing. Nevertheless, there have been a lot of deep investigations in this area and especially important research has been conducted on developing these tools further into purposeful artefacts. Maffei and Mariotti (2011) have investigated how the structure of algebraic expressions can be made explicit by use of the Aplusix CAS (see discussion above). They find that the semiotic potential of the artefact does not emerge spontaneously, but that it needs a specific didactic organisation and that natural language plays a central role in this. The three representation systems that are available to denote mathematical expressions (natural language, standard representation and tree representation) serve different purposes. The role of natural language – beyond communication with the teacher – is to focus on specific features of both the standard representation and the tree representation. Thus, it is the language of a meta-discourse.

Lagrange (2013) illustrates how theories about the organisation of cognitive processes have influenced the design of the Casyopee system and how this helps students to bridge the gap between a real-world situation and its symbolic description by use of dynamic geometry within a computer algebra environment. Similar to Maffei and Mariotti (2011), this work points at the importance of natural language in structuring situations.

Viewing the whole landscape of the use of technology for the learning and teaching of algebra, it is apparent that the orchestration is of crucial importance, but many questions remain unsettled, e.g. the question of how specific or general good teaching software should be. The use of computers parallels, in a sense, the use of symbolic expressions in algebra, as in both cases it is difficult to enable students to use their full power.

4.4 Teachers and teacher education

Teachers and teacher education pervade many of the papers. Yet, surprisingly few tackle the issue directly. This might be due to the very active strands of work in these areas at CERME, which are represented by Chapters 12 and 13 in this book. Nevertheless, several papers do address these issues directly. Ayalon and Even (2010) and Kilhamn (2013) each highlight how teaching materials are enacted differently by different teachers. Novotna and Sarrazy (2006) show how the degree of variation in the problems set by teachers is related to successful student problem solving. Mason (2007) discusses the design of a course for teachers, "Developing Thinking in Algebra", in which pedagogy, didactics and mathematics are "interwoven" – an approach strongly informed by his own research, the "Discipline of Noticing".

5 Evaluation and critique of research on algebraic thinking

The corpus of work on algebraic thinking at CERME is both extensive and impressive. While we intend to celebrate this, our evaluation and critique will identify a number of problematic issues. It is perhaps unsurprising that the Working Group has been dominated by small-scale studies that are difficult to generalise. One of the strengths of the algebraic thinking group has been its focus on work in progress. This has enabled discussions that have influenced ongoing studies theoretically, methodologically and analytically. However, while the results of many of these studies have been reported elsewhere, this inevitably means that collection of papers presents a somewhat partial picture of algebraic thinking work over this period.

Most papers present single-country studies and there have been few bi- or cross-national studies in CERME or beyond. The CERME work as a whole highlights many interesting similarities and differences between different systems. But without specific studies that address this, generalising research findings from one country to another is inherently difficult.

Overall, and again perhaps surprisingly, there is more empirical than theoretical or conceptual research, and the balance appears to be stable over time. Within empirical research, qualitative methods dominate slightly and most quantitative studies are either intervention or cross-section studies. There is a lack of research on the long-term development of algebraic thinking, particularly longitudinal research within cohorts of students. In our view, the quality of quantitative studies has improved over the years. While inferential statistics dominates, there are a few papers that have used more recent modelling methods, such as Rasch analysis or structural equation modelling. (See, for example, Izsák, Remillard and Templin (2016), for a discussion of how such statistical modelling approaches can address critical questions in mathematics education.) We suggest that research on algebraic thinking might draw on methodological (and theoretical) approaches from other fields such as cognitive science. For example, to investigate embodied cognition, Henz, Oldenburg and Schöllhorn (2015) examined the electroencephalographic brain activity of university students while they performed algebraic, geometric, and numerical reasoning tasks. Initial pilot results suggest that bodily movement has a positive effect on the cognitive processing of demanding mathematical tasks.

It is a considerable strength that the papers are focused on algebraic thinking. This allows a depth of discussion and consideration that is valuable. But it is also a potential weakness and there are times when thinking about algebraic thinking could benefit from insights from other strands within CERME, in particular the work on teachers, technology, theory and number. We also consider that there are opportunities to learn from didactical strategies in other areas. Proulx (2013) raises this issue in relation to mental mathematics and argues that too little attention has been devoted to understanding mental mathematics activities with objects other than numbers.

6 Looking forward

The opportunity to review the corpus of CERME work on algebraic thinking is also an opportunity to look forward and it is clear that there are still important issues to be resolved in algebraic thinking.

The papers presented at CERME attest to the fact that students around the world still encounter what Radford (2010) refers to as *legendary difficulties* with algebra and teachers still struggle to overcome these difficulties. This raises important, and we believe urgent, questions for research on algebraic thinking.

An area that has received little attention in research on algebraic thinking is mathematical tasks – we see the need for a more systematic analysis of mathematical tasks used as instruments in classrooms, and in research on teaching and learning of algebra.

Algebraic thinking **43**

We have been somewhat surprised that no substantive literature reviews of algebraic thinking have been presented at CERME. Of course, there are literature reviews elsewhere, but there is a real need to reach a consensus on what is already known and what research questions remain open. We think that literature reviews have considerable potential to help address the question of what algebra and algebraic thinking are, and to help resolve some of the theoretical difficulties that we referred to above. To reiterate this point, research on algebraic thinking draws on a wide and diverse range of theories and there is a real need to examine how these theories interact and align (or not). Without doing this, it is difficult to see how we can build a strong and coherent programme of research.

As we have already noted, CERME is dominated by small-scale studies. For example, many interesting and potentially promising approaches to teaching algebraic thinking have been presented. Indeed, some of these produce very impressive results. But these are almost all conducted in a very small number of settings and often taught by the researchers themselves. Hence, important questions relate to the scaling up of research and to the communication of research beyond the research community – to teachers, policy-makers and others. A similar argument can be made in favour of more longitudinal studies. A feature of algebra is its complexity and this might render teaching interventions that are successful in the short run but unfortunate in the long run because they might lead to misconceptions (e.g. variables as real-world objects). However, according to our experience, the structured discussion process of CERME is optimal for spotting critical issues and inspiring further research that will, hopefully, improve algebra education so that more students can use it as a valuable tool in their everyday and academic lives.

References

Ainley, J., Bills, L., & Wilson, K. (2004). Designing tasks for purposeful algebra. *CERME3*: www.mathematik.uni-dortmund.de/~erme/CERME3/Groups/TG6/TG6_ainley_cerme3.pdf.

Alexandrou-Leonidou, V., & Philippou, G. N. (2011). Can they 'see' the equality? *CERME7* (pp. 410–419).

Ayalon, M., & Even, R. (2010). Offering proof ideas in an algebra lesson in different classes and by different teachers. *CERME6* (pp. 430–439).

Bagni, G. (2006). Inequalities and equations: History and didactics. *CERME4* (pp. 652–662).

Bartolini Bussi, M. G., & Mariotti, M.-A. (2008). Semiotic mediation in the mathematics classroom: Artefacts and signs after a Vygotskian perspective. In L. English, M. G. Bartolini Bussi, G. Jones, R. Lesh, & D. Tirosh (Eds.), *Handbook of international research in mathematics education* (2nd ed., pp. 720–749). Mahwah, NJ: Lawrence Erlbaum Associates.

Bazzini, L. (1999). On the construction and interpretation of symbolic expressions. *CERME1* (pp. 112–122).

Bergsten, C. (1999). From sense to symbol sense. *CERME1* (pp. 123–134).

Block, J. (2015). Flexible algebraic action on quadratic equations. *CERME9* (pp. 391–397).

Bolea, P., Bosch, M., & Gascón, J. (2004). Why is modelling not included in the teaching of algebra at secondary school? *CERME3*: www.mathematik.uni-dortmund.de/~erme/CERME3/Groups/TG6/TG6_bolea_cerme3.pdf.

Brousseau, G. (1997). *The theory of didactical situations in mathematics: Didactique des mathématiques, 1970–1990* (N. Balacheff, M. Cooper, R. Sutherland, & V. Warfield, Eds. & Trans.). Dordrecht: Kluwer.

Cañadas, M. C., Castro, E., & Castro, E. (2011). Graphical representation and generalization in sequences problems. *CERME7* (pp. 460–469).

Caspi, S., & Sfard, A. (2011). The entrance to algebraic discourse: Informal meta-arithmetic as the first step toward formal school algebra. *CERME7* (pp. 470–478).

Chiappini, G. (2011). The role of technology in developing principles of symbolical algebra. *CERME7* (pp. 429–439).

Chua, B. L., & Hoyles, C. (2011). Secondary school students' perception of best help generalising strategies. *CERME7* (pp. 440–449).

Didiş, M. G., Baş, A. S., & Erbaş, K. (2011). Students' reasoning in quadratic equations with one unknown. *CERME7* (pp. 479–489).

Dooley, T. (2011). Using epistemic actions to trace the development of algebraic reasoning in a primary classroom. *CERME7* (pp. 450–459).

Dörfler, W. (2007). Matrices as Peircean diagrams: A hypothetical learning trajectory. *CERME5* (pp. 852–861).

Douady, R. (1999). Relation function/al algebra: An example in high school (age 15–16). *CERME1* (pp. 113–124).

Drouhard, J.-P., Panizza, M., Puig, L., & Radford, L. (2006). Working Group 6: Algebraic thinking. *CERME4* (pp. 631–642).

Gerhard, S. (2011). Investigating the influence of student's previous knowledge on their concept of variables: An analysis tool considering teaching reality. *CERME7* (pp. 490–499).

Gerhard, S. (2013). How arithmetic education influences the learning of symbolic algebra. *CERME8* (pp. 430–439).

Godino, J., Neto, T., Wilhelmi, M., Aké, L., Etchegaray, S., & Lasa, A. (2015). Algebraic reasoning levels in primary and secondary education. *CERME9* (pp. 426–432).

Hadjidemetriou, C., Pampaka, M., Petridou, A., Williams, J., & Wo, L. (2007). Developmental assessment of algebraic performance. *CERME5* (pp. 893–902).

Henz, D., Oldenburg, R., & Schöllhorn, W. (2015). Does bodily movement enhance mathematical problem solving? Behavioural and neurophysiological evidence. *CERME9* (pp. 412–418).

Hewitt, D. (2011). What is algebraic activity? Consideration of 9–10 year olds learning to solve linear equations. *CERME7* (pp. 500–510).

Hodgen, J., Küchemann, D. E., Brown, M., & Coe, R. (2010). Children's understandings of algebra 30 years on: What has changed? *CERME6* (pp. 539–548).

Isler, I., Strachota, S., Stephens, A., Fonger, N., Blanton, M., & Gardiner, A. (2017). Grade 6 students' abilities to represent functional relationships. *CERME10* (pp. 432–439).

Izsák, A., Remillard, J. T., & Templin, J. L. (Eds.). (2016). *Psychometric methods in mathematics education: Opportunities, challenges, and interdisciplinary collaborations*. Reston, VA: National Council of Teachers of Mathematics.

Jones, I. (2010). Presenting equality statements as diagrams. *CERME6* (pp. 549–558).

Kaput, J. (2008). What is algebra? What is algebraic reasoning? In J. J. Kaput, D. W. Carraher, & M. L. Blanton (Eds.), *Algebra in the early grades* (pp. 5–17). New York: Lawrence Erlbaum.

Kieran, C. (2004). The core of algebra: Reflections on its main activities. In K. Stacey, H. Chick, & M. Kendal (Eds.), *The future of the teaching and learning of algebra: The 12th ICMI Study* (pp. 21–33). Dordrecht: Kluwer.

Kilhamn, C. (2013). Hidden differences in teachers' approach to algebra: A comparative case study of two lessons. *CERME8* (pp. 440–449).

Küchemann, D. (1981). Algebra. In K. M. Hart (Ed.), *Children's understanding of mathematics: 11–16* (pp. 102–119). London, UK: John Murray.

Lagrange, J. B. (2013). Covariation, embodied cognition, symbolism and software design in teaching/learning about functions: The case of CASYOPÉE. *CERME8* (pp. 460–469).

Maffei, L., & Mariotti, M. A. (2011). The role of discursive artefacts in making the structure of an algebraic expression emerge. *CERME7* (pp. 511–520).

Malara, N. A., & Iaderosa, R. (1999). The interweaving of arithmetic and algebra: Some questions about syntactic and structural aspects and their teaching and learning. *CERME1* (pp. 159–171).

Mason, J. (2007). Research and practices in algebra: Interwoven influences. *CERME5* (pp. 913–923).

Mestre, L., & Oliveira, H. (2013). Generalising through quasi-variable thinking: A study with grade 4 students. *CERME8* (pp. 490–499).

Novotna, J., & Sarrazy, B. (2006). Model of a professor's didactical action in mathematics education: Professor's variability and students' algorithmic flexibility in solving arithmetical problems. *CERME4* (pp. 696–705).

Oldenburg, R., Hodgen, J., & Küchemann, D. (2013). Syntactic and semantic items in algebra tests: A conceptual and empirical view. *CERME8* (pp. 500–509).

Peacock, G. (1940). *Treatise on algebra*. New York: Dover Publications. (Original work published 1830).

Pittalis, M., Pitta-Pantazi, D., & Christou, C. (2015). The development of student's early number sense. *CERME9* (pp. 446–452).

Postelnicu, V. (2013). Students' difficulties with the Cartesian connection. *CERME8* (pp. 520–529).

Proulx, J. (2013). Mental mathematics and algebraic equation solving. *CERME8* (pp. 530–539).

Radford, L. (2010). Signs, gestures, meanings: Algebraic thinking from a cultural semiotic perspective. *CERME6* (pp. XXXIII–LIII).

Rinvold, R. A., & Lorange, A. (2010). Allegories in the teaching and learning of mathematics. *CERME6* (pp. 609–618).

Schwartz, B., Herschkowitz, R., & Dreyfus, T. (2002). Emerging knowledge structures in and with algebra. *CERME2* (pp. 81–91).

Strømskag, H. (2015). A pattern-based approach to elementary algebra. *CERME9* (pp. 474–480).

Weber, C. (2017). Graphing logarithmic functions: Multiple interpretations of logarithms as a basis for understanding. *CERME10* (pp. 537–544).

Zwetzschler, L., & Prediger, S. (2013). Conceptual challenges for understanding the equivalence of expressions: A case study. *CERME8* (pp. 558–567).

4

RESEARCH ON PROBABILITY AND STATISTICS EDUCATION

Trends and directions

Arthur Bakker, Corinne Hahn, Sibel Kazak and Dave Pratt

Probability and statistics have a special place in the field of mathematics. They are often considered as sub-disciplines, or in the case of statistics, even as a discipline in its own right. Indeed, statistics is a science related to the political, social and economic history of a country, which explains why there are strong cultural differences in teaching practices. In the 17th century, which is considered to be the dawn of modern statistics, two schools were opposed: the descriptive German naturalistic school and the English political arithmetic which developed treatment and extrapolation techniques based on the growing theory of probability. While the latter was adopted very quickly by most countries, the strength of the descriptive tradition depended on the country. Today, researchers in mathematics education agree that it is essential to combine a data-centric perspective with a modelling perspective. Nevertheless, this combination has different emphases from one country to the other.

This history is reflected in the working group at CERME that addressed these topics. It was initially called 'Stochastic Thinking', to emphasise the interdependency between probability and statistics. However, it turned out that the term stochastics was ambiguous. Where its German equivalent (*Stochastik*) captured this interdependency, in most other languages, stochastic has a very particular meaning, as in 'stochastic function' – a function with a random variable. In practice, most papers in the working group focused on either probability or statistics education. For these two reasons, it was renamed 'Probability and Statistics Education'.

This chapter aims to provide an overview of directions and trends within the Probability and Statistics Education Working Group of CERME, but also to compare them with international trends in these areas. We start with a brief history of the working group, and discuss the main themes that recurred throughout the years. These include the relation between probability and statistics, technology, teacher knowledge and the need for interventionist research that goes beyond the

Probability and statistics education **47**

description of problems but offers suggestions of how to improve probability and statistics education. We end with a wish list for the future.

1 Brief history of the working group

The Stochastics Working Group was founded at CERME 3 in Bellaria, Italy. The group has met at each CERME conference since that inaugural year when 17 papers were presented across four themes: probabilistic thinking, statistical thinking, teacher education and computer-based tools. These turned out to be recurrent themes throughout the history of the group.

The theme on probabilistic thinking has resulted in new theoretical perspectives and evidence that recognised the context-sensitive nature of students' probabilistic thinking. In fact, it has been claimed that, even though by nature probability is more mathematical than statistics, the concept of probability is inherently complex and different from other mathematical concepts. There has been concern that the increasing popularity of exploratory data analysis (EDA) has led to the isolation of probability in the curriculum. Below there is discussion on the role that modelling with digital technology might have in reconnecting data and chance. However, there was also a suggestion that an increased emphasis on subjective probability might counter the all-pervasive reference to coins, spinners and dice, which are not so common in children's culture anymore. There has also been research on the understanding of risk, which, although ambiguously defined, does carry some connection with probability and might be an interesting domain for the exploration of subjective probability. Further discussion of this theme can be found below.

Concerns have been expressed about negative attitudes in society toward statistics and statistical thinking. These attitudes are similar to those toward mathematics though, even worse, mathematically minded scholars sometimes reject statistical ways of thinking. There has been research in this theme on the role of language as a mediating tool in learning statistics. This research suggests that students might have good intuitions but often not the statistical language to express these. Research has been reported on how people interpret statistical information from authentic contexts such as newspapers. This research raises the question of what can be considered as statistical as well as the question of the authenticity of the activities. Research has tried to tease out the important role that the construction of a task and the subsequent social interaction has on the quality of observed statistically related discussion. What are students' situated understandings of basic concepts such as average, spread, distribution, determinism, causality, randomness, stochastic and physical dependence vs. independence? This has led to a major theme around the role of context in statistical thinking.

In fact, the role of context is very different in statistics than in mathematics. Mathematics as a discipline typically aims to be de-contextualised whereas statistics typically draws on context. In the mathematics education literature, contexts in word problems are reported to present children with additional difficulties, whereas in statistics the contextual interpretation is important. Nevertheless, if

48 Bakker et al.

tasks lack authenticity by providing students with an artificial context, students are likely to bring in personal knowledge that is not necessarily statistical. Rather than thinking of abstracting, at least in statistics and probability, as a process of decontextualisation, a focus on enriching, disciplining and refining seems to place emphasis on abstracting as a process of generalising. Research on statistical thinking is further elaborated below.

In the theme focused on teachers, CERME authors often report on the impoverished nature of training for teachers of statistics who were not especially knowledgeable in that area. Questions have been raised about how the community might support teacher development in using innovative pedagogies and to become more connectionist in their approach. The key influence of the methodology of teachers on the learning of probability and statistics has been noted. There has been concern that, while research had been finding evidence about what teachers did not know, to design effective teacher education, research would need to identify what teachers know already, including their attitudes, and what they need to know to be effective teachers. One striking observation was that there are many theoretical frameworks for teacher knowledge and so there is a need to clarify the different emphases in the different frameworks and over time reach some convergence in terminology. Details of the teacher education theme are discussed below.

CERME research papers have reported on the importance of computer-based tools in the teaching and learning of probability and statistics, for example in potential for students to appreciate probability distribution as an emergent phenomenon and key concepts such as the mean. Design-related questions have been raised about how research might identify significant affordances of computer-based tools to realise such potential, including the role of microworlds. It has been proposed that digital tools can offer a pathway toward the effective use of modelling to re-connect probability and statistics. Most recently, there was a recognition that the design of learning environments needs to consider the use of computer-based tools alongside the design of the task itself and the nature of classroom interactions.

There is a tendency for some topics to be revisited by each generation (e.g. misconceptions research; problem analysis), possibly because of the challenge of building on research published in diverse disciplines. Another tendency is that CERME papers follow international trends, such as emerging interest in inference or sampling. The Forum for Statistical Reasoning, Thinking and Literacy (SRTL) is one of the international influences in this respect, just like the conferences of the International Association for Statistical Education (IASE) such as the International Conference on Teaching Statistics (ICOTS).

2 The nature of probability and statistics

Probability and statistics have very different historical origins (Stigler, 1986). We see this reflected in how the topics are taught: Probability is highly mathematical (based on combinatorics) while statistics is multidisciplinary. Statistics education is a marginal discipline in the sense that it is at the boundary of many other disciplines including mathematics education, statistics and psychology (Groth, 2015). Statistics education research is carried out by different groups of researchers, who publish in a diversity of journals. The fragmented nature of the field also has an upshot. Statisticians can play in everyone else's backyards. Statistics educators typically draw on many different fields. For a long time, statistics education has lagged behind in terms of theoretical and methodological rigour, but is catching up with, say, mathematics education (Nilsson, Schindler & Bakker, 2018).

The papers presented in the first meeting of the Stochastic Thinking Working Group at CERME 3 tended to focus on statistics and probability as separate topics. However, some studies (Pratt, 2004) brought out the importance of providing students with an experimental situation and computer tools that can help them experience the dual notion of probability (epistemic and statistical) (Hacking, 1975). In the following meetings, several researchers focused on an experimental approach to probability with technology. For instance, Abrahamson and Wilensky (2006) reported on a study with 26 8th-grade students conducting probability experiments using NetLogo models that randomly generated blocks of 3×3 arrays of red and green squares and accumulated the outcomes in columns according to the number of red squares in each block. Researchers designed the task in a way that promoted students' understanding of the connection between the distributions of empirical outcomes from small samples they collected using NetLogo and the distribution of the combined empirical outcomes from all small samples in a collaborative learning environment. Students' probabilistic reasoning was supported by their analysis of distribution of empirical outcomes in this experimental approach to probability. In another study, Prodromou (2007) investigated 15–16-year-old students' coordination of data-centric distribution and modelling distribution as they worked in a microworld environment about throwing basketballs where they could use causality to articulate features of distribution. The paper focused on the work of two pairs of students. Although these students seemed to intuitively understand that the data-centric distribution would converge to the modelling distribution, which was the intended outcome, they had difficulty in viewing the modelling distribution as the generator of the data-centric distribution. In Schnell's (2013) study with students aged 11 to 13, the focus was on the random data generated from chance experiments using a computer simulation tool. Students in pairs evaluated the chance of getting possible outcomes by identifying the patterns and variability in the distribution of outcomes in a bar chart to make predictions in a small number of trials (e.g. n = 20) and a large number of trials (e.g. n = 2000). Eventually, two out of three pairs were able to see the relation between predictability of the random outcomes and the number of trials conducted.

The emphasis on both the frequentist approach and the classical approach to probability in school mathematics curricula also stimulated the discussion of new approaches to teaching probability with the development of new technology. For example, at CERME 7, Henry and Parzysz (2011) provided perspectives on the use of computer simulations for linking the frequentist and the classical approaches

50 Bakker et al.

to probability given the emphasis on teaching both in French high schools. They argued that the use of computer simulations in the classroom as a pseudo-random generator would help learners develop a better understanding of statistical and probabilistic ideas, such as relative frequency and variability.

There are also studies suggesting effective use of a pedagogical approach to make the connection between theoretical and empirical probabilities without the use of computer simulations. For example, the enactive experience of flipping a coin 100 times as a class (Diaz-Rojas & Soto-Andrade, 2015) and students' physical experience of jumping paper frogs (Eichler & Vogel, 2015) were found to be useful in learning situations that involve an experimental approach to probability.

In addition to the frequentist and classical approaches to probability, subjective interpretation is also important and even more intuitive for students when teaching probability. In more recent years, research focused on combining subjective ideas of students with empirical data from random experiments. For instance, Helmerich (2015) studied 8–10-year-old children's use of subjective ideas and empirical data from experimenting with different 'odd dice' in a game context. Moreover, Kazak (2015) reported on 10–11-year-old students' coordination of their subjective ideas and the empirical data in attempting to evaluate the fairness of a chance game. Students also expressed degrees of confidence as they played the game and used different amounts of data generated by TinkerPlots simulations.

With the increasing attention for informal statistical inference at school level, research discussed in the more recent CERME meetings tended to focus on the notion of probability in the context of informal statistical inference and informal inferential reasoning. Different from the focus of aforementioned studies on the data-centric perspective (or experimental approach) in teaching probability, Ben-Zvi, Makar, Bakker and Aridor (2011) brought out the notion of probability within informal statistical inference. They reported on a study of 11-year-old children's reasoning about sampling when making informal statistical inferences in an inquiry-based environment. Engaging students in making informal statistical inferences from samples allowed them to discuss the notions of likelihood, level of confidence and randomness together with statistical notions, such as distribution, spread and average. Moreover, Jacob and Doerr (2013) presented their study about secondary students' informal inferential reasoning as they collected a sample of data in an attempt to draw a conclusion based on the related sampling distribution supported by the use of Fathom software. This study pointed out the importance of probabilistic ideas, such as level of uncertainty and Law of Large Numbers, in making a sound connection between samples and the sampling distribution. Henriques and Oliveira (2015) also studied 8th grade students' informal statistical inference during statistical investigation involving body measurements of students at the school. Students analysed their data using TinkerPlots software. When expressing the uncertainty to make generalisations beyond data, students tended to use probabilistic language, such as "probably", "maybe", "something similar" and "tend to be".

3 Role of context

In 2003, authors raised the question of the effect of context (Monteiro & Ainley, 2004). Then, at CERME 6, Eichler (2010) explicitly drew attention to the role of context in stochastic education, which became the subject of important discussions in CERME 7.

Several authors have raised the question of the authenticity of the problems proposed to pupils, as the context is an integral part of Stochastics, interdisciplinary by nature, and further emphasised by the rise of the EDA perspective (Borovcnik, 2006). Researchers who explored the role of context in the learning of statistics (see e.g. *Educational Studies in Mathematics, 45*(1), 2001; *Mathematical Thinking and Learning, 13*(1&2), 2011) usually recommend the use of real data. Nevertheless, they identified that this use could be a problem, particularly because of the difficulty experienced by students to extricate themselves from the context and the weight of their personal beliefs. What could an authentic situation look like? We identify two main trends through CERME proceedings: Some researchers have recommended the implementation of an inquiry-based process where students collect and handle their own data. Other researchers have proposed activities based on the study of information given by the media. Within the first trend, Ainley, Jarvis and McKeon (2011) designed and experimented with a sequence of science activities which focus on different aspects of exploring flight. Students were led to explore variability through the repeating of measurements. Eckert and Nilsson (2013) conducted an experiment based on farming; they stimulated students to think about the variability of the results of their pumpkin and sunflower plantations. Hauge (2013) proposed to build school activities about risk assessment by drawing inspiration from the management of fisheries. Within the other trend, Sturm and Eichler (2015) used HIV rapid-test information to encourage students to work on Bayes' formula; Plicht, Vogel and Randler (2015) studied students' interpretation of graphs about milk production.

Context may somehow be related to the interest or motivation of the student to engage with statistics. Donati, Primi, Chiesi and Morsanyi (2015) studied 127 psychology students and found that individual interest in statistics has a direct effect on both situational interest and intrinsic motivation. It was also found that the relationship between individual interest and situational interest is mediated by intrinsic motivation and this indirect effect of individual interest on situational interest is regulated by the perceived appeal of the statistical activity.

Alldredge and Brown (2004) investigated two distinct instructional strategies. They found that the association between general confidence toward learning and course performance was more positive for females when using software focused on different statistical methods rather than on the contextualisation of the problems. Perplexingly, for males, it was vice versa. It seems that the role of context is unsettled but we expect research using contexts like the ones above will help to tease out the issues in the future.

52 Bakker et al.

4 Technology

In this section, we discuss how research reported by technology-oriented papers in the CERME conference proceedings has connected to the ideas raised in the handbook chapter by Biehler, Ben-Zvi, Bakker and Makar (2013) on how digital technologies are enhancing statistical reasoning at school level. Biehler et al. described the requirements of digital tools in ways that resonate with an old analysis by a UK quango of the entitlements offered to students by technology (Becta, 2000). These entitlements consisted of 'learning from feedback', 'observing patterns', 'seeing connections', 'working with dynamic images', 'exploring data' and 'teaching the computer'. The requirements laid out by Biehler et al. included a capacity for students not only to practise graphical and data numerical data analysis, engaging in 'exploring data' and 'working with dynamic images', but also to create new methods, such as by programming or similar activity involving 'teaching the computer'. Biehler et al. also recognised the importance of using embedded microworlds, thus 'learning from feedback', and constructing models, which would also involve 'teaching the computer'. It is not difficult to see how 'observing patterns' and 'making connections' are fundamental to all of the requirements of tools as described by Biehler et al.

Among the main themes identified by Biehler et al. (2013) was the issue of when students should use software as opposed to working manually. Without exception, the technology-oriented papers in CERME proceedings have involved direct use of technology as a tool for learning by students and there has not been research specifically addressing when the use of technology might be beneficial.

A second theme pointed out the tension between adopting technology that might be beneficial for learning statistics and probability, but which requires considerable effort to master before such learning might become evident. There is of course an argument that the length of time needed to learn a new technology is time that could have been spent making sense of difficult statistical and probabilistic ideas. It is the teacher under pressure to cover a large curriculum who has to manage this tension (and at school level it is often the mathematics teacher working with a curriculum that contains relatively little content of statistics and probability). Researchers compete for scarce research funds and so encounter a similar tension because they are unlikely to afford the time for the gradual integration of technology into a classroom. Hence, research using software that is more narrowly focused can be easier to manage than research on the use by students of more general educational softwares such as TinkerPlots and Fathom. Nevertheless, over the years, CERME has reported examples of both.

There have been five studies that have used specially designed microworlds to study student learning of specific key concepts: 'fairness' (Paparistodemou & Noss, 2004); the Law of Large Numbers (Paparistodemou, 2006); randomness (Pratt & Prodromou, 2006) and distribution (Prodromou & Pratt, 2009). These studies demonstrated how digital resources can be harnessed to explore specific research questions, throwing light upon students' understanding of these key concepts.

Reasoning with key concepts is a theme identified by Biehler et al. (2013) but, whereas that report discussed the use of technology in the teaching and learning of key concepts, these CERME studies illustrated the use of technology to research perturbations in students' statistical reasoning about key concepts. Indeed, at a higher level of abstraction, one other study, Pratt (2004) deployed a microworld to propose a general theory for how probabilistic knowledge emerges. The designed microworlds in these studies were sufficiently narrowly focused on the specific concept in question that the data collection could take place over a relatively short time span without a considerable time commitment for the students to learn the tool. Most of the other technologically oriented studies reported in CERME conferences used more general educational software, such as NetLogo, Fathom and TinkerPlots. These studies needed strategies for embracing the challenge of enabling students to master the software sufficiently to elaborate their research aims and so were either closely related to the researchers' teaching activity or were part of a wider long-term study.

Two studies by Abrahamson and Wilensky (2004, 2006) deployed the programming language NetLogo to explore how students learned through design and collaboration. In the terminology of Biehler et al. (2013), these studies immersed the students in a setting where they might create their own methods of solution, such as building for themselves the Normal distribution and engaging in collaborative activity around the Law of Large Numbers, thus emphasising the development of aggregate thinking, an important theme in the Biehler et al. (2013) report.

There has been an increasing emphasis on studies involving model construction (one out of the two technologically oriented papers in each of CERME 5, 7 and 8 and all four in CERME 9). These studies have focused on students' reasoning with key concepts and aggregates as they: follow a schema for simulation (Maxara & Biehler, 2007); reason informally about sampling using TinkerPlots (Martins, Monteiro & Carvalho, 2015); reason about probability and randomisation tests using TinkerPlots (Frischemeier & Biehler, 2013); compare groups (Frischemeier & Biehler, 2015); reason about uncertainty while playing a game and using a TinkerPlots simulation (Kazak, 2015); and use simulations for informal inference (Lee, Tran & Nickel, 2015).

The move from research that studies either probability or statistics exclusively to studies of these two connected domains in an integrated way has been discussed in the earlier section on the nature of statistics and probability. It is worth mentioning here that this transition is reflected in the changing nature of those studies involving technology. Indeed, it is a reasonable conjecture that the innovation of educational software such as TinkerPlots and Fathom has been a significant trigger for that change. At one time, experimental studies of probability were restricted in the main to students using familiar random generating devices such as coins, urns, dice and playing cards. Some studies then began to exploit additional affordances of technology by simulating those devices. At the same time, early uses of technology focused on exploratory data analysis, arguably intentionally avoiding the difficulty

54 Bakker et al.

of probabilistic thinking. Innovations in educational software have facilitated the re-connection of statistics and probability in research studies. The integration into TinkerPlots of samplers that allow the simulation of familiar random devices into exploratory data analysis software marks the move among CERME researchers to conduct research that considers both key concepts and aggregate thinking, and allows students to create new methods and models. Without doubt, such research makes additional demands because of the increased time commitment for the students to learn such tools. There is a danger perhaps that the pressure to conduct research with these tools will lead to a narrowing of research settings to those few situations where long-term research is being conducted or where the subjects of the research are in fact closely connected to the researchers, for example through the teaching commitments of the researchers. There continues, therefore, to be a place for research that is more narrowly focused alongside the exciting research now being conducted with larger educational software packages.

5 Teacher knowledge

As the statistics and probability topics have become part of the mainstream mathematics curricula in various countries since the late 1990s, teachers' knowledge on these topics (i.e. their conceptions of statistical and probabilistic concepts and ideas) became an ongoing interest to mathematics education researchers. Discussions about the following main issues related to teachers began in CERME 3 and seem still to be relevant: (1) impact of teachers' strong beliefs about the nature of mathematics on teaching and learning of statistics (deterministic vs. uncertain), (2) insufficient training of teachers both in terms of content knowledge and pedagogical knowledge related to statistics and probability.

Much of the research with pre-service and in-service teachers has focused on teacher knowledge of statistics and probability, involving conceptions, competencies and reasoning in various topics: Probability (e.g. Contreras, Batanero, Díaz & Fernandes, 2011), randomness (Paparistodemou, Potari & Pitta, 2007), graphs (e.g. Batanero, Arteaga & Ruiz, 2010), risk (Pratt, Levinson, Kent & Yogui, 2011), sampling distribution (Doerr & Jacob, 2011), statistical literacy (Koleza & Kontogianni, 2013), uncertainty (e.g. Frischemeier & Biehler, 2013), variability and sampling variability (Gonzales, 2013; Jacob, Lee, Tran & Doerr, 2015), and measures of central tendency (Santos & De Ponte, 2013). A general implication from these studies seems to be the need to improve teachers' knowledge of specific content that they are expected to teach. Indeed, it seems that competence in statistics also supports general interest in the subject. Batanero, Estrada, Díaz and Fortuny (2006), after studying 367 pre-service teachers in different subject areas, suggested that teachers' attitudes toward statistics were highly correlated with their cognitive competence in statistics.

Having adequate knowledge of content is not sufficient for developing students' understanding of statistical concepts and procedures. For instance, Eichler (2007) pointed out that students' difficulties in understanding dependence, conditional probability, and Bayes' theorem might be due to the teacher's use of tree diagrams in a traditional way, rather than with natural frequencies. This finding suggests the importance of teachers' pedagogical content knowledge which involves "an understanding of what makes the learning of specific topics easy or difficult" (Shulman, 1986, p. 9) and links content and pedagogy-related aspects of knowledge for teachers. However, only five of the CERME papers focusing on teachers' knowledge dealt with pedagogical content knowledge of teachers, while there were 23 papers on teachers' knowledge/conceptions in statistics and probability. For example, one of the studies conducted with pre-service teachers (Paparistodemou et al., 2007) indicated some difficulties in combining pedagogical practices (e.g. group work, use of concrete tools, games and time management) and the mathematical content in their teaching and in considering students' intuitive ideas, possible student responses and how they would think during the implementation of their lesson plans on the idea of randomness for children of ages 4–5.5. Given that these pre-service teachers were doing their teaching practice in pre-primary schools and had completed courses on statistics and probability as well as teaching mathematics, there is a need to address pedagogical content knowledge of teachers in teaching statistics and probability in teacher education programmes. The other four papers focused on in-service teachers' knowledge, in particular pedagogical content knowledge, with regard to teaching topics such as statistical graphs (González Astudillo & Pinto Sosa, 2011), probability (Eckert & Nilsson, 2013) and variability (Gonzales, 2013; Quintas, Oliveira & Tomás Ferreir, 2013) at different grade levels, from primary school to university. Although these studies tended to report on findings from a very small number of teachers, their attempt to identify characteristics of teachers' pedagogical content knowledge or knowledge for teaching of different statistical and probabilistic topics seems to be promising for our understanding of what knowledge teachers should have in order to promote students' understanding of these topics.

The use of technology to support teachers' statistical understanding also seems to be getting more attention in recent years. For example, the role of using technology tools in training future mathematics teachers has been studied in the context of modelling a random experiment within Fathom (Maxara & Biehler, 2007) and reasoning about uncertainty during randomisation tests with TinkerPlots (Frischemeier & Biehler, 2013). Research with in-service teachers included teachers' understanding of sampling distribution with the use of Fathom software (Doerr & Jacob, 2011) and teachers' models of simulation processes in the context of informal statistical inference through the use of TinkerPlots software (Lee et al., 2015).

6 Types of research conducted and needed

It is possible to identify four types of research needed to improve probability and statistics education: (1) Descriptive or evaluative research that focuses on education as it currently is. Often such studies involve a problem analysis, baseline study, or needs analysis. (2) Research that identifies sensible learning goals, for example discussions of statistical literacy or risk. In such cases, scholars analyse what would be good learning goals given today's or tomorrow's society. (3) Research that offers suggestions or advice on how to promote particular learning. These are typically design-based interventions. New technology is often used to foster desirable ways of learning. But there are also creative ideas such as using random walks (Soto-Andrade, 2013). (4) Effect studies and evaluations of interventions, which are closely connected to (3) and focus on what was actually learned. A closely related question is how to assess student or teacher knowledge or skills validly and reliably.

There has always been much descriptive research in the CERME group, typically about students' statistical or probabilistic knowledge, or lack of it. Such studies are important, for example to flag up a problem in a country and underpin the need for improvement or redesign of the curriculum. At CERME, however, several commentators (e.g. Per Nilsson in 2013) have observed that the research community knows pretty well what the problems are so that we need more design-based, prescriptive or advisory research: didactical ideas about how to improve probability and statistics education. This implies that, in their view, the field asks for more research of the third type. However, to know how effective and efficient these approaches are, we also need more systematic evaluation of new interventions (type 4).

We end with a wish list of research we think is needed:

- A large proportion of the papers focused on student learning. However, because most mathematics teachers have little knowledge of statistics, more research on teachers and teaching is needed. Many teachers try to teach statistics like other mathematical topics, focusing on only the results, procedures, graphs, etc., rather than on statistical thinking and reasoning processes. What is it that teachers need to know? The concept of Statistical Knowledge for Teaching (SKT) might be fruitful here (Groth, 2007). And, also relevant: How can teachers be supported to develop this SKT?
- In line with the previous point, teachers also need better familiarity with how to use technology. The notion of Technological Pedagogical Content Knowledge (TPCK) has been suggested as a theoretical lens on this issue.
- In mathematics and science education, modelling is coming up as an important learning goal, but also a means of supporting learning. As some of the CERME papers indicated, modelling can also act as a bridge between statistics and probability in an era when probability is becoming isolated. Technology offers new possibilities, as numerous CERME papers have shown (cf. difference between dice games and computer games), but what are effective ways to promote students' understanding?
- The rare studies on vocational and professional usage of statistics emphasise that statistics in its many contextual manifestations is becoming more and more important for the workplace. More interest from educational researchers for this domain would be welcome.
- A large proportion of CERME papers focused on rather basic probability and statistics. Most welcome would be attention for more difficult concepts and techniques.

References

Abrahamson, D., & Wilensky, U. (2004). The quest of the bell curve: A constructionist designer's advocacy of learning through designing. *CERME3*: www.mathematik.uni-dortmund.de/~erme/CERME3/Groups/TG5/TG5_abrahamson_cerme3.pdf.

Abrahamson, D., & Wilensky, U. (2006). Problab goes to school: Design, teaching, and learning of probability with multi-agent interactive computer models. *CERME4* (pp. 570–579).

Ainley, J., Jarvis, J., & McKeon, F. (2011). Designing pedagogic opportunities for statistical thinking within inquiry based science. *CERME7* (pp. 705–714).

Alldredge, J. R., & Brown, G. (2004). Association of course performance with student beliefs: an analysis by gender and instructional software environment. *CERME3*: www.mathematik.uni-dortmund.de/~erme/CERME3/Groups/TG5/TG5_alldredge_cerme3.pdf.

Batanero, C., Arteaga, P., & Ruiz, B. (2010). Statistical graphs produced by prospective teachers in comparing two distributions. *CERME6* (pp. 368–377).

Batanero, C., Estrada, A., Díaz, C., & Fortuny, J. M. (2006). A structural study of future teachers' attitudes towards statistics. *CERME4* (pp. 508–517).

Becta (2000). *Secondary mathematics with ICT: A pupil's entitlement to ICT in secondary mathematics*, downloaded on December 6, 2016 from: www.stem.org.uk/elibrary/resource/29209.

Ben-Zvi, D., Makar, K., Bakker, A., & Aridor, K. (2011). Children's emergent inferential reasoning about samples in an inquiry-based environment. *CERME7* (pp. 745–754).

Biehler, R., Ben-Zvi, D., Bakker, A., & Makar, K. (2013). Technology for enhancing statistical reasoning at the school level. In M. A. (Ken) Clements et al. (Eds.), *Third international handbook of mathematics education* (pp. 643–689). New York: Springer Science and Business Media. doi: 10.1007/978-1-4614-4684-2_21.

Borovcnik, M. (2006). Probabilistic and statistical thinking. *CERME4* (pp. 484–506).

Contreras, J. M., Batanero, C., Díaz, C., & Fernandes, J. A. (2011). Prospective teachers' common and specialized knowledge in a probability task. *CERME7* (pp. 776–775).

Diaz-Rojas, D., & Soto-Andrade, J. (2015). Enactive metaphoric approaches to randomness. *CERME9* (pp. 629–635).

Doerr, H. M., & Jacob, B. (2011). Investigating secondary teachers' statistical understandings. *CERME7* (pp. 776–786).

Donati, M. A., Primi, C., Chiesi, F., & Morsanyi, K. (2015). Interest in statistics: Examining the effects of individual and situational characteristics. *CERME9* (pp. 740–745).

Eckert, A., & Nilsson, P. (2013). Contextualizing sampling: Teaching challenges and possibilities. *CERME8* (pp. 766–776).

Eichler, A. (2007). The impact of a typical classroom practice on students' statistical knowledge. *CERME5* (pp. 722–731).

Eichler, A. (2010). The role of context in statistics education. *CERME6* (pp. 378–387).

Eichler, A., & Vogel, M. (2015). Aspects of students' changing mental models when acting within statistical situations. *CERME9* (pp. 636–642).

Frischemeier, D., & Biehler, R. (2013). Design and exploratory evaluation of a learning trajectory leading to do randomization tests facilitated by Tinkerplots. *CERME8* (pp. 798–808).

Frischemeier, D., & Biehler, R. (2015). Preservice teachers' statistical reasoning when comparing groups facilitated by software. *CERME9* (pp. 643–650).

Gonzales, O. (2013). Conceptualizing and assessing secondary mathematics teachers' professional competencies for effective teaching of variability-related ideas. *CERME8* (pp. 809–818).

González Astudillo, M. T., & Pinto Sosa, J. E. (2011). Instructional representations in the teaching of statistical graphs. *CERME7* (pp. 797–806).

58 Bakker et al.

Groth, R. E. (2007). Toward a conceptualization of statistical knowledge for teaching. *Journal for Research in Mathematics Education, 38*(5), 427–437.

Groth, R. E. (2015). Working at the boundaries of mathematics education and statistics education communities of practice. *Journal for Research in Mathematics Education, 46*(1), 4–16.

Hacking, I. (1975). *The emergence of probability: A philosophical study of early ideas about probability, induction and statistical inference.* Cambridge: Cambridge University Press.

Hauge, K. I. (2013). Bridging policy debates on risk assessment and mathematical literacy. *CERME8* (pp. 819–828).

Helmerich, M. (2015). Rolling the dice: Exploring different approaches to probability with primary school students. *CERME9* (pp. 678–684).

Henriques, A., & Oliveira, H. (2015). Student's informal inference when exploring a statistical investigation. *CERME9* (pp. 685–691).

Henry, M., & Parzysz, B. (2011). Carrying out, modelling and simulating random experiments in the classroom. *CERME7* (pp. 864–874).

Jacob, B., & Doerr, H. M. (2013). Students' informal inferential reasoning when working with the sampling distribution. *CERME8* (pp. 829–839).

Jacob, B., Lee, H., Tran, D., & Doerr, H. (2015). Improving teachers' reasoning about sampling variability: A cross institutional effort. *CERME9* (pp. 692–699).

Kazak, S. (2015). A Bayesian inspired approach to reasoning about uncertainty: 'How confident are you?' *CERME9* (pp. 700–706).

Koleza, E., & Kontogianni, A. (2013). Assessing statistical literacy: What do freshmen know? *CERME8* (pp. 840–849).

Lee, H., Tran, D., & Nickel, J. (2015). Simulation approaches for informal inference: Models to develop understanding. *CERME9* (pp. 707–714).

Martins, M., Monteiro, C., & Carvalho, C. (2015). How teachers understand sampling when using Tinkerplots. *CERME9* (pp. 715–721).

Maxara, C., & Biehler, R. (2007). Constructing stochastic simulations with a computer tool: Students' competencies and difficulties. *CERME5* (pp. 762–771).

Monteiro, C., & Ainley, J. (2004). Developing critical sense in graphing. *CERME3*: www.mathematik.uni-dortmund.de/~erme/CERME3/Groups/TG5/TG5_monteiro_cerme3.pdf.

Nilsson, P., Schindler, M., & Bakker, A. (2018). The nature and use of theory in statistics education. In D. Ben-Zvi, K. Makar, & J. Garfield (Eds.), *International handbook of research in statistics education* (pp. 359–386). Cham: Springer.

Paparistodemou, E. (2006). Young children's expressions for the law of large numbers. *CERME4* (pp. 611–618).

Paparistodemou, E., Potari, D., & Pitta, D. (2007). Looking for randomness in tasks of prospective teachers. *CERME5* (pp. 791–800).

Paparistodemou, E., & Noss, R. (2004). Fairness in a spatial computer environment. *CERME3*: www.mathematik.uni-dortmund.de/~erme/CERME3/Groups/TG5/TG5_paparistodemu_cerme3.pdf.

Plicht, C., Vogel, M., & Randler, C. (2015). An interview study on reading statistical representations in biology education. *CERME9* (pp. 734–739).

Pratt, D. (2004). The emergence of probabilistic knowledge. *CERME3*: www.mathematik.uni-dortmund.de/~erme/CERME3/Groups/TG5/TG5_pratt_cerme3.pdf.

Pratt, D., & Prodromou, T. (2006). Towards the design of tools for the organization of the stochastic. *CERME4* (pp. 619–626).

Pratt, D., Levinson, R., Kent, P., & Yogui, C. (2011). Risk-based decision-making by mathematics and science teachers. *CERME7* (pp. 875–884).

Prodromou, T. (2007). Making connections between the two perspectives on distribution. *CERME5* (pp. 801–810).

Prodromou, T., & Pratt, D. (2009). Students' causal explanations for distribution. *CERME6* (pp. 394–403).

Quintas, S., Oliveira, H., & Tomás Ferreir, R. (2013). The didactical knowledge of one secondary mathematics teacher on statistical variation. *CERME8* (pp. 860–869).

Santos, R., & De Ponte, J. P. (2013). Prospective elementary school teachers' interpretation of central tendency measures during a statistical investigation. *CERME8* (pp. 870–879).

Schnell, S. (2013). Coping with patterns and variability: Reconstruction of learning pathways towards chance. *CERME8* (pp. 880–889).

Shulman, L. S. (1986). Those who understand: Knowledge growth in teaching. *Educational Researcher, 15*(2), 4–14.

Soto-Andrade, J. (2013). Metaphorical random walks: A royal road to stochastic thinking? *CERME8* (pp. 890–900).

Stigler, S. M. (1986). *The history of statistics: The measurement of uncertainty before 1900*. Cambridge, MA: Harvard University Press.

Sturm, A., & Eichler, A. (2015). Changing beliefs about the benefit of statistical knowledge. *CERME9* (pp. 761–767).

5

RESEARCH ON UNIVERSITY MATHEMATICS EDUCATION

Carl Winsløw, Ghislaine Gueudet, Reinhard Hochmuth and Elena Nardi

1 Introduction

Mathematics is an ancient scholarly discipline, with distinct and related practices of research and education. CERME is concerned with one such practice: research on mathematics education (RME), which is the study of mathematics teaching and learning at all levels, including university mathematics education (UME). And that, research on university mathematics education (RUME), is the topic of this chapter. [Note: the abbreviation RUME is also used by the American association SIGMAA-RUME, but with a slightly different meaning: undergraduate mathematics education.]

Internationally, RUME has emerged along with RME in the course of the 20th century, beginning with the efforts of Klein (1908) to develop stronger ties between UME and mathematics education in schools. In terms of systematic research, early milestones for RUME appear in the volume on *Advanced Mathematical Thinking* (Tall, 1991) and the 11th ICMI study (Holton, 2001). The first six CERMEs each had 10–20 papers on RUME in different thematic working groups (TWG); at CERMEs 4, 5 and 6, most RUME papers appeared in the TWG on 'Advanced Mathematical Thinking' (AMT). A group explicitly focusing on RUME has been present since CERME 7 in 2011, and the number of accepted papers has increased from 24 in 2011 to 47 in 2017. In addition, papers on RUME also appear in other groups, e.g. 'Applications and Mathematical Modelling' and 'Proof'.

In this chapter, we present selected main problems and results from RUME which have continuously and fruitfully appeared at CERME, with a deliberate focus on current and emerging trends. The chapter as a whole, and in particular the selection of papers from CERME to illustrate the points we make in each section, is naturally shaped and limited by the perspectives of the authors.

The chapter has two main parts: research into current practices of UME (with no direct intervention), and developmental or experimental research,

University mathematics education **61**

where an intervention design is part of the research project. We have thus chosen to concentrate on *problems* and *results* and leave *methods* and *theoretical paradigms* more in the background (for the latter, we refer to the special issue edited by Nardi, Biza, González-Martín, Gueudet and Winsløw (2014), largely based on CERME work on RUME). In the concluding section, we draw up some possible directions for the future of RUME and, in particular, its potential contributions to UME.

2 Mathematics education at university: what is it?

This first section focuses on *descriptive* research, i.e. studies of UME 'as it is' (no intervention). The section is structured so that it *foregrounds research objects*: contents; methods and resources; transition phenomena; student experiences; and teaching non-mathematics specialists.

2.1 Mathematical content in university programmes

As in all areas of RME, it is important within RUME to be attentive to the variations of cultural and institutional contexts of the phenomena we study. We notice, however, that the *mathematical contents* taught and learned appear remarkably stable and recognisable in many studies. A large portion involve Calculus and Linear Algebra, which are indeed commonly taught during the first year of university programmes, in the mathematical sciences, natural sciences, engineering and business (cf. Adam, 2002). Many papers study specific challenges students face with notions such as limits and derivatives (e.g. Hähkiöniemi, 2006) or infinite series (González-Martín, 2013); these and other studies find that students may succeed with institutional (computational) requirements related to such notions, and still have little idea about their theoretical and practical significance.

Studies involving more advanced domains from pure mathematics, such as Abstract Algebra or Topology, slowly but surely begin to appear in CERME papers (e.g. Hausberger, 2015). We also find more and more studies carried out in settings where the content is not one of the above classical topics in mathematics, such as an analysis of modelling within an introductory course on electrical engineering, and the ways in which knowledge from high school mathematics acts as a resource or obstacle for students in such a course (Biehler, Kortemeyer & Schaper, 2015). An important insight from such research is that the uses and meanings of mathematics at university are not simply confined to courses with a mathematics label.

2.2 Methods and resources in UME teaching

The investigation of teaching methods, both new and 'traditional', is a constant theme in RUME. For instance, lectures are a common format in UME, unlike other levels, and are a recurring focus of RUME studies. For example, Bergsten (2011) studied Swedish students' views of this format. He observes that the students value the lectures, for instance because they consider them as a helpful indicator of what is expected at the exam. They also value the explanations given by the lecturer, both intuitive and more formal, and consider them useful to learn proofs. They appreciate their lecturer's pedagogical awareness. These results, obtained in a single context, seem to go against the prevailing pedagogical criticism of the lecture format (e.g. Sikko & Pepin, 2013); but more research is needed to understand when and why students might appreciate lectures.

The effects of new technology in UME have been studied rather extensively, just as at other educational levels. In some cases, the technology considered is specific to the university level, for example, concerning distance learning (Misfeldt & Sanne, 2007). Recent works also treat technology as belonging to sets of resources of a different nature (as is also observed in Chapter 11). Gueudet (2013), for example, analyses how a teacher in a technological institute designs his own resources, using a computer algebra system (Scilab) and develops a structured resource system, according to his professional knowledge and beliefs.

Recent works have also taken a closer look at assessment practices in UME. Thoma and Iannone (2015) analysed three examination tasks by applying the 'Mathematical Assessment Task Hierarchy' by Smith et al. (1996), that focuses on knowledge and skills, and a discourse-orientated framework by Morgan and Tang (2012) that is based on systemic functional linguistics (Halliday, 1978) and Sfard's (2008) Theory of Commognition. Both approaches allow for identifying various complementary and implicit characteristics of tasks and, potentially, the 'learning approaches' they promote. We note that a surge in assessment-focused works at UME as well as other levels has led to the emergence, from CERME 10, of a related TWG.

A special characteristic of UME is that many of the teachers are research mathematicians, and research presented at CERME conferences has drawn on different forms of engagement with mathematicians, often in the form of intimate interviews that explore their epistemological and pedagogical perspectives. Burton (1999) interviewed 70 mathematicians about their views on mathematics and on their teaching: this study is an example of a CERME paper that was later developed considerably, to result in an influential book (Burton, 2004). Burton's results drew a worrying picture. For example, mathematicians acknowledged the existence of different *thinking styles* in mathematics, and that they themselves had a dominant one. However, they did not seem to consider that their own thinking style influenced their teaching, and their accounts of teaching did not include many references to efforts toward catering for students' different thinking styles. Subsequent works drew a more nuanced – and in some ways less alarming – picture, with participants demonstrating substantial pedagogical awareness. For example, Nardi and Iannone (2006) demonstrate that mathematicians can have a rich reflection on their students' mathematical practices, and be interested in engaging in teaching innovation. Similarly, Mesa and Cawley (2015) analyse the learning processes of university teachers who engaged in a project on Inquiry Based Learning. Such projects seem to share the aspiration of Burton (1999) for

University mathematics education **63**

mathematics taught at university with "approaches not too dissimilar from those of research mathematicians" (p. 98). The tendency that UME teachers are increasingly engaging in efforts to this end – and that these efforts are now part of an increasing number of studies of teaching *practice*, not only of teachers' *perspectives* – is reinforced by the bulk of recent CERME work in this area.

2.3 The transition from school to university

The transition from school to university has been an important topic in all CERME right from the start. We notice a key evolution in the theoretical frames of transition studies presented at CERME: developmental approaches, which were very present in the TWG on AMT, seem in recent years less frequent, while sociocultural, discursive and semiotic approaches have been steadily growing. We claim that this evolution has two main consequences.

The first consequence concerns the focus of the studies. Many changes happen during the secondary–tertiary transition; the choice of a specific change as focus of a study is strongly linked with the study's theoretical perspective. In the AMT group, many papers focused on *cognitive aspects of this change*. Biza, Souyoul and Zacharides (2006) showed that novice students create synthetic models, mixing personal beliefs and scientific theory, and stressed the need for conceptual change. Studying these models in the case of the concept of tangent to a curve, they observed that the students developed a synthetic model incorporating the properties of the circle tangent, hindering the development of an adequate concept image of the tangent. Based on an epistemological-institutional approach (the Anthropological Theory of the Didactic (ATD) see e.g. Chevallard, 2002), Winsløw (2006) chose a different focus: the changes in the *praxeologies* of school and university mathematics. He identified two kinds of such changes: (1) from praxeologies at secondary school, centred on tasks and techniques, to full praxeologies including technology and theory at university; (2) from more to less familiar tasks and techniques to tasks and techniques of a more theoretical nature. This second kind of transition can happen later, for example from the first to the second university year. De Vleeschouwer and Gueudet (2011) studied related changes in the *didactic contract*, between secondary school and university. They noticed different kinds of changes: general changes, concerning, for example, personal work; changes related to mathematics as a discipline, such as expectations in terms of rigour; and changes related to specific mathematical content, in their case duality in Linear Algebra. Other authors studied changes in the mathematical discourse. Petterson, Stadler and Tambour (2013) analysed students' understanding of function as a 'threshold concept' (an initially troublesome concept which leads to a new understanding), using the commognitive approach (Sfard, 2008). They identify specific cases of evolution and stability in the discourses of the students. In these and other cases, we can observe how theoretical framings of transition phenomena, initiated at one CERME, are subsequently taken up and developed by other authors in later conferences, and beyond.

64 Winsløw et al.

The second, strongly related, consequence concerns a broader view on secondary–tertiary transition conveyed by CERME research studies. The view on transition has evolved from a local cognitive change, taking place at the beginning of university, to a long-term social and cultural process of change developing over one or several years. Most of the early studies concentrate on the difficulties with particular mathematical contents met by students entering university. For Linear Algebra, for example, Dorier, Robert, Robinet and Rogalski (1999) argued that novice students fear formalism. Nevertheless, according to their historical-epistemological analysis, formalism should not be avoided: on the contrary it must be put forward and introduced as the answer to a problem. For example, Vandebrouck (2011) notes three perspectives that can be adopted on functions: a point-wise perspective, a global perspective and a local perspective; the first two are generally introduced in school, but at university it is necessary to master, and be able to combine, all three perspectives. Vandebrouck reported that incoming students still have difficulties with the point-wise and the global perspectives, and struggle to combine them with the local one.

The sociocultural approaches mentioned above study the transition from a more global and long-term perspective. For example, Stadler, Bengmark, Thunberg and Winberg (2013) followed the approaches of students to learning mathematics along the first university year. They showed that students progressively rely less on the teacher, and more on peers or Internet resources.

2.4 Students' experience of UME

Over the years, research into students' experience of university mathematics presented at CERME has gained scope, substantively (in terms of the areas this research examines) and theoretically (in terms of the theoretical underpinnings of this research). In what follows, we trace this growth from studies of student learning of particular mathematical topics and specific forms of mathematical activity such as proof and proving, to include also studies of student affect – including attitudes, motivations and emotions – as well as studies of broader institutional and social issues, such as recruitment and retention. Much along the lines we described in Section 2.3 in relation to studies of transition, the broadening of substantive scope has occurred in tandem with a broadening of theoretical scope. We dare say that *it is in the study of student experience of university mathematics that the shift from the AMT TWG* (dominated by what we labelled in Section 2.3 as developmental studies of mathematical learning) *to the UME TWG* (currently populated with a very diverse set of studies that endorse cognitive as well as sociocultural, discursive and anthropological perspectives) *has occurred more tellingly.*

In the earlier conferences, there is a proliferation of well-known binaries in the way students' mathematical learning is described and explored. These include: concept image/concept definition (Tall & Vinner, 1981), procedural/conceptual understanding (Hiebert, 1986), process/object (Sfard, 1991) and informal/formal modes of reasoning (Fischbein, 1994), often along with related developmental frameworks such as APOS theory (Asiala, Cottrill, Dubinsky & Schwingendorf, 1997). These studies are typically conducted in the context of a relatively limited set of mathematical topics.

Typical investigations and findings on Calculus or Analysis (e.g. Hähkiöniemi, 2006) concern students' procedural or conceptual knowledge about the limiting process, their semiotic and mental representations, and their use and combination of such representations. Similar investigations into elementary Linear Algebra provide insight into the processes and conditions for students' construction of schemas. For instance, Trigueros, Oktaç and Manzanero (2007) demonstrate, through structured interviews with six students in a Linear Algebra course, how the learning of equation systems depends crucially on the versatility of previously constructed schemas for the notion of variable. In this and several other studies, APOS-based genetic decompositions are proposed as tools to both examine and to support students' learning.

With regard to proof and proving, a main theme is what Inglis and Simpson (2006) identify as the two parts of dual process theory: intuition and formalism/abstraction. Students are frequently uneasy with the latter and uncertain about the validity of the former. In these studies, there is also a tendency to juxtapose novice and expert approaches and to explore ways in which novices can learn to emulate experts. One way, explored in several studies, is to assist students toward constructing rich and meaningful example spaces. For example, Meehan (2007) proposes a systematic sequence of example generation activities in an Analysis course: these activities gradually add extra conditions, alert students to the significance of each condition, and ask students to continuously review and validate their responses.

While the aforementioned binary and stage theories remain part of the vernacular in the more recent studies (e.g. Breen, Larson, O'Shea & Pettersson, 2015), numerous recent works are based on more dynamic and fluid theoretical constructs such as the RBC Model (Recognising, Building with and Constructing: Tabach, Rasmussen, Hershkowitz & Dreyfus, 2015) and constructs originating in discourse analysis, such as those from the Theory of Commognition (Sfard, 2008; Pettersson, Stadler & Tambour, 2013).

Finally, the number of studies with a self-proclaimed *exclusive* focus on the student experience, has decreased over the years (for example, from eight of the 12 papers accepted for publication in CERME 4 to also eight, but out of 45, papers accepted for publication in CERME 9). While we see the distinction between student-centred and teacher-centred studies as slightly artificial, it is at times necessary for pragmatic reasons – as the structure of this part of our chapter suggests. However, since the early days of CERME, there have been some alerts to the importance of focusing on the student–teacher, student–resource, and student–institution interfaces (Cazes, Gueudet, Hersant & Vandebrouck, 2006). The number and quality of studies presented in the more recent conferences on these interfaces is promising in its coverage of a range of pedagogical (Sikko & Pepin, 2013), institutional (Farah, 2015), social (Bergsten & Jablonka, 2013) and affective (Liebendörfer & Hochmuth, 2015) influences on the student experience.

66 Winsløw et al.

For example, Sikko and Pepin (2013) present a survey of university students in their second year, which explored the teaching and study methods from their first year that the students found most effective. They report the proliferation of active and collaborative ways of working in the student responses, and a limited appreciation for lectures (see our comment on contrasting evidence on this in Section 2.2). Farah (2015) focuses on students from preparatory classes for French business schools – and, especially, their "autonomous study and the gestures involved in it" (p. 2097). The study combines qualitative and quantitative methods for answering questions concerning the evolution of individual activities in terms of quantity and forms of study, and how these activities relate to student performance, social relationships, study track and so on.

Similarly, but in a different context, Liebendörfer and Hochmuth (2015) identify different factors that support or hinder the perceived autonomy of students in a first year Analysis course. In particular, they observe that students who plan to become teachers might not be convinced about the need for university mathematics as preparation for teaching at school, leading to a perceived gap between their personal goals and their study activity. They also experience the transition from school to university mathematics as an often perplexing mixture of familiar and novel experiences (see also Stadler, 2011), where they might lack strategies, e.g. for autonomous problem solving. Retaining a theoretical perspective that networks ATD and the construct of didactic contract from the Theory of Didactic Situations, González-Martín (2013) studies how textbooks and teachers' practices shape the institutional didactic contract, and in particular its rules about specific parts of the mathematical content. He presents cases where students, confronted with tasks that do not obey these rules, produce inappropriate answers. In these studies, student experience appears as deeply entangled with (and often depending on) given institutional conditions, such as regulations, pedagogies, resources and prescribed contents.

Another way in which research on student experience has changed over the years is an increasing and deliberate emphasis on courses and programmes other than in pure mathematics. We turn to these developments next.

2.5 UME in study programmes of other disciplines

The studies cited in the preceding sections *may* concern UME in a variety of study programmes. In this section, we consider UME research focused exclusively on students who are not specialising in pure or applied mathematics. We highlight mostly research that places emphasis on this difference of institutional context. Many of these papers choose an institutional approach, based on ATD.

First, we observe that CERME papers recently took increased interest in the teaching and learning of mathematics in study programmes of other disciplines. This may be linked with the growth of research in the teaching and learning of STEM disciplines at university level and an increasing concern about the quality and perceived relevance of 'service mathematics' courses in study areas such as business or engineering. It could also be the increasing role played by connections with other disciplines in the development of mathematics itself, and thus an increasing awareness of the weakenesses in dealing with service courses in typical university courses. In particular, many papers document and analyse the isolation of mathematics within such programmes, and a lack of connectedness of the curricula integrating mathematics and other disciplines.

For example, Barquero, Bosch and Gascòn (2011) investigated institutional restrictions on teaching modelling in first year 'mathematics for natural science' courses at ten different Spanish universities. They identified a dominant epistemology, 'applicationism', concerning the role attributed to mathematics, which is characterised by "strict separation between mathematics and other disciplines" (p. 1940). Together with other restrictions, applicationism makes it difficult for students to relate mathematics to the discipline of their course.

With regard to modelling tasks, the separation between mathematical and non-mathematical parts, and how it is represented in 'modelling cycles', has been questioned by Biehler et al. (2015). A construed student-expert solution for a task from a German first year course on 'Foundations of Electrical Engineering' shows that "a division into separate phases [. . .] is not adequate" (p. 2066), as the electrical engineering task requires a 'non-separated' knowledge. Similarly, Xhonneux and Henry (2011) conducted a praxeological analysis of how Lagrange's multiplier method is presented in mathematics and economics textbooks, showing that, in mathematics, one begins with Lagrange's Theorem and its justification, which is then applied to particular cases – whereas, in economics, students encounter the technique of Lagrange's multipliers as an integrated part of solving economic optimisation problems.

Naturally, isolated mathematics courses may influence non-specialist students' beliefs and attitudes toward mathematics. Bergsten and Jablonka (2013), drawing on interview data from students of five different engineering Masters level programmes at two Swedish universities, demonstrate that students experience mathematics and their non-mathematical major courses as two separate worlds. Students mostly associate mathematics with merits such as "enhancing individual problem solving abilities" (p. 2294), but not with skills needed in the discipline that is the major focus of study in their programme.

While building our knowledge of current practices and challenges for UME is certainly a main endeavour of RUME, experimenting and developing UME based on RUME research is a growing focus. We now turn to this.

3 Mathematics education at university: what could it be?

Proposing and discussing new possibilities for improving the teaching of university mathematics are important tasks for RUME. In Section 3.1 we focus on UME interventions in specific courses or programmes and experiments with specific ideas for UME, and in Section 3.2 on issues pertaining to the preparation of UME teachers.

68 Winsløw et al.

3.1 Research on and for innovation in UME

Many research results evoked in the previous sections have been followed by interventions tested in class. In this section we present examples of such research-driven experiments in UME. As almost all of those reported on in CERME papers to date, the interventions we highlight in this section are mostly explorative and small-scale (concerning only one university and frequently just one course unit).

Some interventions focus on a particular mathematical topic or aspect of mathematical activity. In such papers, an epistemological analysis is presented as a necessary prerequisite. Hausberger (2015), for example, studied the case of a Year 3 course on Abstract Algebra. He developed a series of activities regarding so-called 'banquets'. These start with a logical investigation of an underlying 'toy' axiomatic system, and the classification of adequate models of that system. This is followed by "the elaboration of an abstract theory of tables and structure theorem for banquets" (p. 2148) and concludes with a link to permutations. This proposal is rooted in the methodology of didactic engineering (Artigue, 2015) and is based on a complex net of frameworks concerning abstraction, semiotics, relations between syntax and semantics, and algebraic structuralism. These are intended as support for the conceptualisation of algebraic structures. The observation of students participating in this intervention evidences difficulties with this conceptualisation, due in particular to an insufficient syntax-semantic dialectic.

Some interventions concern mathematical modelling. Drawing on Barquero et al. (2011) mentioned earlier, and more generally on the ATD, Barquero, Serrano and Serrano (2013) proposed a 'Study and Research Course' – more recently and now commonly known as SRP, 'Study and Research Path' – to first year business and administration students. The teaching started with a single question (e.g. 'How does the population of users of a social network evolve over time?') which was studied during the whole year in workshops. The students had access to some pre-established answers, the *media*, and used these to build their own answers. The course also provided *milieus* to check the correctness of their answers. These questions/answers and media/milieu dialectics are proposed as crucial in avoiding applicationism.

Other interventions aim to address the secondary–tertiary transition (Section 2.3) through support for novice students at their point of entry into the university institution toward meeting institutional expectations. A typical example is the VEMA project, involving several German universities (Biehler, Fischer, Hochmuth & Wassong, 2011). Its design involves a blended bridging course which offers opportunities for self-diagnostic and self-regulated learning. Student evaluations show that students find it useful; yet its impact on students' actual mathematical practices would require further study.

Still, more explorative papers go beyond the secondary–tertiary transition. One example is the 'transition' from a science to a social sciences paradigm needed by postgraduate students in mathematics education whose undergraduate studies are typically in mathematics. This issue is taken up in (Nardi, 2013), with a Masters level intervention based on activities born out of pedagogical principles that foster participation, cultural sensitivity, creativity and critical thinking and with a focus on specific aspects of said paradigm shift, such as 'engaging with research literature in mathematics education', 'forming the conceptual and theoretical framework of a mathematics education research project' and 'choosing and applying data analysis methods'.

In her CERME 5 plenary, Artigue (2007) combines general considerations about educational interventions based on digital technologies with a discussion of theoretical issues concerning RME. She argues that "research dealing with digital technologies reflects the general trends and major evolutions of the field but it is also a source of inspiration for these" (p. 69). For instance, Dreyfus, Hillel and Sierpinska (1999) proposed and analysed a teaching of Linear Algebra using digital geometry software (Cabri-Géomètre). Testing it with undergraduate students, they evidenced that the choices made in the software, for example for the representation of vectors, strongly influenced students' conceptions.

Deploying the ATD construct of media/milieu dialectics already evoked above, Grønbæk and Winsløw (2015) designed an interactive self-study module on complex numbers for non-mathematics students in an introductory mathematics course. The authors explicitly questioned the extent to which students "develop a critical and autonomous relationship to the 'answers' found in the media" (p. 2133). To this aim, these authors intended

> to create a media-milieu dialectics in three ways: as a generator of dynamic text (media with embedded milieu), as a generator of drills for techniques to solve specific tasks with feedback to students' solution proposals (milieu with embedded media), and to explore and exhibit phenomena (media with embedded milieu)
>
> *(p. 2134)*

Whereas students criticised the lack of feedback in an anonymous on-line evaluation, there are indicators that they appreciated the idea of 'drill'.

3.2 Professionalisation of UME practice

As at other educational levels, RUME has increasingly turned its attention to mathematics teachers, to investigate their practices, beliefs, knowledge and preparation, but also to involve them as partners in research – both exploratory and developmental. That the interest in teachers' practices and beliefs has also materialised in RUME was exemplified in Section 2.2. Much less research exists when it comes to university teachers' pedagogical knowledge and its development through formalised education. University teachers are often experts in a field they do research on, and this might be a reason why their knowledge is not much investigated, let alone questioned. Studies of university teachers' knowledge of specific teaching methods (Mesa & Cawley, 2015) and of their practices in problematic

70 Winsløw et al.

or 'new' situations, such as in dealing with digital resources (Gueudet, 2013), contribute toward investigating their professional development as *teachers*, and the contribution RUME may supply to this aim. But organised, deliberate development of UME teachers, based on RUME, is still rare. An exception is a report from Jaworski and Matthews (2011) which presents experiences from a seminar for mathematicians and educators that began as a voluntary and explorative activity, and has since become part of the professional development programme for new mathematics lecturers at their university. But in most other universities we know of, subject-specific educational offers are not available to UME teachers.

4 The future of RUME in CERME

At CERME 10, where a first draft of this paper was presented, the TWG on RUME had more than doubled since its first edition in 2009. With more than 80 participants, the group had to be split in two 'subgroups' (each dealing with about half of the communications) in order to facilitate communication and collaboration. At the final session of the TWG in at CERME 10, the whole group discussed the possibility of creating new groups to cater for the growth. The leader team suggested two emerging themes that could inspire this process: 'mathematics as a service subject' and 'general questions about teaching and learning at university level' (Gonzáles-Martín et al., 2017). Other specialisations could become viable as RUME develops.

We conclude with some general observations and directions for work on UME and RUME which the previous look at the past seem to suggest.

Studies of the current practices and results of UME (Section 2) show that the field is faced with a number of considerable challenges. At CERME as well as in RUME more generally, we still have mostly relatively local knowledge of these challenges, based on small-scale studies; the same goes for studies based on developmental and experimental 'intervention' (Section 3.1). International collaboration, as increasingly fostered and appearing in CERME, should help to establish our knowledge of the challenges, as well as the viability of possible solutions, on a firmer and more substantial basis.

The motivation (and funding!) of RUME is largely connected to the purpose of *solving (some) problems faced by UME*. As in pre-university levels, this naturally brings the teachers into focus, along with the questions of professional development and formal university teacher education (Section 3.2), which we foresee will gain importance both in research questions and as rationale for RUME.

At the same time, there is an increasing realisation of the importance of teachers as not merely 'audiences' or 'objects', but foremost as 'partners' in effecting change. In RUME, one cannot talk of 'teachers' and 'researchers' without recalling that many university mathematics teachers are also researchers, although not typically with RUME as their research specialty. Moreover, in the CERME papers reviewed for this chapter, especially those from the last ten years, we find still more authors and coauthors who, besides (as such) being involved in RUME, are also teachers of mathematics at university, frequently in the contexts their study investigates. This also means that the RUME group at CERME – and conceivably RUME more generally – includes scholars with a variety of backgrounds and experience in research, including pure and applied mathematics. For a number of authors with their PhD in mathematics, including several of those quoted in this chapter, CERME has functioned as a venue for learning about and induction into RUME – not least thanks to its focus on international *collaboration* and constructive *communication*. INDRUM, the International Network for Didactic Research in University Mathematics, which was launched at the end of CERME 9, further adds to the venues serving this purpose. The first congress took place in Montpellier in 2016 and the second is to be held in 2018 in Kristiansand (https://indrum2018.sciencesconf.org/).

Interfaces between UME teachers and RUME can be conceived as especially important to create and sustain research-based innovations in UME. As mentioned in Section 3.1, these currently tend to be quite local and small scale often concerning one course (or part of it) at one university. In the future, we call out for more RUME studies that stretch across institutions, courses and even national borders – and we note that CERME, and INDRUM, should be a promising framework to enable this. We also believe that such developments will help to make results and projects from RUME contribute to more systematic UME teacher education, as briefly touched in Section 3.2.

This, finally, raises the question of *impact* of RUME on UME. Besides collaboration and communication with teachers through professional development and research, we should mention also the potential, but currently quite limited, impact at the level of innovation of curricula and policy. Some of the more global problems identified by RUME, such as transition problems between secondary and tertiary institutions (Section 2.3), the isolation of mathematics courses in non-mathematics majors (Section 2.5) or the need for more systematic and research-based UME teacher development and UME practice innovation (Section 3), clearly call for research and impact at this level.

References

Adam, S. (2002). Towards a common framework for Mathematics degrees in Europe. *Newsletter of the European Mathematical Society* 4526–4528.

Artigue, M. (1994). Didactical engineering as a framework for the conception of teaching products. In R. Biehler, R. W. Scholz, R. Strässer, & B. Winkelmann (Eds.), *Didactics of mathematics as a scientific discipline* (pp. 27–39). Dordrecht: Kluwer.

Artigue, M. (2007). Digital technologies: A window on theoretical issues in mathematics education. *CERME5* (pp 68–82).

Artigue, M. (2015). Perspectives on design research: The case of didactical engineering. In: A. Bikner-Ahsbahs, C. Knipping, & N. Presmeg (Eds.), *Approaches to qualitative research in mathematics education* (pp. 467–496). Dordrecht: Springer.

Asiala, M., Cottrill, J., Dubinsky, E., & Schwingendorf, K. (1997). The development of students' graphical understanding of the derivative. *Journal of Mathematical Behavior, 16*(4), 399–431.

Barquero, B., Bosch, M., & Gascòn, J. (2011). 'Applicationism' as the dominant epistemology at university. *CERME7* (pp. 1937–1948).

Barquero, B., Serrano, L., & Serrano, V. (2013). Creating the necessary conditions for mathematical modelling at university. *CERME8* (pp. 950–959).

Bergsten, C. (2011). Why do students go to lectures? *CERME7* (pp. 1960–1970).

Bergsten, C., & Jablonka, E. (2013). Mathematics as 'meta-technology' and 'mindpower': Views of engineering students. *CERME8* (pp. 2284–2293).

Biehler, R., Fischer, P., Hochmuth, R., & Wassong, T. (2011). Designing and evaluating blended learning bridging courses in mathematics. *CERME7* (pp. 1971–1980).

Biehler, R., Kortemeyer, J., & Schaper, N. (2015). Conceptualizing and studying students' processes of solving typical problems in introductory engineering courses requiring mathematical competences. *CERME9* (pp. 2060–2066).

Biza, I., Souyoul, A., & Zachariades, T. (2006). Conceptual change in advanced mathematical thinking. *CERME4* (pp. 1727–1736).

Breen, S., Larson, N., O'Shea, A., & Pettersson, K. (2015). Students' concept images of inverse functions. *CERME9* (pp. 2228–2234).

Burton, L. (1999). Mathematics and their epistemologies – and the learning of mathematics. *CERME1* (pp. 87–102).

Burton, L. (2004). *Mathematicians as enquirers: Learning about learning mathematics*. Netherlands: Springer.

Cazes, C., Gueudet, G., Hersant, M., & Vandebrouck, F. (2006). Problem solving and web resources at tertiary level. *CERME4* (pp. 1737–1747).

Chevallard, Y. (2002). Organiser l'étude, 3: Écologie et régulation. In J.-L. Dorier et al., *XIe école d'été de didactique des mathématiques* (pp. 41–56). Grenoble: La Pensée Sauvage.

De Vleeschouwer, M., & Gueudet, G. (2011). Secondary-tertiary transition and evolutions of didactic contract: the example of duality in linear algebra. *CERME7* (pp. 2113–2122).

Dorier, J.-L., Robert, A., Robinet, J., & Rogalski, M. (1999). Teaching and learning linear algebra in first year of science university in France. *CERME1* (pp. 103–112).

Dreyfus, T., Hillel, J., & Sierpinska, A. (1999). Cabri-based linear algebra: Transformations. *CERME1* (pp. 209–221).

Farah, L. (2015). Students' personal work in mathematics in French business school preparatory classes. *CERME9* (pp. 2096–2102).

Fischbein, E. (1994). The interaction between the formal, the algorithmic and the intuitive components in a mathematical activity. In R. Biehler, R. W. Scholz, R. Strässer, & B. Winkelmann (Eds.), *Didactics of mathematics as a scientific discipline* (pp. 232–245). Dordrecht: Kluwer Academic Publishers.

González-Martín, A. (2013). Students' personal relationship with series of real numbers as a consequence of teaching practices. *CERME8* (pp. 2326–2335).

González-Martín, A., Biza, I., Cooper, J., Ghedamsi, I., Hausberger, T., Pinto, A., et al. (2017). Introduction to the papers of TWG14: University mathematics education. *CERME10* (pp. 2073–2080).

Grønbæk, N., & Winsløw, C. (2015). Media and milieus for complex numbers: An experiment with Maple based text. *CERME9* (pp. 2131–2137).

Gueudet, G. (2013). Digital resources and mathematics teachers' professional development at university. *CERME8* (pp. 2336–2345).

Hähkiöniemi, M. (2006). Is there a limit in the derivative? Exploring students' understanding of the limit of the difference quotient. *CERME4* (pp. 1758–1767).

Halliday, M. (1978). *Language as social semiotics. The social interpretation of language and meaning*. London, UK: Edward Arnold.

University mathematics education   **73**

Hausberger, T. (2015). Abstract algebra, mathematical structuralism and semiotics. *CERME9* (pp. 2145–2151).

Hiebert, J. (Ed.) (1986). *Conceptual and procedural knowledge: The case of mathematics*. Hillsdale, NJ: Erlbaum.

Holton, D. (Ed.) (2001). *The teaching and learning of mathematics at university level*. New ICMI Study Series, Vol. 7. New York: Kluwer.

Inglis, M., & Simpson, A. (2006). Characterising mathematical reasoning: Studies with the Wason selection task. *CERME4* (pp. 1768–1777).

Jaworski, B., & Matthews, J. (2011). How we teach mathematics: Discourses on/in university teaching. *CERME7* (pp. 2022–2032).

Klein, F. (1908). *Elementharmathematik vom höherem Standpunkte aus*. Leipzig: B. G. Teubner.

Liebendörfer, M., & Hochmuth, R. (2015). Perceived autonomy in the first semester of mathematics studies. *CERME9* (pp. 2180–2186).

Meehan, M. (2007). Student generated examples and the transition to advanced mathematical thinking. *CERME5* (pp. 2349–2358).

Mesa, V., & Cawley, A. (2015). Faculty knowledge for teaching inquiry based mathematics. *CERME9* (pp. 2194–2200).

Misfeldt, M., & Sanne, A. (2007). Flexibility and cooperation: Virtual learning environments in online undergraduate mathematics. *CERME5* (pp. 1470–1479).

Morgan, C., & Tang, S. (2012). Studying changes in school mathematics over time through the lens of examinations: The case of student position. In T. Y. Tso (Ed.), *Proceedings of the 36th Conference of the International Group for the Psychology of Mathematics Education* (Vol. 3, pp. 241–248). Tapei, Taiwan: PME.

Nardi, E. (2013). Shifts in language, culture and paradigm: The supervision and teaching of graduate students in mathematics education. *CERME7* (pp. 2396–2405).

Nardi, E., & Iannone, P. (2006). To appear and to be: Mathematicians on their students' attempts at acquiring the 'genre speech' of university mathematics. *CERME4* (pp. 1737–1747).

Nardi, E., Biza, I., González-Martín, A., Gueudet, G., & Winsløw, C. (Eds.) (2014). Institutional, sociocultural and discursive approaches to research in university mathematics education. *Research in Mathematics Education, 16*(2) [special issue].

Pettersson, K., Stadler, E., & Tambour, T. (2013). Transformation of the students' discourse on the threshold concept of function. *CERME8* (pp. 2406–2415).

Sfard, A. (1991). On the dual nature of mathematical conceptions: Reflections on processes and objects as different sides of the same coin. *Educational Studies in Mathematics, 22*, 1–36.

Sfard, A. (2008). *Thinking as communicating: Human development, the growth of discourses, and mathematizing*. New York: Cambridge University Press.

Sikko, S., & Pepin, B. (2013). Students' perceptions of how they learn best in higher education mathematics courses. *CERME8* (pp. 2446–2455).

Smith, G., Wood, L., Coupland, M., Stephenson, B., Crawford, K., & Ball, G. (1996). Constructing mathematical examinations to assess a range of knowledge and skills. *International Journal of Mathematical Education in Science and Technology, 27*(1), 65–77.

Stadler, E. (2011). The secondary-tertiary transition: A clash between two mathematical discourses. *CERME7* (pp. 2083–2092).

Stadler, E., Bengmark, S., Thunberg, H., & Winberg, M. (2013). Approaches to learning mathematics: Differences between beginning and experienced university students. *CERME8* (pp. 2435–2445).

Tabach, M., Rasmussen, C., Hershkowitz, R., & Dreyfus, T. (2015). First steps in re-inventing Euler's method: A case for coordinating methodologies. *CERME9* (pp. 2249–2255).

Tall, D. (1991). *Advanced mathematical thinking*. Dordrecht: Kluwer.

Tall, D., & Vinner, S. (1981). Concept image and concept definition in mathematics with special reference to limits and continuity. *Educational Studies in Mathematics, 12*, 151–169.

Thoma, A., & Iannone, P. (2015). Analysing university closed book examinations using two frameworks. *CERME9* (pp. 2256–2262).

Trigueros, M., Oktaç, A., & Manzanero, L. (2007). Understanding of systems of equations in linear algebra. *CERME5* (pp. 2359–2368).

Vandebrouck, F. (2011). Students' conceptions of functions at the transition between secondary school and university. *CERME7* (pp. 2093–2102).

Winsløw, C. (2006). Research and development of university level teaching: The interaction of didactical and mathematical organisations. *CERME4* (pp. 1821–1830).

Xhonneux, S., & Henry, V. (2011). A didactic survey of the main characteristics of Lagrange's theorem in mathematics and in economics. *CERME7* (pp. 2123–2133).

6

ARGUMENTATION AND PROOF

Maria Alessandra Mariotti, Viviane Durand-Guerrier and Gabriel J. Stylianides

1 Introduction

The theme of this chapter, argumentation and proof, has been the theme of a CERME working group since the very beginning and is still one of the main themes of research in mathematics education. However, for a long time, proof and proving, and their relation with argumentation, have been themes of debate not only in the community of mathematics educators, but also in the community of mathematicians and in the communities of researchers in history and philosophy of mathematics (Thurston, 1994; Hanna, 1989; Hanna, Jahnke & Pulte, 2010). Proof represents a very special case in respect to other mathematical topics, such as fractions or decimals, and though some might consider it to be a mathematical topic in itself, it is intimately and specifically related to any field of mathematics. In the past years, the debate among the researchers has been very passionate, sometimes reflecting great divergences. This makes this theme fascinating but also testifies to its complexity, mainly in respect to the objective of outlining didactical implications that can be useful in school practice.

Therefore, this chapter will start by addressing some epistemological issues which, since the beginning of CERME, have been at the core of the participants' discussions and that have led to the elaboration of specific epistemological stances concerning the possible tension between argumentation and proof. The following part will be devoted to one specific direction of research that emerged in the group, namely the role of logic in argumentation and proof. Finally, the third part will deal with questions of teaching proof.

2 Historical, epistemological and theoretical issues

The variety of approaches and the diversity of the positions in the working group are related to deep and partly implicit epistemological stances. Thus, epistemological, as

76 Mariotti, Durand-Guerrier & Stylianides

well as historical, issues have been addressed not only in the group discussions, but also explicitly elaborated in the papers presented at CERME. This section offers a critical synthesis of these contributions and attempts to describe how specific issues emerged and evolved within the group.

2.1 The issue of terminology in search of common ground

Since the beginning of the working group, it has been clear that there is a need to communicate common meanings for the terminology we were using in our contributions. However, the first explicit approach to discussing terminological issues, and consequently epistemological issues, started at CERME 4 when the discussion on "The meaning of proof in mathematics education" was opened by Reid (2005) who raised a key question "Is there any prototype of proof?" In line with the analysis carried out in Reid's paper, when confronted with specific examples, participants generally rejected certain arguments as proofs and accepted others, even though they were generally not able to define the characteristics of a prototype proof. Thus, it seemed that, regardless of the different epistemological positions that participants might have held, when asked to judge arguments as proofs or non-proofs, experts find common ground. Nevertheless, such a consensus seemed not to correspond to any specific shared set of explicit characteristics, and not surprisingly it also became clear that there was no common ground concerning teaching approaches toward proof. Thus, the key question arose: Do researchers' epistemologies have any significant influence on their teaching approaches to proof? Such an issue was also mentioned in Mariotti (2006), and was later explicitly addressed by Balacheff (2008), who questioned the existence of a shared meaning of the term 'proof' and consequently of any other term or expression related to it:

> Currently, the situation of our field of research is quite confusing, with profound differences in the ways to understand what is a mathematical proof within a teaching–learning *problématique* but differences which remain unstated.
>
> *(Balacheff, 2008, p. 501)*

The acknowledgement of the crucial influence of researchers' epistemology was perhaps one of the first shared achievements of the group's discussions: this shared point led to a fundamental principle for constructive discussion, namely asking the researchers to make their epistemology explicit. On the basis of this principle, in the following years a fruitful discussion developed within the group.

2.2 Tension between epistemological and didactical issues

The introduction of an educational perspective induces a tension between epistemological and didactical issues that must be resolved. An interesting example of a coherent position in this respect is the 'genetic approach to proof' presented by

Jahnke (2006). Starting from a specific and explicitly stated epistemological position, "to treat geometry in an introductory period as an empirical theory", Jahnke expounds his teaching approach. The relationship between empirical evidence and proof gains a new meaning in the 'genetic teaching approach', consistent with the objective of constructing a theory stemming from empirical observations and deriving new facts from shared stated principles, facts that can be checked by observation. Jahnke claims that there is a kind of dialectic or mutual support between intellectual proofs (when arguments are detached from action and experience) and pragmatic proofs (when arguments are based on empirical data). Such a dialectic sheds a new and refreshing light on the relationship between proof and empirical data and offers a new educative perspective that overcomes well-known conceptions about the role of empirical evidence and counterexample reported by the classic findings of Fischbein and Kedem (1982) and Vinner (1983). We will come back to Jahnke's proposal in the section on 'Teaching of proof' where we will outline the evolution of issues arising within the working group from a teaching and learning perspective.

In addressing different research foci, it became clear that making explicit one's own epistemological positions was not sufficient for making communication and discussion possible. Thus, it was necessary to elaborate specific suitable theoretical tools based on explicit epistemological stances, but also suitable to describe and analyse proof and proving processes as they occur in the school practice.

2.3 The elaboration of specific epistemological stances

The shared concern about making explicit one's own perspectives and assumptions, related to the use of a certain terminology, was at first aimed at improving mutual understanding; however, it was progressively joined by the development of specific epistemological stances into theoretical tools for framing research studies on teaching and learning proof. Such tools were borrowed from different research fields.

First examples can be found in some of the papers presented at CERME 5. For instance, Pedemonte (2007a) and Antonini and Mariotti (2007) exploit specific notions, such as that of Cognitive Unity and that of Theorem, emerging from the elaboration of Toulmin's model for argumentation (Toulmin, 1969) to describe and compare argumentation processes occurring in the production of conjectures and proofs of the conjectured statements. While Stylianides and Stylianides (2007) brought to the attention of mathematics education researchers a rich body of psychological research on deductive reasoning related to the well-known paradigm of mental models (e.g. Johnson-Laird, 1983), other contributions concerned the specific role of tools of analysis coming from logic. Because of the development of this trend of research within the working group, the next section will be devoted to the specific topic of logic.

We intend to focus on two specific cases that we consider exemplars of the development of the group discussion: the case of Cognitive Unity which introduced

78 Mariotti, Durand-Guerrier & Stylianides

Toulmin's model for argumentation as one of its parts and thus was at the origin of the spreading of the use of this model among some of the members of the group; and the case of Habermas' notion of rationality (Habermas, 2003).

The case of the notion of Cognitive Unity

The notion of Cognitive Unity, first introduced in the discussions of the group by Pedemonte at CERME 2 (Pedemonte, 2002), offers one possible research direction to resolve the tension between epistemological and didactical issues concerning the relationship between argumentation and proof. Elaborating on a previous definition (Boero, Garuti & Mariotti, 1996), the main objective of this theoretical construct was that of providing a tool of analysis to investigate the possible gap between argumentation and proof. This first and subsequent contributions presented by Pedemonte at CERME 3 (2004) and CERME 5 (2007a), and later published in extended form (Pedemonte, 2007b), were centred on the structural analysis of arguments, where the main theoretical tool used was borrowed from Toulmin and adapted to the specific case of producing conjectures as answers to open problems.

The effectiveness of Toulmin's model in analysing the complex structure of an argumentative chain is not limited to highlighting the possible gap between argumentation and proof. It also provides a powerful tool to discriminate between different types of arguments. As a matter of fact, in some papers presented at the subsequent CERMEs, we can observe the appearance of an explicit reference to using Toulmin's model, though not by referring to the notion of Cognitive Unity, nor by referring to the specific case of producing conjectures. For example, at CERME 7, a number of contributions integrated this model into their theoretical framework. This is the case, for example, of Cramer (2011) who develops a methodology combining Toulmin's scheme and a collection of topical schemes with an epistemic action model in order to shed light on the relations between argumentation and knowledge construction.

While Toulmin's model has shown its usefulness, its limits have emerged as well. It has become clear that, in some cases, students meet difficulties inherent in the lack of 'structural continuity' when they have to move from creative arguments supporting the validity of a statement to the organisation of these same arguments in a deductive chain to produce an acceptable proof. Such difficulties can be clearly described in the frame of Toulmin's model, but their origin cannot be fully explained by the model. Thus, beyond awareness of these difficulties, the need arose of interpreting their origins, in order to investigate how they could be overcome.

The case of Habermas' model of rationality

At CERME 6, Morselli and Boero (2010) presented the analysis of some examples applying the model of rationality introduced by Habermas (2003), showing

Argumentation and proof **79**

the viability and usefulness of such a model in the special case of conjecturing and proving. Habermas' model, conceiving rational behaviour on the basis of the relationships between teleological, epistemic and communicative rationality, seemed to offer a fruitful tool for analysing the dynamics of students' arguments. The suggestion of integrating other models with Habermas' model was followed by some of the researchers attending CERME 7. For instance, Perry, Molina, Camargo and Samper (2011) presented a paper where they analysed the proving activity of a group of three university students solving a geometrical problem in a dynamic geometry system. Solving the problem requires formulation of a conjecture and justification in a theoretical system. For their analyses, the authors refer to the integration of Toulmin's and Habermas' models and define specific components of a successful performance.

Similarly, at CERME 7, two other papers on research from this perspective were presented and discussed. One is the paper by Arzarello and Sabena (2011) which draws explicitly on the integration of the models of Toulmin and Habermas. In particular, Arzarello and Sabena used the idea of 'meta-cognitive unity' in order to give reasons for success and difficulties in indirect proofs. The other is a paper by Boero (2011), which constitutes a continuation of his work on analysing classroom discussions through combining the Toulmin and the Habermas models. According to the different foci of their research studies, the different authors make use of both models.

However, it is only more recently, in a paper presented at CERME 9 by Boero (2015), that the need for, and the potentialities of, integrating the two theoretical tools was explicitly discussed, and Cognitive Unity was analysed through both Toulmin's model and Habermas' approach based on the three components of rationality. Boero's paper starts with the story of a protocol and the problems that the complexity of its interpretation posed to the researcher: how to interpret the apparently chaotic argumentation process supporting the validation of a quite obvious statement. A mere description in terms of its structure according to Toulmin's model was not enough to do justice to the complexity of the student's mental process. As the author writes

> The [first] need suggested us to try and adapt Habermas' construct of rational behaviour to Ivan's problem solving – as a process driven by intentionality to get a correct result by enchaining correct steps of reasoning, and to communicate it in an understandable way in a given community.
>
> *(op. cit., p. 96)*

3 The role of logic in argumentation and proof

The role of logic in argumentation and proof is a rather controversial issue in the international mathematics education research community (Durand-Guerrier, Boero, Douek, Epp & Tanguay, 2012). From CERME 3, there were regularly papers referring explicitly to issues of logic. We present below the two main aspects that were discussed over the years.

3.1 Quantified logic as a tool for analysing proof and proving

In the 1980s, following psychological studies seeming to show that logic was not relevant for understanding human reasoning (e.g. Johnson-Laird, 1983), the idea that formal logic was not a model for how people make inferences was rather popular among mathematics educators and researchers. Nevertheless, as underlined by Stenning and van Lambalgen (2008, p. xiii), "the picture of logic current in psychology and cognitive science is completely mistaken". As a matter of fact, these studies criticised by Stenning and van Lambalgen generally rely on analysis in the frame of the propositional calculus in which the sentence is the unit of analysis, while human reasoning involves an on-going interaction between syntax and semantics (Durand-Guerrier et al., 2012, p. 372).

Durand-Guerrier (2004) claimed that from a didactic perspective, it was necessary to consider not only sentences but also objects, properties and relationships in order to analyse students' reasoning. She brought into consideration the model-theoretic approach introduced by Tarski (1936), which distinguishes three dimensions: the syntactic (the linguistic form), the semantic (the reference objects), and the pragmatic (the context, and the subject's knowledge in the situation), considering these distinctions as important in order to foster argumentation and proving processes of students. In her paper at CERME 4, Durand-Guerrier (2006) claimed that 'natural deduction' is a powerful tool to analyse proofs from a didactic perspective, for teachers as well as researchers, and argued that the model-theoretic approach introduced by Tarski calls for continuity between argumentation and proof, in contrast with the discontinuity seen by researchers working in a cognitive approach. This point has been discussed anew in the paper by Barrier, Mathe and Durand-Guerrier (2010) at CERME 6, opening a discussion with papers whose main reference was Toulmin, such as Pedemonte's (2007a) at CERME 5. Barrier et al. supported the importance of taking into account the distinction between truth and validity, and this issue emerged again in CERME 8 with the paper by Mesnil (2013).

The importance of taking quantification issues into account is highlighted in the paper by Deloustal-Jorand (2007) presented at CERME 5. It is well known that students face strong difficulties when dealing with implication, while the complexity of this concept is often underestimated by teachers. In her paper Deloustal-Jorand analyses mathematical implication from three different points of view: formal logic, deductive reasoning, and sets. A didactical engineering, based on the assumption of the necessity of making these points of view interact, is carefully described and its implementation is discussed, showing how a suitable situation can raise the issue of implication.

3.2 Relationships between logic and language in proof and proving

The relationship between logic and language in proof and proving has also been a recurrent theme that has been widely discussed, for example at CERME 8 with five papers focusing on this theme (Azrou, 2013; Chellougui & Kouki, 2013; Cramer, 2013; Mesnil, 2013; Müller-Hill, 2013).

The main points that emerge from the discussion are:

- identification, in the relationship between logic and language, of aspects that are likely to be an obstacle for developing proof and proving skills, and of aspects that are likely to favour it;
- the value of teaching logic for fostering proof and proving competencies;
- the usefulness for teachers, of logical analysis in mathematical discourse, and how to do it; and
- the relationships between logic and formalisation.

The following questions were discussed during the sessions without reaching a consensus:

- Should we consider logical competencies and/or logic as a body of knowledge, logic as a theory modelling human reasoning and/or as a theory aiming to control validity of proof?
- Should logical proof be considered both in terms of a final product and as a process in action, or should we consider logical proof only as a final product, considering that human reasoning does not require logical competences?
- Is it relevant to teach logic as a separate subject at secondary school or not?

Two main lines of research emerged from our discussions. The first concerns the role of logic as a possible tool for researchers and the implication that research findings could have for teachers. The second concerns the relevance, or not, of teaching logic in order to foster proof competencies and, in case of a positive answer, how to do this.

As far as the first line is concerned, the logical analysis of a statement appears to be a fruitful means to deepen and enrich an *a priori* analysis of a task, for example by revealing possible unexpected ambiguities that could impede the understanding of the mathematical statement being considered, or by favouring the identification of didactical variables in order to enlighten possible choices for the study of a given concept. In addition, logical analysis of mathematical discourse can be used to analyse students' productions in order to better understand them and, in some cases, open new interpretations of their genesis. Last, but not least, logical systems such as natural deduction (Durand-Guerrier, 2006) provide researchers with powerful tools to analyse proofs both, *a priori* and *a posteriori*, in order to identify possible invalid steps that could be hidden when written in natural language.

This emergence of logical issues in European research on argumentation and proof discussed at CERME drew the interest of non-European researchers involved in the 19th ICMI Study on Proof and Proving in Mathematics Education, where several papers on this topic were submitted. As a consequence, a chapter was devoted explicitly to this question in the Study book (Durand-Guerrier et al., 2012).

82 Mariotti, Durand-Guerrier & Stylianides

Three active members of the CERME working group on 'Argumentation and Proof' were involved, as were colleagues from the USA and Canada. In this chapter, the relevance of, and interest in, including some instruction in logic in order to foster competence with proof in the mathematics classroom is examined. A contradiction between two assumptions is discussed: on the one hand, doing mathematics at the secondary level in itself seems sufficient to develop logical abilities; on the other hand, many tertiary students lack the logical competence needed to learn advanced mathematics, especially proof and other mathematical activities that require deductive reasoning. The authors support the claim that it is necessary and possible to introduce in the school curriculum activities aiming at explicitly developing the logical competences required by advanced mathematical activity.

4 Teaching of proof

In the previous section on logic, we already touched upon issues of teaching proof, such as, for example, the issue of including some instruction in logic in order to foster competence with proof in the mathematics classroom. This section will further consider issues of teaching proof in the classroom, focusing more on issues of epistemology and terminology that we introduced at the beginning of the chapter, considering those issues both from the point of view of the teacher and from the point of view of the students. We use the term 'classroom' broadly to denote formal learning settings at both the school and the university levels, including mathematics teacher education.

Issues of teaching proof started to receive more attention in the discussions of the working group during the most recent decade of CERME. For example, in their introduction to the papers from CERME 6, the working group leaders noted that the discussions did not focus much on didactic issues related to proof, with the only exception being the teaching intervention discussed in Douek's (2010) paper. After the discussions that took place during the working group sessions at CERME 7, the group concluded that more research and discussions on the teaching of proof were needed during upcoming CERMEs, especially in relation to designing activities that could be used to foster proof skills across the levels of education.

The epistemological and terminological issues we discussed previously in this chapter received more attention in the earlier discussions of the working group. This is not surprising as it was sensible for the group to try to first clarify epistemological and terminological issues before focusing on the design of classroom interventions to promote school and university students' understandings of proof. We do not claim, of course, that epistemological or terminological issues have now been clarified completely, and certainly not that they have been resolved, in the context of CERME or in the field of mathematics education more broadly. Yet the evolution of the issue of the design of classroom interventions in the area of proof within the working group mirrors its general evolution outside the group in the mathematics education community and was influenced by developments in the field in relation to (1) terminological and epistemological issues and (2) theoretical constructs and

Argumentation and proof **83**

frameworks that helped explain classroom phenomena. We exemplify each of these points in the next two subsections.

4.1 Terminological and epistemological issues and the design of classroom interventions

In relation to terminological issues, as presented above, the discussion about the meaning of proof was taken up by Andreas Stylianides who presented a conceptualisation of proof at CERME 5 that appeared also in a journal article later that year (Stylianides, 2007). This conceptualisation, while not comprehensive, is sufficiently flexible to allow description of proof across different levels of education and integrates different perspectives on proof discussed in the literature. These perspectives include, for example, the view of proof as a logical deductive chain of arguments and other views that highlight the cognitive or social aspects of proof. The conceptualisation has been used in various classroom-based research studies in the area of proof. For example, it constituted the basis for designing a classroom intervention at the university level (Stylianides & Stylianides, 2009). It was also used and elaborated by other researchers, some of whom presented papers at CERME. For instance, the conceptualisation informed the theoretical framework that guided Morselli's teaching intervention with lower secondary school students aimed at introducing proof by using algebraic language as a proving tool (Morselli, 2013) and, more recently at CERME 10, it was used to describe the development of a 'toolbox' in Reid and Vargas' (2017) proof-based teaching intervention that was used to help third-grade students develop their knowledge of division of natural numbers.

In relation to theoretical constructs, as mentioned above, at CERME 4 Jahnke (2006) used the metaphor of a *theoretical physicist* to explain common student thinking about the meaning of empirical verification in the proving process, arguing that this student thinking occurs with a certain necessity and is a consequence of a meaningful behaviour. Jahnke then went on to propose a didactical approach, which he further elaborated in Jahnke (2007) and illustrated in the domain of geometry. In this approach, geometry is treated in the early stages as an empirical theory that bears similarity with the natural sciences, such as physics. The didactical approach is

> centred around the idea that inventing hypotheses and testing their consequences is more productive for the understanding of the epistemological nature of proof than forming elaborate chains of deductions [. . .] In this approach proving and forming models get in close contact.
>
> *(Jahnke, 2007, p. 79)*

This didactical approach, together with other similar approaches in which experiments at the school level are combined with theoretical aims, were discussed at CERME (e.g. Bartolini-Bussi, 2010) and beyond (e.g. Boero, Garuti & Lemut, 2007) and informed the classroom-based intervention study by Jahnke and Wambach (2013) which we discuss below.

The analysis of the proving process through the notion of Cognitive Unity and its elaboration through Habermas' model of rationality, discussed earlier, also has interesting implications for teaching and learning. Indeed, Morselli and Boero (2010) claim the possibility of dealing with the approach to theorems and proving in school as a process of scientific enculturation consisting of the development of a special kind of rational behaviour, according to that described by Habermas' model.

4.2 Theoretical constructs for explaining classroom phenomena

In this section, we discuss two studies that aimed to promote students' understandings of proof. The first is Douek's (2010) study presented at CERME 6, as an example of an early CERME paper that focused on issues of teaching in the area of proof. The other is the intervention study we mentioned earlier by Jahnke and Wambach (2013), as an example of a study that built on ideas discussed in earlier CERMEs.

Douek (2010) presented a theoretical approach to deal with theorems for which according to the notion of Cognitive Unity a serious gap appeared between argumentation and proof. Such an approach could be used with lower secondary students to help them develop an awareness of some important features of proving theorems. The approach, exemplified in a task sequence specific to the Pythagorean theorem, is based on the key idea of engaging students in conjecturing activities and guiding their work to proof construction, finally helping them through discussion and 'story making'. The latter can help students make sense of the links between statements and arguments in the proof, to focus on important characteristics of the organisation of a proof.

Jahnke and Wambach (2013) conducted an intervention involving eighth-grade students in Germany, which aimed to help develop students' understanding that proofs are based upon certain assumptions. The intervention took place during eight geometry lessons and was situated in the attempts of ancient Greeks to model the so-called 'anomaly of the sun'. The students were asked to assume that the methods and tools that were known to ancient astronomers at the time were available to them. These restrictions were similar to the restrictions imposed on the Cabri tools available to the students in another classroom-based intervention discussed by Mariotti (2011) in CERME 7 and were an important factor that contributed to students becoming more conscious of the role of assumptions in building a deductive theory.

5 Concluding remarks

In the short space of this chapter we have tried to give a snapshot of the progress of the 'Proof and Argumentation' working group. Of course, reading the original

Argumentation and proof **85**

papers gives a deeper and more analytic perspective on the different topics that were presented, discussed and elaborated within the group, and for this reason we hope that this overview has triggered the curiosity of the reader. The initial disputes about terminology, the gradual emergence of new theoretical tools and new thematic issues, as well as the contribution of specific perspectives such as that of logic has certainly contributed to the progress of research on argumentation and proof.

Before presenting some of the emergent issues and open questions that we would like to propose to the research community, let us mention some of what we consider good practices that, over the years, have been implemented in the organisation of the working group and that we think have been effective for the development of the group as a research community.

5.1 Practices in the organisation of the working group over the years

Participation in the group has been rather stable, which means that a number of participants attended the working group activities for many years, so that we can really think of the group as a community where researchers know each other's work and not only support but also integrate the different theoretical perspectives that are presented and discussed. Of course, the size of the working group has increased over time, but in spite of the possible organisational difficulties, this has allowed for highly rewarding debates, and definitely contributed to building up a community of researchers.

For some of the youngest among us, participation started during or immediately after their doctoral studies, and continues up to the present. Some of these have by now served as group co-leaders, such as Samuele Antonini, Bettina Pedemonte, Kirsti Hemmi and Christine Knipping. Some of the contributions that were first presented at CERMEs were subsequently published in international journals, so that we can reasonably think of the positive effect of the group discussion on both specific and general issues explored in these studies. As a matter of fact, the ZDM special issue on the theme of 'Argumentation and Proof', edited by Balacheff and Mariotti (2008), though not explicitly mentioned as sprouting from CERME, collected contributions that, in the majority of the cases, were expanded versions of earlier papers presented at the CERME group.

It is a general policy of the working group that participants read the papers in advance in order to minimise the time needed for presentations and maximise the time for the discussion. In addition, the leader and the co-leaders carry out organisational work – sometimes particularly complex – in order to group the presentations, and consequently the discussions, according to specific research perspectives. Sometimes, the success of the discussion was fostered by asking specific participants to act as discussants, preparing specific questions to stimulate the debate. This has been particularly effective in creating a collaborative attitude. Similarly, when time allows, after the discussion of the papers, the working group participants split into small groups, with the objective of deepening specific theoretical or

5.2 Emergent issues and open questions

As stated above, it is only recently that issues of teaching proof have received attention, following the suggestion that emerged explicitly at CERME 7 to develop design-based studies focused on introducing students to proof, including at the early grades of school. In this new trend of investigation, we find fresh issues emerging.

The first has already been addressed at CERME 9 and concerns the analysis of textbooks. A shared assumption is that textbooks play a major role in school practice, orienting and shaping teachers' approaches to specific subjects, and influencing, through tasks and texts, students' ways of thinking about mathematics. On the basis of this assumption it becomes a focus to investigate how argumentations and proofs are presented in textbooks. A range of papers have addressed this issue in European countries and elsewhere: Israel (Silverman & Even, 2015), Spain (Conejo, Arce & Ortega, 2015), Sweden and Finland (Bergwall, 2015), as well as in the USA (Thompson, Senk & Johnson, 2012). This new direction of research is very promising, though the originality of the theme gives rise to a number of methodological problems, for instance, how to select the unit of analysis (e.g. task, lesson, chapter, etc.) and how to identify proof and argumentation tasks in textbooks (e.g. looking for keywords like 'prove' or 'show', as it was done in most of the above-mentioned studies, is only one possible way).

A second direction of research is strictly related to promoting design-based investigation in the classroom; as a matter of fact, when looking at the classroom we are led to move the focus from the learners to the teacher. This shift of focus raises different issues concerning the analysis of the teacher's role in designing and managing didactic situations concerning argumentation and proof, and concerning the identification of specific teacher competencies requested for playing this role. Argumentation and proof are related to specific competences that must be differently articulated in different mathematical domains, and also related to disciplines other than mathematics. Not much has been done to characterise these competences and consequently to outline possible elements on which teacher education can be based. All this calls for new research. Unlike other topics, not many studies have been carried out concerning teacher education on argumentation and proof.

A third direction of research, partially intersecting the others, concerns a specific aspect relating logic and proof. As said above, consensus has not been reached on the relationship between logic and proof; nevertheless, recent contributions concerning advanced mathematical thinking have highlighted interesting new aspects concerning the meta-theoretical level that call for further investigation. In spite of a shared opinion that formalisation is essential for mathematical work, in particular to control correctness of both definitions and proofs, it appears that for many students, formalisation, and specifically symbolic expressions, is a challenging obstacle, as shown in Azrou's paper (2013). The fact that at this level of education formalisation is an underestimated issue is shown by the variability of formalisation in text books, as presented by Chellougui and Kouki (2013) in the case of continuity. This opens a research area questioning the relationship between conceptualisation and formalisation at the advanced mathematics level.

Acknowledgement

We are deeply indebted to all the colleagues who read and commented on the draft version of the manuscript. A very special thanks to David Reid who kindly accepted to revise an advanced draft; his precious work definitely improved our text.

References

Antonini, S., & Mariotti, M. A. (2007). Indirect proof: An interpreting model. *CERME5* (pp. 541–550).

Arzarello, F., & Sabena, C. (2011). Meta-cognitive Unity in indirect proof. *CERME7* (pp. 99–109).

Azrou, N. (2013). Proof in Algebra at the university level: Analysis of students' difficulties. *CERME8* (pp. 76–85).

Balacheff, N. (2008). The role of the researcher's epistemology in mathematics education: An essay on the case of proof. *ZDM. The International Journal on Mathematics Education*, *40*, 501–512.

Balacheff, N., & Mariotti, M. A. (2008). Introduction to the special issue on didactical and epistemological perspectives on mathematical proof. *ZDM. The International Journal on Mathematics Education, 40*, 341–344.

Barrier, T., Mathe, A. C., & Durand-Guerrier, V. (2010). Argumentation and proof: A discussion about Toulmin's and Duval's models. *CERME6* (pp. 191–200).

Bartolini-Bussi, M. G. (2010). Experimental mathematics and the teaching and learning of proof. *CERME6* (pp. 221–230).

Bergwall, A. (2015). On a generality framework for proving tasks. *CERME9* (pp. 86–92).

Boero, P. (2011). Argumentation and proof: Discussing a 'successful' classroom discussion. *CERME7* (pp. 120–130).

Boero, P. (2015). Analysing the transition to epsilon-delta Calculus: A case study. *CERME9* (pp. 93–99).

Boero, P., Garuti, R., & Lemut, E. (2007). Approaching theorems in grade VIII. In P. Boero (Ed.), *Theorems in schools: From history, epistemology and cognition to classroom practice* (pp. 249–264). Rotterdam: Sense Publishers.

Boero, P., Garuti, R., & Mariotti, M. A. (1996). Some dynamic mental processes underlying producing and proving conjectures. *Proceedings of the 20th PME Conference, Valencia, Spain, 2*, 121–128.

Chellougui, F., & Kouki, R. (2013). Use of formalism in mathematical activity – case study: The concept of continuity in higher education. *CERME8* (pp. 96–105).

Conejo, L., Arce, M., & Ortega, T. (2015). A case study: How textbooks of a Spanish publisher justify results related to limits from the 70s until today. *CERME9* (pp. 107–113).

Cramer, J. (2011). Everyday argumentation and knowledge construction in mathematical tasks. *CERME7* (pp. 141–150).

Cramer, J. C. (2013). Possible language barriers in processes of mathematical reasoning. *CERME8* (pp. 116–125).

Deloustal-Jorrand, V. (2007). Relationship between beginner teachers in mathematics and the mathematical concept of implication. *CERME5* (pp. 601–610).

Douek, N. (2010). Approaching proof in school: From guided conjecturing and proving to a story of proof construction. *CERME6* (pp. 332–342).

Durand-Guerrier, V. (2004). Logic and mathematical reasoning from a didactical point of view: A model-theoretic approach. *CERME3*: www.erme.tu-dortmund.de/~erme/CERME3/Groups/TG4/TG4_Guerrier_cerme3.pdf.

Durand-Guerrier, V. (2006). Natural deduction in Predicate Calculus: A tool for analysing proof in a didactic perspective. *CERME4* (pp. 409–419).

Durand-Guerrier, V., Boero, P., Douek, N., Epp, S., & Tanguay, D. (2012). Examining the role of logic in teaching proof. In G. Hanna & M. de Villiers (Eds.), *ICMI Study 19 Book: Proof and proving in mathematics education* (pp. 369–389). New York: Springer.

Fischbein, E., & Kedem, I. (1982). Proof and certitude in the development of mathematical thinking. *Proceedings of the sixth international conference for the psychology of mathematics education* (pp. 128–131). Antwerp, Belgium: Universitaire Instelling Antwerpen.

Habermas, J. (2003). *Truth and justification.* Cambridge, MA: MIT Press.

Hanna, G. (1989). More than formal proof. *For the Learning of Mathematics, 9*(1), 20–25.

Hanna, G., Jahnke, H. N., & Pulte, H. (Eds.) (2010). *Explanation and proof in mathematics: Philosophical and educational perspectives.* New York: Springer.

Jahnke, H. N. (2006). A genetic approach to proof. *CERME4* (pp. 428–437).

Jahnke, H. N. (2007). Proofs and hypotheses. *ZDM–The International Journal on Mathematics Education, 39*(1–2), 79–86.

Jahnke, H. N., & Wambach, R. (2013). Understanding what a proof is: A classroom-based approach. *ZDM – The International Journal on Mathematics Education, 45*, 469–482.

Johnson-Laird, P. N. (1983). *Mental models: Towards a cognitive science of language, inference, and consciousness.* Cambridge, MA: Harvard University Press.

Mariotti, M. A. (2006). Proof and proving in mathematics education. In A. Gutiérrez & P. Boero (Eds.), *Handbook of research on the psychology of mathematics education* (pp. 173–204). Rotterdam: Sense Publishers.

Mariotti, M. A. (2011). Proof and proving as an educational task. *CERME7* (pp. 61–89).

Mesnil, Z. (2013). New objectives for the notions of logic teaching in high school in France: Complex request for teachers. *CERME8* (pp. 166–175).

Morselli, F. (2013). Approaching algebraic proof at lower secondary school level: Developing and testing an analytical toolkit. *CERME8* (pp. 176–185).

Morselli, F., & Boero, P. (2010). Proving as a rational behaviour: Habermas' construct of rationality as a comprehensive frame for research on the teaching and learning of proof. *CERME6* (pp. 211–220).

Müller-Hill, E. (2013). The epistemic status of formalizable proof and formalizability as a meta-discursive rule. *CERME8* (pp. 186–195).

Pedemonte, B. (2002). Relation between argumentation and proof in mathematics: Cognitive unity or break. *CERME2* (Vol. 2, pp. 70–80).

Pedemonte, B. (2004). What kind of proof can be constructed following an abductive argumentation? *CERME3*: www.erme.tu-dortmund.de/~erme/CERME3/Groups/TG4/TG4_Pedemonte_cerme3.pdf.

Pedemonte, B. (2007a). Structural relationships between argumentation and proof in solving open problems in algebra. *CERME5* (pp. 643–652).

Pedemonte, B. (2007b). How can the relationship between argumentation and proof be analysed? *Educational Studies in Mathematics, 66*(1), 23–42.

Perry, P., Molina, O., Camargo, L., & Samper, C. (2011). Analyzing the proving activity of a group of three students. *CERME7* (pp. 151–160).

Reid, D. (2005). The meaning of proof in mathematics education. *CERME4* (pp. 458–467).

Reid, D., & Vargas, E. V. (2017). Proof-based teaching as a basis for understanding why. *CERME10* (pp. 235–242).

Silverman, B., & Even, R. (2015). Textbook explanations: Modes of reasoning in 7th grade Israeli mathematics textbooks. *CERME9* (pp. 205–212).

Stenning, K., & van Lambalgen, M. (2008). *Human reasoning and cognitive science*. Cambridge, UK: Bradfors Books.

Stylianides, A. J. (2007). Proof and proving in school mathematics. *Journal for Research in Mathematics Education, 38*, 289–321.

Stylianides, A. J., & Stylianides, G. J. (2007). The mental models theory of deductive reasoning: Implications for proof instruction. *CERME5* (pp. 665–674).

Stylianides, G. J., & Stylianides, A. J. (2009). Facilitating the transition from empirical arguments to proof. *Journal for Research in Mathematics Education, 40*, 314–352.

Tarski, A. (1936). *Introduction to logic and to the methodology of deductive sciences* (4th ed., 1994). New York: Oxford University Press.

Thompson, D. R., Senk, S. L., & Johnson, G. J. (2012). Opportunities to learn reasoning and proof in high school mathematics textbooks. *Journal for Research in Mathematics Education, 43*, 253–295.

Thurston, W. P. (1994). On proof and progress in mathematics. *Bulletin of The American Mathematical Society, 30*(2), 161–177.

Toulmin, S. E. (1969). *The uses of arguments*. Cambridge, UK: Cambridge University Press.

Vinner, S. (1983). The notion of proof: Some aspects of students' views at the senior high level. *Proceedings of the Seventh International Conference for the Psychology of Mathematics Education* (pp. 289–294).

7

THEORY–PRACTICE RELATIONS IN RESEARCH ON APPLICATIONS AND MODELLING

Morten Blomhøj and Jonas Bergman Ärlebäck

1 Modelling and application as a field of research

From 2006, CERME has included a Thematic Working Group on applications and modelling (WGAM). Since then, the WGAM has been part of the infrastructure of the international research community on the teaching and learning of mathematical modelling and applications (ICTMA; see www.ictma15.edu.au/). The WGAM organising teams have been recruited among the European members of the international community, and the WGAM has attracted experienced researchers as well as welcomed new researchers to the field, primarily from Europe, but also from outside Europe.

In general, the research field has developed in close interplay with the development of the practices of teaching, and as surveyed, summarised, and documented in the ICMI-study on modelling and applications (Blum, Galbraith, Henn & Niss, 2007), research has clarified arguments for including models and modelling in general mathematics education, conceptualised modelling competency at different levels, identified teaching and learning obstacles related to models and modelling, and exemplified the potential for enhancing students' learning through modelling activities. Frameworks for justifying, designing and implementing modelling and applications at different levels in educational systems, as well as methodologies supporting systematic collaboration between researchers and teachers, are continuously being developed, refined and applied in the research field. The WGAMs continue to contribute to these important lines of research.

In line with the ERME spirit, the WGAMs have framed presentations and discussions of research related to current challenges and ongoing developments in practices of teaching modelling and application in, primarily, European countries. Accordingly, the relationship between the development of teaching practices and theories on the teaching and learning of modelling and applications has been one

Research on applications and modelling **91**

of the dominant themes throughout the WGAMs. Moreover, different educational aims for including modelling and applications lead to different needs for theoretical underpinnings of teaching practices, and thus also to different types of research.

In his ICME-12 plenary address, Werner Blum surveyed the research achievements in the field from the perspective of the quality of teaching of applications and modelling at the secondary level. Based on empirical findings, Blum identified ten important aspects of successful and productive teaching methodologies. However, he ends with the following:

> I would like to emphasise that all these efforts will not be sufficient to assign applications and modelling its proper place in curricula and classrooms and to ensure effective and sustainable learning. The implementation has to take place systemically, with all system components collaborating: curricula, standards, instruction, assessment and evaluation, and teacher education.
>
> *(Blum, 2015, p. 87)*

Thus, despite the progress made in research and its influence on curricula, there are still major challenges concerning the development of practices of teaching modelling and applications as an integrated element in mathematics teaching in general and in tertiary education. A necessary, but of course not sufficient, condition for overcoming these challenges is to further strengthen the interplay in research between the development of theory and practice. This challenge is clearly reflected in the WGAMs. Therefore, we have decided to focus our analysis, on the relationship between theory and teaching practice as manifested in the WGAM research.

2 Characterising the contributions to the WGAMs

In the seven congresses CERMEs 4–10 (2006–2017) there have been in total 122 papers (not including five shorter introductory papers) and ten posters presented in the WGAMs as shown in Figure 7.1. The contributors are from 25 different countries from four continents, although 84% of the contributors are affiliated with European countries. Twenty-four per cent of the contributors are affiliated with Germany, 18% with Spain, and around 5% with each of France, Sweden, the Netherlands and the United Kingdom. The US (5%) and Mexico (4%) are the two major non-European contributors.

The WGAM contributions display a broad shared understanding of the basic notions and the research achievements, although the field is still being researched within different theoretical frameworks and perspectives. Some of these differences seem to be founded in national traditions and preferences. The most dominant ones are the German and Scandinavian holistic and project-oriented approaches for developing and researching modelling competency (Blomhøj & Jensen, 2003; Frejd, 2013); the French and Spanish ATD-based (Anthropological Theory of the Didactics) approach for theorising the teaching and learning of mathematics in general (García, Gascón, Higueras & Bosch,

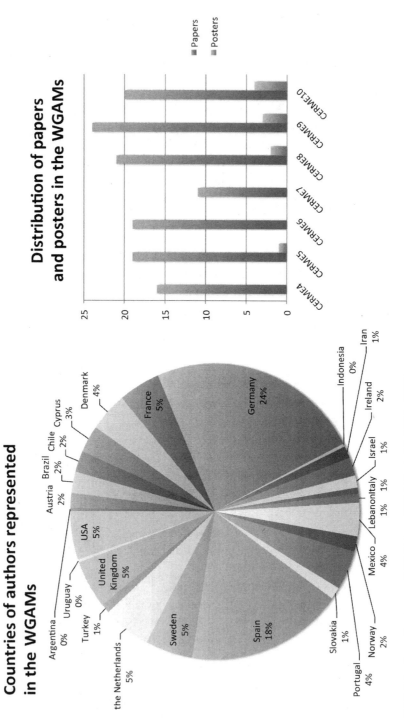

FIGURE 7.1 Contributions to WGAM at CERMEs 4–10 (2006–2017)

Research on applications and modelling **93**

·2006; Barquero, Serrano & Serrano, 2015); the Realistic Mathematics Education (RME) perspective on modelling applied by many Dutch researchers and others (Treffers, 1987; Vos, 2006). However, other theoretical perspectives across the national trends can also be identified. The cognitive perspective on modelling (Borromeo Ferri, 2006; Roorda, Vos & Goedhart, 2007) and the critical mathematics education approach toward the teaching and learning of applications and modelling (Skovsmose, 1994; Barbosa, 2007) are two examples of other approaches. In addition, recently an increasing number of WGAM contributions have drawn on the models and modelling perspective developed by Lesh and colleagues (Lesh & Doerr, 2003; Sriraman, 2006). For each perspective, we give a reference defining the perspective and a WGAM paper exemplifying it.

Already at CERMEs 4 and 5, the diversity of theoretical perspectives and approaches used in the contributions to the WGAMs and in the research field more generally were identified and discussed. This work resulted in a characterisation of the different research perspectives found in the field. A first characterisation was presented in the introduction to the WGAM proceedings from CERME 5 (Kaiser, Sriraman, Blomhøj & García, 2007). The work, initiated at CERME 4 was elaborated and used to structure and discuss the status of and developments in field in two thematic issues of *ZDM* in 2006 (nos. 38 and 39) (Kaiser, Blomhøj & Sriraman, 2006). This type of general characterisation of the research facilitates the formation of a shared understanding of the field and can help in moving the discussions in directions of challenges to be addressed and away from pro and con discourses on particular theoretical perspectives or frameworks. In our opinion, the WGAMs have contributed to the research field in exactly that way.

While representing a large diversity in theoretical perspectives, the WGAM contributions have many commonalities. To a great extent, there is a shared understanding in the papers of basic notions such as mathematical model, modelling process and modelling competency, and of the principal difficulties related to the teaching and learning of modelling and applications. Many papers report research on the development and testing of didactical designs for teaching modelling and applications in specific educational contexts. Over 75% of the 122 papers analysed have an empirical element connected to the teaching practice mainly at secondary level or at the first year of tertiary mathematics.

To some degree, the focus on the practices of teaching modelling and applications found in WGAM reflects the general situation in the research field. However, the format for CERME contributions in itself encourages research papers reporting on small-scale projects or parts of larger projects in progress. Some of the papers – especially among those related to tertiary education – are reporting developmental projects related to the author(s)' own teaching environment. Quite a few of the papers –around 20% – are related to PhD projects. In general, the close connection to teaching practice has, to a large degree, framed the discussions at the WGAMs. In our view, these features and a general inclusive and supporting atmosphere has made the WGAM accessible for newcomers and in some cases has served as an entrance point to the international research community.

94 Blomhøj & Ärlebäck

3 Structuring the analyses according to the theory–practice relation

Returning to the need for a more systemic approach to the integration of applications and modelling in mathematics education as pinpointed by Blum (2015), we note that in many European countries secondary education serves both the purpose of general education for citizenship in a democracy and the purpose of preparation for further education and professions. The same fundamental duality is found in the aims for including applications and modelling in mathematics teaching: e.g. the duality between, on the one hand, the aim of developing students' competences to set up, apply, analyse and critique models and their applications in contexts, and, on the other hand, the aim of motivating and supporting the students' learning of mathematics through modelling activities. Especially at the secondary level, this duality is important because both aims are often being pursed at the same time in classrooms.

The role of, and relationship between, theory and practice differs depending on which of these two educational aims are the focus of the research. When surveying the WGAM papers, we found, although often not stated explicitly, that most of them have a clear focus on one of these aims. In nearly all the papers, there is an emphasis on either the development of teaching practice in a particular educational context or on the use or development of theory capturing potentials and difficulties concerning the teaching and learning of applications and modelling.

Therefore, in the following we analyse the WGAM contributions according to the four categories defined by the matrix shown in Figure 7.2, with the goal of characterising the theory–practice relationship in the research in terms of the two general aims and purposes we have identified. Our analysis will also point to potential areas for further developments.

Modelling as a means for / Focusing on	the learning of mathematics	developing modelling competency
the practices of teaching mathematical modelling **(80%)**	(1) Integrating modelling in the teaching of mathematics **(33%)**	(2) Developing, implementing and analysing modelling projects and activities in teaching practice **(41%)** — 6%
theories on the teaching and learning of modelling **(20%)**	(3) Using and developing theories on the learning of mathematics connected to modelling **(7%)** — 9%	(4) Developing theories on the teaching and learning of modelling competency **(4%)**

FIGURE 7.2 Categorisation of the WGAM papers with the percentages, see the discussion

4 Method

We read all the WGAM contributions and tried to place each into one of the four categories in the matrix. The focused reading of the papers helped us to clarify and better articulate the differences between and within the two dimensions in the matrix.

For the reader to gain some insight into our conceptualisation of the four categories and how we have characterised the WGAM papers, we will now briefly explain and elaborate on these differences.

The divide between the two rows in the matrix is related to the existence and role of empirical data. Most of the papers categorised as belonging in the top row (practices of teaching) include qualitative or quantitative empirical data typically used to evaluate a didactical design in a particular educational context. Often this context is closely related to the authors' own teaching practice or to the authors' collaboration with teachers. Empirical data in papers placed in the second row (theories on teaching and learning) have a more global or general function, such as being used for illustrative purposes to give meaning to theoretical notions or as a basis for developing theory. Pure theoretically based designs without empirical data on its implementation are also placed in the second row.

Concerning the divide between the two columns, we note that the theoretical elements used or developed are different in nature, reflecting either modelling as a means for the learning of mathematics, or modelling as a means for developing modelling competency in its broadest sense. Hence, in the left column, the theoretical elements are typical general theories on the teaching and learning of mathematics, e.g. theoretical frameworks such as ATD, RME, or theories on students learning of more or less specified mathematical concepts. While in the right column, the theories used or developed are related to the teaching and learning of modelling and applications, e.g. theories defining modelling competency and/or addressing teaching and learning challenges related to developing students' modelling competency.

In our analysis, we found some of the WGAM papers ($n = 21$, 17%) challenging to place in only one category. Some papers ($n = 8$) clearly had a focus related to teaching practices but were hard to place with respect to categories (1) and (2). Similarly, there were theoretical oriented papers ($n = 13$) that posed the same conundrum for categories (3) and (4). One type of paper that was categorised as belonging to one of these mixed categories is meta-studies. It should be noted, however, that we did not find this ambiguity with respect to the category-pairs (1)–(3) or (2)–(4). The contributions analysed were either empirical in nature, researching the development of practices of teaching modelling, or theoretical with the goal to build or further develop theories and their meaning. We now turn to present the result of our analysis.

5 Results: the contributions in the four categories

We present a brief general characterisation of the papers in each of the four categories with descriptions of some illustrative papers for each category. In the discussion,

96 Blomhøj & Ärlebäck

we pinpoint challenges for further research on the relationship of practice and theory with respect to the educational aims of teaching models and modelling.

5.1 Integrating modelling in the teaching of mathematics

The 40 contributions (33%) categorised as belonging to category (1) often focus on using a modelling approach to provide a learning environment for the learning of specific mathematical concepts, content areas, or more general ideas or methods. The following examples illustrate the characteristics and variation within this category.

Carriera and Baioa (2015) used experimental activities involving modelling the design of a 'convenient' staircase to study how two classes of 14–15-year-old students conceptualise slopes of linear functions. The students went out into the city to experience and make measurements of a number of different stairs displaying great diversity both between them and internally (having inhomogeneous measures of risers (vertical length) and treads (horizontal length)). Based on the data, the students developed models for designing stairs with a constant overall slope and homogeneous steps. Carriera and Baioa found that the students calculated the overall slope by either using the steps' total rise over total tread or averaging over the slopes of the steps.

Zell and Beckmann (2010) studied German 6th-grade students' work on one of three physical experiments focusing on what aspects of the concept of variable the experiments elicited. Using a framework conceptualising variables by Malle (1986), the research concluded that all of the aspects of variable in the framework (e.g. object, placeholder, and calculation aspects and nuances of these) surfaced in the students' work and were articulated in post-experimentation interviews. Although variables were used and expressed mostly on a descriptive level, Zell and Beckmann found that the students especially noted the functional relationship between two measured quantities and concluded that the physical experiment, though cognitively challenging, motivated the students and provided a venue to elicit and introduce different aspects of the concept of variable.

Blomhøj and Kjeldsen (2007), drawing on data from a course in mathematical modelling structured around six mini-projects, used modelling to challenge first-year university students' conceptions of integrals. The paper reported on the students' perceptions of the integral concept before and after engaging in working with the mini-project CO_2-*balance of a lake*. This project was "designed to challenge students' understanding of the concepts of the definite integral and the anti-derivative, the significance of the constant, and the interpretations of these concepts in different problem situations" (p. 2073). The authors found that by engaging in the mini-project the students developed their understanding of the connections between derivatives, anti-derivative functions, and definite integrals. More concretely, the arbitrary constant C in the formula for the anti-derivative was demystified for many students through the mini-project. The modelling contexts provided new opportunities for students to learn mathematics in different, non-traditional, ways.

An even more general mathematical idea (average rate of change) is the focus of a series of papers by Doerr and colleagues (Doerr & O'Neil, 2011; Ärlebäck, Doerr & O'Neil, 2013; Doerr, Ärlebäck & O'Neil, 2013) who report on research using a models and modelling perspective (c.f. Lesh & Doerr, 2003) to design and evaluate teaching and learning of average rate of change as the major structuring theme of a six-week summer programme for beginning university engineering students.

Another example of a broad perspective on the learning of mathematics through modelling activities is found in the paper by Grigoras (2010). A task where students explore and investigate a pattern of craters on Mars due to meteor impacts is used to research 13–14-year-old students' mathematisation processes. Analysing transcripts of video recordings of the students' group work using the notions of fundamental ideas (c.f. Schweiger, 2006), Grigoras reported that the students repeatedly engaged in approximation and geometrisation when conceptualising the craters as more abstract mathematical objects: locating and measuring when representing the positions and distributions of the craters using coordinate systems; counting and optimising as they organised and investigated the potential pattern of the craters. However, some of these fundamental ideas were only implicit in the students' work and although elicited, the students could not build productively on them due to limitations in their previous knowledge and experiences.

In a somewhat related study, Siller, Kuntze, Lerman and Vogl (2011) investigated in what way 159 pre-service teachers from Germany and Austria understood modelling as a fundamental idea. Using an exploratory study design, their analysis of the pre-service teachers' answers on a questionnaire initially put modelling on a par with other fundamental ideas, such as functional dependence, argumentation/proof, and generalising/specialising. The data analysis revealed that a large portion of the pre-service teachers saw modelling as relatively insignificant compared to other big ideas.

5.2 Developing, implementing and analysing modelling projects and activities

The 50 papers (41%) categorised as researching aspects of modelling competency as an educational goal focus primarily on how to develop the practices of teaching. These papers showed great diversity in approach, scope and educational context. Some contributions study issues related to students' activity, engagement and learning, whereas others investigate the role of the teacher. In addition, on the one hand, one can find papers that are inherently empirical and rich on data, and, on the other hand, more methodological contributions, reporting on the design of research instruments and modelling activities. One set of papers address the issue raised by Burkhart (2006), namely that modelling by and large is not an integrated part of everyday mathematics classroom practices even though it is included in the curricula.

Focusing on teachers as key stakeholders in everyday mathematics classroom practices, Ärlebäck (2010) presents a case study on teachers' beliefs about mathematical models and modelling as understood in terms of a belief structure consisting of five sub-belief objects: the nature of mathematics, the real-world, problem solving, school mathematics, and applications of mathematics. Interviews with two teachers centred around five mathematical problems serving as a basis for discussion and reflection (three standard textbook problems, one Fermi Problem (c.f. Sriraman & Lesh, 2006), and one modelling-eliciting activity (Lesh & Doerr, 2003). The interviews were analysed using a contextual sensitive categorisation scheme based on the five belief objects. The study found that the two teachers did not have any well-formed beliefs about mathematical models and modelling. The belief structure of the teachers contained inconsistencies that were made explicit within the framework. Ramirez (2017) found results similar for US and Chilean teachers using an online 10-item questionnaire and open-ended questions and gave examples of the participating teachers' experiences with modelling and their envisioned teaching of modelling in a mathematics class.

Also focusing on teachers' beliefs about models and modelling, the study by Bautista, Wilkerson-Jerde, Tobin and Brizuela (2013) explored the relationship between mathematics teachers' educational backgrounds and their expressed perceptions about mathematical models of a real-world phenomenon and the relationship between models and real data. A content analysis of written responses to three open-ended questions from 56 US in-service teachers (grades 5 to 9) in a professional developmental programme revealed, similar to the results of Ärlebäck (2010), that the teachers did not hold a unitary understanding of the notion of mathematical models. Factoring in the participating teachers' educational background, Bautista et al. (2013) concluded that teachers with backgrounds in science disciplines and mathematics education tended to tone down the notion of models being exact and, rather, stressed the flexible utilitarian aspect of models as tools. Teachers with backgrounds in other disciplines exhibited a more rigid view of models as exactly right or completely arbitrary and stressed the importance of producing an exact result.

Focusing on another aspect of teachers as key stakeholders, Schmidt (2010), working in the LEMA-project,[1] reported on the design and first results from a questionnaire aimed at assessing teachers' arguments against and for modelling in terms of obstacles and motives (Burkhart, 2006). The paper describes the development of the questionnaire, resulting in an instrument containing 120 five-level Likert questions organised in 23 categories. These categories were intended to capture areas where obstacles and motives are likely to surface relative a given offer-and-use model (i.e., a conceptualisation of how to think about the effectiveness of a lesson as dependent on various inputs and on the quality of teaching). The research instrument was evaluated using data from 240 teachers, resulting in rich descriptions of some of the 23 categories of obstacles and motives.

The work presented by Schmidt (2010) was followed up by Borromeo Ferri and Blum (2013) who focused on teachers at the primary level. After adapting the final version of the questionnaire from the LEMA study to be suitable for primary teachers, Borromeo Ferri and Blum recruited 71 primary teachers to respond to

Research on applications and modelling **99**

the adapted questionnaire that was comprised of 43 items distributed over 14 scales and one additional open item. The analysis showed that the three greatest barriers for primary teachers to implement modelling were a lack of material, followed by time pressure (lack of time), and concerns related to assessment.

5.3 Using and developing theories on the learning of mathematics via modelling

In this category (with eight papers, 7%) the papers focused on using and developing theory that can support the integration of modelling and applications as a means for teaching and learning mathematics. These theories could be on different ecological levels of the educational system such as curriculum, teacher education, textbooks, assessment systems, classroom interactions or on students' learning. We illustrate the scope of the category by means of two examples from CERMEs 7 and 5, respectively.

In the paper by Wake (2011), different theoretical elements from pedagogical science and from science and mathematics education research are combined to form a theoretical framework for understanding the possible roles of modelling in interdisciplinary mathematics and science teaching. The framework is illustrated in the design and analysis of an interdisciplinary upper secondary project on the phenomenon of floating. The aim of the paper is to develop a theoretical framework that can support interdisciplinary teaching in mathematics and science by means of mathematical modelling and applications were the learning goals for the students' are both key concepts in science and mathematics as well as developing their modelling competency. In this latter sense, the paper is close to category (4). However, Wake described the research by saying that "This type of exploratory environment, therefore, might be classified as being of the perspective: Educational modelling type (b) conceptual modelling (focusing on conceptual introduction and learning) in the classification system as proposed by Kaiser et al. (2007)" (p. 1007). The modelling activities in the interdisciplinary context are primarily seen as a means for students to communicate with and make sense of the science and mathematics involved. So the focus is on supporting the students' conceptual understanding through modelling.

Wake's paper is anchored in the EU project COMPASS and is hereby also representative of a strong current trend in the research field. During the latest decade, The European Union has launched numerous large research and developmental projects within the field of mathematics and science education. Many of these projects include mathematical modelling and applications, with a large number of projects contributing didactical designs for teaching and professional development activities for teachers (see www.scientix.eu).

Borromeo Ferri and Mousolides' (2017) paper is coded (3) and (4). The paper represents a continuing interest in integrating the teaching of mathematical modelling in interdisciplinary approaches to science teaching (e.g. in Science, Technology, Engineering and Mathematics (STEM) education). The paper makes a theoretical contribution by developing a model for interdisciplinary mathematics education aiming both at developing the students' modelling competency and supporting their learning of important concepts in mathematics and science. Hence, the paper is coded as belonging to both categories.

Ruiz, Bosch and Gascón (2007) provide an example of a research paper entirely framed within a single theoretical framework, namely Anthropological Theory of Didactics (ATD). The paper analysed the conditions needed to teach and learn functional-algebraic modelling in an experimental activity designed for and implemented at the end of the secondary school level. In addition, the constraints that hindered the development of such a teaching practice were also analysed theoretically. The focus is on the theoretical underpinnings within ATD of the didactical design of a modelling situation. The didactical design takes its point of departure in the question of how to earn money from the production and sale of T-shirts. The authors discussed in detail how to construct a series of praxeologies that can support the students' learning of algebra and functional relationships through modelling activities using this design. In ATD, modelling in and with mathematics is seen as an essential activity for the learning of mathematics. As a consequence, research within this framework inherently contributes to the development of theories on the learning of mathematics through modelling activities. However, ATD-based research is not concerned with the development of the students' modelling competency as an aim in its own right. The mathematical elements in the modelling process and the students' critical reflections related to the validity of models and/or the model's role and function in societal contexts are not a focus in ATD. Therefore, typically ATD-based research focuses on modelling as a means for learning mathematics and thus qualifies to belong in category (3). In relation to the integration of modelling into existing practices of mathematics teaching, it is a challenge for the ATD approach that the theory is not easy for teachers to understand and that its implementation requires rather thorough and radical changes in the organisation of mathematics teaching.

5.4 Developing theories on the teaching and learning of modelling competency

The papers in category (2) draw on theory on the teaching and learning of mathematical modelling, often in the form of connected notions and viewpoints developed in the field of research on modelling and applications. This includes, among other things, different versions of the modelling process, ideas about the difficulties students might encounter in different sub-processes, and ways of conceptualising progress in the individual student's development of modelling competency. Although the theoretical developments in the field were surveyed by Blum (2015), it is characteristic of these theoretical developments that they are dynamic and still gaining new meaning as they are used in different contexts for research and the development of teaching. Therefore, in quite a few cases papers placed in category (2) can be said to contribute also to the development of theory. However, in order to be categorised in category (4) the main focus of the paper should be development of theory. We found only five papers (4%) fulfilling this criterion, and we will showcase two of these for illustration.

The paper by Henning and Keune (2006) presented at CERME 4, and later used in their research, is an example of a theoretical contribution regarding modelling competency as an educational aim. In this study, the authors develop a model with three levels of students' modelling competency, namely: Level 1: Recognise and understand modelling; Level 2: Independent modelling; and Level 3: Meta-reflection on modelling. Each level is specified in terms of which competences the students should have at that particular level, and these are illustrated through examples of modelling tasks and what types of challenges students should be able to deal with at each competence level.

In the paper by Cabassut (2010), the concept of didactical transposition within ATD was used to analyse the mathematisation process at the primary school level. Through an analysis of different examples, the author developed this notion: "The mathematisation teaching is the place of a double didactic transposition, one from real world into the classroom and the other one from the mathematical world into the classroom" (p. 2157). The theoretical noting of the double transposition in mathematisation captured a fundamental difficulty for the teacher when teaching mathematical modelling. The notion was used for analysing the PISA model for mathematisation. The research draws on the EU project LEMA for professional development. Accordingly, in the paper, the author proposed generic questions to be addressed in professional development to help teachers deal with the challenge of the double transposition:

> In a mathematisation task, what knowledge of real world and of mathematical world has to be transposed? What techniques, justifications and validations from both worlds have to be used? How different knowledge, techniques, justifications and validations in the two worlds are articulated and interfering? What effects on teachers' practice, on pupils' learning and on class didactical contract have these articulations and interferences?
>
> *(p. 2164)*

Such questions are relevant for developing further theories on the teaching and learning of modelling competency.

6 Discussion

Our analysis with respect to the four categories shows that around 80% of the WGAM papers are researching the development of practices of teaching in various educational contexts rather than focusing on developing theory on the teaching and learning of applications and modelling. These papers are close to evenly distributed between focusing on modelling and applications as a means for learning mathematics (category (1), 33%) and as a means for developing modelling competency (category (2), 41%) with 6% placed on the border between the two.

102 Blomhøj & Ärlebäck

Only about 20% of the WGAM papers have their main focus on using or developing theory. Seven per cent were placed in category (3) and 4% in category (4). Nine per cent of the papers were characterised as having a theoretical focus, but not placed in either (3) or (4). See Figure 7.2 for an overview.

Concerning educational levels, we only found a few cases directed toward primary education. The vast majority of the papers were found to be focusing on secondary education, with about half the categorised papers belonging to category (1) (primarily papers involving the lower secondary level) and half belonging to category (2) (primarily papers involving the upper secondary level). Around 15% of the papers addressed teaching at tertiary level and these were found to be evenly distributed between categories (1) and (2). Most of these papers addressed modelling in teacher education or in professional development projects for teachers. Even though the format and the tradition developed in the WGAMs invites research reports on developmental projects rather than theoretical syntheses, we see the picture emerging in our analysis as a reflection of the situation within the research field more generally. The majority of the research in the field internationally is driven by a wish to develop and improve the practices of mathematics teaching by means of integrating applications and modelling. In many European countries, applications and modelling is already part of the mathematics curricula, especially at the secondary level, and in several other countries there is a quest for reforms. Although reforms have been influenced by the research in the field of applications and modelling, there still is a general need for research providing a basis for, and supporting, development at curricular level, especially regarding formats of assessment including applications and modelling. This, together with professional development for teachers, is crucial for the integration of applications and modelling in the actual practices of mathematics teaching (Blum, 2015).

We note that in this past decade the EU has strongly promoted and supported a political agenda for the development of mathematics and science education in a direction that emphasises inquiry-based teaching, applications and modelling and the integration of information technology. This educational policy is seen as instrumental for the socioeconomic development needed in Europe as established, for instance, in the 'Rocard report' by the EU Commission (Rocard et al., 2007). Accordingly, a number of large projects, typically with 6–12 countries involved, focusing on mathematics teachers' professional development, have been launched. Several of these projects have applications and modelling in mathematics teaching as part of their programmes, and related research and developmental work have been presented at the WGAMs. In fact, the WGAMs have served as a basis for recruiting researchers from different European countries to these projects. Issues related to the organisation of and the role played by research in these projects has been discussed at many WGAMs. Often in these projects, it is difficult to allocate sufficient resources for theoretical research. In line with this, we find that the strong focus on developing the practices of teaching applications and modelling in the WGAMs is relevant in relation to the European situation as well as to the research field internationally.

Research on applications and modelling **103**

However, our analysis has documented a scarcity of theoretical research. In particular in category (3), we see a need for research developing theory that can establish connections between the potentials for learning mathematics through modelling activities and what is known empirically, and theoretically explained about the learning of mathematical concepts. There are a few cases of such theory-driven research within the ATD and Realistic Mathematics Education frameworks, but no cases of theory development explicitly addressing this missing connection.

Concerning category (4) we mentioned that the basic theoretical notions in the field are still being explored and gaining further meaning and mutual connections through use and re-contextualisation in concrete teaching and learning situations. In that respect, some of the papers in category (2) are contributing to the development of theory about the teaching and learning of modelling competency. However, the few cases placed in categories (3) and (4) display a scarcity of research striving to develop our theoretical understanding of the teaching and learning of applications and modelling in mathematics teaching. In our opinion, the WGAM papers offer a rich resource for such theory development.

In general, we found that the papers presented at the WGAMs represent a multifaceted body of research exhibiting a close connection to the development of practices of teaching applications and modelling at various levels and forms of mathematics teaching. The element of theory development can be strengthened through the future conferences by initiating research using the WGAM papers as a resource for theory development. Thereby, the interplay between research and teaching practice can be enhanced in the continuing work of the WGAM.

Acknowledgement

We thank Despina Potari, Helen Doerr and Werner Blum for valuable comments, suggestions and corrections.

Note

1 LEMA stands for *Learning and Education in and through Modelling and Applications* and was an EU Comenius funded collaborative project between Cyprus, Germany, Hungary, France, Spain and the United Kingdom aiming at producing material for professional development (see www.lema-project.org/).

References

Ärlebäck, J. B. (2010). Towards understanding teachers' beliefs and affect about mathematical modelling. *CERME6* (pp. 2096–2105).
Ärlebäck, J. B., Doerr, H. M., & O'Neil, A. H. (2013). Students' emerging models of average rate of change in context. *CERME8* (pp. 940–949).
Barbosa, J. C. (2007). Mathematical modelling in parallel discussion. *CERME5* (pp. 2101–2109).
Barquero, B., Serrano, L., & Serrano, V. (2015). Creating necessary conditions for mathematical modelling at university level. *CERME9* (pp. 950–959).

Bautista, A., Wilkerson-Jerde, M. H. Tobin, R., & Brizuela, B. M. (2013). Diversity in middle school mathematics teachers' ideas about mathematical models: The role of educational background. *CERME8* (pp. 960–969).

Blomhøj, M., & Højgaard Jensen, T. (2003). Developing mathematical modelling competence: Conceptual clarification and educational planning. *Teaching Mathematics and its Applications, 22*(3), 123–139.

Blomhøj, M., & Kjeldsen, T. H. (2007). Learning the integral concept through mathematical modelling. *CERME5* (pp. 2070–2079).

Blum, W. (2015). Quality teaching of mathematical modelling: What do we know, what can we do? In S. J. Cho (Ed.), *The Proceedings of the 12th International Congress on Mathematical Education: Intellectual and Attitudinal Challenges* (pp. 73–96). New York: Springer.

Blum, W., Galbraith, P. L., Henn, H.-W., & Niss, M. (Eds.) (2007). Modelling and applications in mathematics education. *The 14th ICMI study*. New York: Springer.

Borromeo Ferri, R. (2006). Theoretical and empirical differentiations of phases in the modelling process. *Zentralblatt für Didaktik der Mathematik, 38*(2), 86–95.

Borromeo Ferri, R., & Blum, W. (2013). Barriers and motivations of primary teachers for implementing modelling in mathematics lessons. *CERME8* (pp. 1000–1009).

Borromeo Ferri, R., & Mousolides, N. (2017). Mathematical modelling as a prototype for interdisciplinary mathematics education? *CERME10* (pp. 900–907).

Burkhart, H. (2006). Modelling in mathematics classrooms: Reflections on past developments and the future. *Zentralblatt für Didaktik der Mathematik, 38*(2), 178–195.

Cabassut, R. (2010). The double transposition in mathematisation at primary. *CERME6* (pp. 2156–2165).

Carriera, S., & Baioa, A. M. (2015). Assessing the best staircase: Students' modelling based on experimentation with real objects. *CERME9* (pp. 834–840).

Doerr, H. M., & O'Neil, A. H. (2011). A modelling approach to developing an understanding of average rate of change. *CERME7* (pp. 937–946).

Doerr, H. M., Ärlebäck, J. B., & O'Neil, A. H. (2013). Teaching practices and modelling changing phenomena. *CERME8* (pp. 1041–1050).

Frejd, P. (2013). An investigation of mathematical modelling in the Swedish national course tests in mathematics. *CERME8* (pp. 947–956).

García, F., Gascón, J., Higueras, L., & Bosch, M. (2006). Mathematical modelling as a tool for the connection of school mathematics. *Zentralblatt für Didaktik der Mathematik, 38*(3), 226–246.

Grigoras, R. (2010). Modelling in environments without numbers: A case study. *CERME6* (pp. 2206–2015).

Henning, H., & Keune, M. (2006). Levels of modelling competencies. *CERME4* (pp. 1666–1674).

Kaiser, G., Blomhøj, M., & Sriraman, B. (2006). Towards a didactical theory for mathematical modelling. Editorial. *Zentralblatt für Didaktik der Mathematik, 38*(2), 82–85.

Kaiser, G., Sriraman, B., Blomhøj, M., & Garcia, F. J. (2007). Report from the working group modelling and applications: Differentiating perspectives and delineating commonalties. *CERME5* (pp. 2035–2041).

Lesh, R. A., & Doerr, H. M. (Eds.). (2003). *Beyond constructivism: Models and modeling perspectives on mathematics problem solving, learning, and teaching*. Mahwah, NJ: Erlbaum.

Malle, G. (1986). Variable: Basisartikel mit Überlegungen zur elementaren Algebra. *Mathematik Lehren, 15*, 2–8.

Ramirez, P. (2017). Teachers' beliefs about mathematical modelling. *CERME10* (pp. 972–979).

Rocard, M., Csermely, P., Jorde, D., Lenzen, D., Walberg-Henriksson, H., & Hemmo V. (2007). *L'enseignement scientifique aujourd'hui: une pédagogie renouvelée pour l'avenir de l'Europe*. Brussels: Commission Européenne, Direction générale de la recherche, Science, économie et société.

Roorda, G., Vos, P., & Goedhart, M. (2007). Derivatives in applications: How to describe students' understanding. *CERME5* (pp. 2160–2169).

Ruiz, N., Bosch, M., & Gascón, J. (2007). The functional algebraic modelling at Secondary level. *CERME5* (pp. 2170–2179).

Schmidt, B. (2010). Modelling in the classroom: Motives and obstacles from the teachers' perspective. *CERME6* (pp. 2066–2075).

Schweiger, F. (2006). Fundamental ideas: A bridge between mathematics and mathematics education. In J. Maazs & W. Schloeglmann (Eds.), *New mathematics education research and practice* (pp. 63–73). Rotterdam: Sense Publishers.

Siller, H.-S., Kuntze, S., Lerman, S., & Vogl, C. (2011). Modelling as a big idea in mathematics with significance for classroom instruction: How do pre-service teachers see it? *CERME7* (pp. 990–999).

Skovsmose, O. (1994). *Towards a philosophy of critical mathematical education*. Dordrecht: Kluwer.

Sriraman, B. (2006). Conceptualizing the model-eliciting perspective of mathematical problem solving. *CERME4* (pp. 1686–1695).

Sriraman, B., & Lesh, R. (2006). Modeling conception revisited. *Zentralblatt für Didaktik der Mathematik, 38*(3), 247–254.

Treffers, A. (1987). *Three dimensions: A model of goal and theory descriptions in mathematics instruction – the Wiskobas Project*. Dordrecht, NL: Kluwer.

Vos, P. (2006). Assessment of mathematics in a laboratory-like environment: The importance of replications. *CERME4* (pp. 1696–1705).

Wake, G. (2011). Modelling in an integrated mathematics and science curriculum: Bridging the divide. *CERME7* (pp. 1000–1009).

Zell, S., & Beckmann, A. (2010). Modelling activities while doing experiments to discover the concept of variable. *CERME6* (pp. 2116–2225).

8

EARLY YEARS MATHEMATICS[1]

Esther S. Levenson, Maria G. Bartolini Bussi and Ingvald Erfjord

1 Introduction

The Early Years Mathematics (EYM) Working Group met for the first time in 2009. As such, it is a relatively 'young' group, vibrant, and still growing. Perhaps the first question that must be dealt with at the onset is the necessity for this working group, or in other words, what makes early years mathematics special enough to warrant its own working group. Thus, this chapter begins with a discussion of what it means to develop young children's mathematics knowledge and the importance of learning mathematics during the early years. The second part of this chapter deals with themes that have continuously interested researchers in this group, as well as ideas that have been developed over the last four CERME conferences. The third part of this chapter takes a look at connections between this working group and other CERME working groups. Finally, the last part reviews some of the challenges this group faces as well as future research directions.

2 Mathematics for young children

In their review of early childhood mathematics learning, Clements and Sarama (2007) noted that "researchers have changed from a position that young children have little or no knowledge of or capacity to learn mathematics . . . to theories that posit competencies that are either innate or develop in the first years of life" (p. 462). But, what is the mathematics that children are learning at such a young age? And what age is considered young?

Historically, the construct of 'early years' has been associated with informal learning prior to formal schooling. However, current curricula and standards for preschool mathematics (e.g. in England the Curriculum Guidance for the Foundation Stage (DfEE/QCA, 2000)), offer several suggestions for practitioners in how to specifically foster children's knowledge of counting, calculations, shapes and measures. In addition, several curricula also advocate the promotion of mathematical processes such as problem solving, reasoning and justifying conjectures (e.g. Israel National Preschool Mathematics Curriculum [INPMC], Ministry of Education, Israel, 2008, p. 8). In the EYM group at CERME, we have discussed children's development of mathematical concepts (e.g. Koleza & Giannisi, 2013), as well as how to encourage children's ability to explain and justify their reasoning, and promote early generalisation and abstraction (e.g. Vighi, 2013). Regarding the age of 'young' learners, in the past, we referred to children between the ages of 3 and 8 years.[2] This takes into account the different transition ages between preschool and primary school in different countries. It also fosters cooperation between mathematics education researchers of different age groups. At the 2015 CERME conference, there were nine papers which reported on studies of primary school children (up to age 8). Although those papers may have also fit in with other groups (e.g. the Arithmetic and Number Systems group), the approach in the EYM group is to emphasise the development over time of early mathematical concepts, focusing on the transition between informal and more formal learning environments. (See also Section 3 of this chapter.) In 2017, Björklund presented a paper regarding toddlers' encounters with aspects of numbers. Toddlers refer to children ages 1 to 3 years old, younger than the age group which had thus far been included in the EYM group. This was a welcome addition and exemplifies the inclusive spirit of CERME.

There are several important reasons for focusing on EYM. First, focusing on the transition to formal schooling, not all children come to first grade with the same knowledge. For example, low-income children often come to first grade with less mathematical knowledge than higher-income children (Starkey & Klein, 2000). Tsamir, Tirosh, Levenson, Tabach and Barkai (2011) found significant differences between the knowledge of abused and neglected children and other children. In addition, research has found that early knowledge of mathematics may be seen as a predictor of later school success (Duncan et al., 2007). Finally, when the EYM group convened for the first time, the introduction noted how several countries, including Germany, Finland, Cyprus, Denmark and others, have become increasingly aware of the need to support mathematics learning during the early years. Yet, not all countries have a mandatory or even recommended curriculum for this age, nor do all have compulsory or financially supported education for young children. By supporting the work of the EYM group, the mathematics education community is sending a message that this research is vital and can be used to inform countries who are developing curricula for this age. In the next section, we highlight a few themes that have continuously run through CERME meetings at the EYM group.

3 Sustained themes

There are at least three particularly important issues related to EYM education. The first is the importance of what the child learns at home, before or alongside formal instruction at school. At times, children learn mathematical concepts at home that are not necessarily utilised in the most appropriate way (Meaney, 2010). In Germany, studies have investigated mathematical learning embedded in family discourse and how family members can support mathematical learning (e.g. Brandt & Tiedemann, 2010).

From the adults at home, we turn toward the adults in the classroom, namely the teachers. At the first EYM meeting, one paper dealt with preschool teachers' mathematics beliefs (Benz, 2010) and showed that although teachers might agree with a constructivist approach to learning, they might agree more with an acquisitionist approach. The author suggested that professional development could help teachers in supporting children's own constructions. It was also shown that kindergarten teachers' views of learning and teaching mathematics differ from primary school teachers' beliefs, due, in part, to different curricula as well as different trainings (Schuler, Kramer, Kröger & Wittmann, 2013). For these and other reasons, several countries have initiated professional development programmes for practising preschool teachers. In Italy, for example, teacher educators, education committees, policy makers and teachers combined their efforts to help teachers implement a new mathematics curriculum for preschool children (Bartolini Bussi, 2013). At the last CERME, studies dealt with the roles of the preschool teacher and their views of how to guide children's learning (e.g. Delacoeur, 2015).

Perhaps the most discussed and debated issue in the early mathematics education community regards the roles of instruction and guided learning versus the roles of play and types of play in early years mathematics (e.g. Alpaslan & Erden, 2015). We face differences linked to our cultures and the roles of kindergarten and early schooling. According to the OECD (2006), kindergartens in the Nordic and Central European countries are situated within a social pedagogical tradition, as an educational institution where upbringing, care, play and learning are the core enterprises and where free play has a key role. The OECD mentions countries in the Western part of Europe (UK, Ireland, France and the Netherlands) as having a pre-primary approach, meaning they "tend to introduce the contents and methods of primary schooling into early education" (p. 61). Thus, kindergartens in the Western part of Europe, compared to Nordic and Central Europe have stronger similarities with formal schooling and less focus on play.

The notion and interpretation of play are problematic and vary between contexts, countries and theoretical frameworks. Learning can be difficult to capture in play-based activities compared to a more structured classic school type of activity, where more is expressed orally than in written form. Thus, there is also a need to discuss methodological perspectives on data collection and data analysis of play situations and exploration (Vogel & Jung, 2013). In this chapter we add to the discussion of play and learning by introducing the term *playful learning* (Hirsh-Pasek, Golinkoff, Berk & Singer, 2009). This term takes into account that for a child, play and learning are one and the same. Playful learning builds on developmental research which has documented children of all ages experimenting with mathematical concepts through play. In their outline of playful learning, Hirsh-Pasek et al. distinguish between *free play* and *guided play*. In guided play, a teacher typically offers activities and materials for the children to engage with, where the aim is to learn some mathematical concept. Several studies in the EYM group explored characteristics of guided play such as when children and parents play board games together (Tubach, 2015), or where children play with games designed to develop their conceptions of number (e.g. Sinclair & SedaghatJou, 2013). In these guided play situations, the teacher typically considers opportunities to interact in ways that support children's learning of these concepts. Free play, on the other hand, takes part without any input or structure from the teachers and, with the exception of Flottorp's (2011) study, which captured two boys' verbal and non-verbal expressions in an outside free play situation, free play has rarely been studied in our group.

An issue related to the balance between children's free play, guided play and the role of formal instruction and teaching materials, is the notion of agency. In free play situations, children's own choices and explorations take part in their natural setting, and the mathematics might be hidden rather than explicit. In guided play, children work in a more structured way supported by materials, where the mathematics is made visible and the 'play' is structured by the teacher. In the latter situation, children may become passive 'receivers' of ready-made structures. In a free play situation, the children have the agency; in a guided play situation, the teacher and the material being offered have the main agency and children's agency is limited to a few options (e.g. Erfjord, Carlsen & Hundeland, 2015). The learner needs to have some agency in order to learn. However, full agency might be problematic from a mathematical point of view, where concepts are crucial and build on each other.

4 The relevance of EYM research to other research groups

By its very nature, the EYM group overlaps with other groups. Although other working groups at CERME discuss arithmetic, geometry and algebraic thinking, these conceptualisations begin before primary school and develop over time. The transition from preschool to primary school is another concern of the EYM group.

Alpaslan and Erden (2015) presented a review of EYM research papers from 2000 to 2013 published in seven peer-reviewed research journals. The research topic most investigated was number systems and arithmetic, a topic addressed in many EYM papers at CERME. We are also aware of the fact that several papers in the Arithmetic and Number Systems group study this topic among children aged 6–8 years. The importance of these papers to the EYM group is seeing how numerical concepts develop over time, from a very young age to the first years of school. Thus, although number concepts are dealt with in the Arithmetic group, we address these concepts in the EYM group, not just from a preschool point of view, but also with a look toward the first years of schooling, addressing transitions within educational systems (e.g. Vennberg, 2015).

One of those transitions includes developing a sense for arithmetic operations by building on key number concepts. For example, Sinclair and SedaghatJou (2013) showed how a touchscreen application may engage children with the concept of cardinality, which can then lead them to use cardinality when joining (i.e. adding) two groups. In another study, Maffia and Mariotti (2015) described how second graders worked with an artefact to learn about the distributive property. Notable was the movement from personal meanings, to mathematical meanings, a movement that is of particular interest to the EYM group, where it can be adapted to other mathematical properties. Rational number concepts, usually developed during the elementary school years, might have their roots in early fraction concepts such as how children conceptualise the notion of a half (Tirosh, Tsamir, Tabach, Levenson & Barkái, 2011).

Another content strand discussed at the EYM group is geometrical thinking, which is the main theme of the Geometrical Thinking group. The difference between the groups is that at the preschool level, geometrical learning is mostly related to shapes, whereas at the upper levels, geometrical thinking encompasses a wider perspective including transformations and proof schemes. This does not mean that some overlap does not occur. In 2015, a paper was presented at the Geometrical Thinking group regarding 6-year-old students' knowledge of intuitive triangles (Rodrigues & Serrazina, 2015). This paper would have been interesting for the EYM group and in fact, Rodrigues and Serrazina suggested that some of their findings might be due to students' previous experiences with shapes. Those authors, as well as the participants of the EYM group, might have benefitted from the presentation of that paper to the EYM group. In the opposite direction, a geometry-related paper presented at the EYM group dealt with defining rectangles and squares in first grade (Bartolini Bussi & Baccaglini-Frank, 2015). This paper exemplified an EYM approach to geometrical definition by taking into consideration young children's natural tendency to place these figures into distinct categories, a tendency reinforced by everyday language.

A more recent development within our group is the integration of Information and Communication Technologies (ICT) with preschool mathematics, connecting to the 'Technologies and Resources' group in mathematics education (TWG15). The issue of ICT first appeared in the EYM group in 2011 (Ladel & Kortenkamp, 2011) investigating the linking of different forms of representation (hands, fingers and multi-touch technologies). This area was further explored in 2013. Once again, cultural differences were noted along with the previously mentioned dilemma between free play and directed learning in preschool. In Sweden (Lange & Meaney, 2013), one study explored the mathematics that might be hidden in popular (not necessarily mathematical) games played on the iPad. On the other hand, Sinclair and SedaghatJou (2013), from Canada, investigated a game (TouchCounts) specifically designed to promote young children's knowledge of number concepts. Hundeland, Erfjord and Carlsen (2013) explored the roles of the teacher when employing ICT-related mathematics activities and how ICT can support mathematical learning processes through exploration and discovery of mathematical relations. This area of research led to additional research questions, such as the difference between engaging children with concrete versus virtual manipulatives.

In 2015, a short activity with TouchCounts was analysed, focusing on the audible, the visible and the tangible (Pimm & Sinclair, 2015), seeking to gain insight into the nature of number, in particular the ordinal aspect, in this complex assemblage. Demetriou (2015) compared the use of concrete and virtual manipulatives in three symmetry tasks. Bartolini Bussi and Baccaglini-Frank (2015) presented fragments of a first-grade experiment where the seeds were sown for a mathematical definition of rectangles that includes squares, by means of programming a very simple robot. It is important to note that all ICT-related papers in our group investigated the potential use of ICT in its own right, and not merely as a replacement for concrete manipulatives. We expect this to be a growing concern for our group.

5 Looking back and looking ahead – challenges and future research directions

5.1 Challenges

From the first meeting of the EYM group, scholars from different countries brought with them research traditions and related theoretical frameworks that appeared to be strictly linked to cultural contexts. At CERME 8, there was a growing interest in exploring differences among the various cultures, sparked by two papers, one from China (Sun, 2013) and one from Italy (Ramploud & Di Paola, 2013). Within the context of addition and subtraction, these papers explored task design in different countries, different epistemologies, pedagogies and beliefs related to cognitive processes. This issue of different research traditions was raised again in CERME 9, where the presence of several theoretical traditions was noted in the introduction to this group's papers:

> In most cases, . . . theoretical frameworks are chosen with reference to . . . authors from the same country; this choice could be misunderstood as patriotism, but it is not necessary the case. For instance, the semiotic mediation theory is useful in countries where the focus is on long term studies, which in turn depends on the institutional role of a teacher working for more than one year with the same group of pupils . . . The relationship between contexts and theoretical frameworks is a challenge for the diffusion of findings and the possibility of exploiting findings from different cultural contexts.
>
> *(pp. 1886–1889)*

Another challenge is conducting mathematics education research with toddlers, specifically children who are preverbal. Yet, in several countries (e.g. Italy, Finland and Sweden), nearly all 2 year olds, and even 1 year olds, are enrolled in preschools, sometimes following the same curriculum as 3–5 year olds. In other countries (e.g. Israel), although most 1 and 2 year olds are enrolled in day-care or preschool, there are no curricular guidelines. In addition, as noted by Palmér

112 Levenson, Bartolini Bussi & Erfjord

and Björklund (2017), preschool teachers are mostly generalists, lacking necessary insights into mathematics learning. These issues pose additional challenges for the EYM group.

5.2 Future directions

The EYM Working Group faces many difficult, but important and interesting questions. In Section 3, we discussed balancing agency, free play, guided play, structured materials and teacher guidance. Some of our differences are contextual and not something we, as researchers, can influence directly. As previously noted, in northern Europe there is a strong emphasis on play, whereas in the Western part of Europe there is a tendency for more structured mathematical activities. Bringing together these cultures at CERME allows Northern European researchers to explore more structured activities and for Western European researchers to explore more play-based activities. Exploring together play and learning from a developmental and research perceptive could link members of EYM at CERME. The question of how to approach the 'education' of children was a key question at CERME in 2011. One concrete result of this issue was a decision by participants in Germany to organise a workshop-based conference during the spring of 2012. This was the start of POEM – A Mathematics Education Perspective on Early Mathematics Learning between the Poles of Instruction and Construction. Participants from the EYM group at CERME, as well as others, met, and have continued to meet every second spring (in between the CERME congresses). The overall focus for POEM has been the seemingly distant poles of instruction and construction, with some arguing that children's play should be the starting point, and playful activities should contain some elements of instruction (e.g. van Oers, 2014). Another conference for EYM is the Children's Mathematical Education conference (CME) taking place in Poland every second year. We believe strong links between these conferences strengthens each conference and group. At CERME in 2017, the EYM group had leaders centrally engaged in POEM and CME.

Although previous researchers (e.g. Clements & Sarama, 2007) outlined key early conceptions of number and geometry, we believe that additional research is necessary. Such research could help illuminate for practitioners ways in which to develop rich mathematical concepts for children in different settings, using different kinds of activities, and tools (manipulative as well as virtual). As researchers, we consider both where we stand and where we wish to go. As such, we believe that at future EYM Working Group sessions, it would be mutually beneficial for researchers from different countries to discuss new avenues of research and questions. We end by giving an example. We have seen several CERME research papers (e.g. Saebbe & Mosvold, 2015) that have taken the teacher's perspective and studied, for example, the role of questions in mathematical activities in EYM. However, there is little research into the nature and task design of mathematical activities and teacher's orchestration that might foster children's questioning and children's own investigations. This can be a next step.

Notes

1 DEDICATION. Our young friends Zişan Güner Alpaslan and her husband Mustafa Alpaslan passed away in a terrible car accident on July 31, 2015. Both were in CERME 9 in Prague: in particular, Zişan was in TWG 13 (Early Years Mathematics). They both were brilliant researchers and active members of the ERME community. This text is dedicated to both of them.
2 Preschool in many countries (e.g. Sweden) is from 1–5 years old, but our concern has been for the ages 3–5 years. Some countries label the first years as kindergarten, and some label kindergarten as the year prior to primary school. In this chapter, we use preschool and kindergarten almost interchangeably (mainly in the way the authors we refer to use them).

References

Alpaslan, Z., & Erden, F. (2015). The status of early childhood mathematics education research in the last decade. *CERME9* (pp. 1933–1939).

Bartolini Bussi, M. G. (2013). *Bambini Che Contano*: A long term program for preschool teachers' development. *CERME8* (pp. 2088–2097).

Bartolini Bussi, M. G., & Baccaglini-Frank, A. (2015). Using pivot signs to reach an inclusive definition of rectangles and squares. *CERME9* (pp. 1891–1897).

Benz, C. (2010). "Numbers are actually not bad": Attitudes of people working in German kindergarten about mathematics in kindergarten. *CERME6* (pp. 2547–2556).

Björklund, C. (2017). Aspects of numbers challenged in toddlers' play and interaction. *CERME10* (pp. 1821–1828).

Brandt, B., & Tiedemann, K. (2010). Learning mathematics within family discourses. *CERME6* (pp. 2557–2566).

Clements, D. H., & Sarama, J. (2007). Early childhood mathematics learning. In F. K. Lester (Ed.), *Second handbook of research on mathematics teaching and learning* (pp. 461–555). New York: Information Age Publishing.

Delacoeur, L. (2015). How the role of the preschool teacher affects the communication of mathematics. *CERME9* (pp. 1905–1910).

Demetriou, L. (2015). The use of virtual and concrete manipulatives. *CERME9* (pp. 1911–1917).

DFEE, QCA (2000). *Curriculum guidance for the foundation stage*. London: DFEE.

Duncan, G. J., Dowsett, C. J., Claessens, A., Magnuson, K., Huston, A. C., Klebanov, P., et al. (2007). School readiness and later achievement. *Developmental Psychology, 43*(6), 1428–1446.

Erfjord, I., Carlsen, M., & Hundeland, P. (2015). Distributed authority and opportunities for children's agency in mathematical activities in kindergarten. *CERME9* (pp. 1918–1924).

Flottorp, V. (2011). How do children's classification appear in free play? A case study. *CERME7* (pp. 1852–1861).

Hirsh-Pasek, K., Golinkoff, R. M., Berk, L. E., & Singer, D. G. (2009). *A mandate for playful learning in preschool: Presenting the evidence*. New York: Oxford University Press.

Hundeland, P. S., Erfjord, I., & Carlsen, M. (2013). Use of digital tools in mathematical learning: Kindergarten teachers' approaches. *CERME8* (pp. 2108–2117).

Koleza, E., & Giannisi, P. (2013). Kindergarten children's reasoning about basic geometric shapes. *CERME8* (pp. 2118–2127).

Ladel, S., & Kortenkamp, U. (2011). Implementation of a multi-touch environment supporting finger symbol sets. *CERME7* (pp. 2278–2287).

Lange, T., & Meaney, T. (2013). iPads and mathematical play: A new kind of sandpit for young children? *CERME8* (pp. 2138–2147).

Maffia, A., & Mariotti, M. A. (2015). Introduction to arithmetical expressions: A semiotic perspective. *CERME9* (pp. 1947–1953).

Meaney, T. (2010). Only two more sleeps until the school holidays: Referring to quantities of things at home. *CERME6* (pp. 2617–2626).

Ministry of Education, Israel (2008). *Israel national mathematics preschool curriculum* (INMPC). Retrieved April 7, 2009, from http://meyda.education.gov.il/files/Tochniyot_Limudim/KdamYesodi/Math1.pdf.

OECD (2006). *Starting Strong II: Early childhood education and care.* Paris: OECD Publishing. Retrieved April 7, 2016, from www.oecd-ilibrary.org/education/starting-strong-ii_9789264035461-en.

Palmér, H., & Björklund, C. (2017). How do preschool teachers characterize their own mathematics teaching in terms of design and content? *CERME10* (pp. 1885–1892).

Pimm, D., & Sinclair, N. (2015) "How do you make numbers?": Rhythm and turn-taking when coordinating ear, eye and hand. *CERME9* (pp. 1961–1967).

Ramploud, A., & Di Paola, B. (2013). [Shuxue: Mathematics] Take a look at China. A dialogue between cultures to approach arithmetic at first and second Italian primary classes. *CERME8* (pp. 2188–2197).

Rodrigues, M., & Serrazina, L. (2015). Six years old pupils' intuitive knowledge about triangles. *CERME9* (pp. 578–583).

Saebbe, P., & Mosvold, R. (2015). Asking productive mathematical questions in kindergarten. *CERME9* (pp. 1982–1988).

Schuler, S., Kramer, N., Kröger, R., & Wittmann, G. (2013). Beliefs of kindergarten and primary school teachers towards mathematics teaching and learning. *CERME8* (pp. 2128–2137).

Sinclair, N., & SedaghatJou, M. (2013). Finger counting and adding with TouchCounts. *CERME8* (pp. 2198–2207).

Starkey, P., & Klein, A. (2000). Fostering parental support for children's mathematical development: An intervention with Head Start families. *Early Education and Development, 11*(5), 659–680.

Sun, X. (2013). The structures, goals and pedagogies of "variation problems" in the topic of addition and subtraction of 0–9 in Chinese textbooks and its reference books. *CERME8* (pp. 2208–2217).

Tirosh, D., Tsamir, P., Tabach, M., Levenson, E., & Barkai, R. (2011). Can you take half? Kindergarten children's responses. *CERME7* (pp. 1891–1902).

Tsamir, P., Tirosh, D., Levenson, E., Tabach, M., & Barkai, R. (2011). Investigating geometric knowledge and self-efficacy among abused and neglected kindergarten children. *CERME7* (pp. 1902–1911).

Tubach, D. (2015). "If she had rolled five then she'd have two more": Children focusing on differences between numbers in the context of a playing environment. *CERME9* (pp. 2617–2624).

van Oers, B. (2014). The roots of mathematising in young children's play. In U. Kortenkamp, B. Brandt, C. Benz, G. Krummheuer, S. Ladel, & R. Vogel (Eds.), *Early mathematics learning: Selected papers of the POEM 2012 conference* (pp. 111–123). New York: Springer.

Vennberg, H. (2015). Preschool class: One year to count. *CERME9* (pp. 2045–2046).

Vighi, P. (2013). Game promoting early generalization and abstraction. *CERME8* (pp. 2238–2247).

Vogel, R., & Jung, J. (2013). Videocoding: A methodological research approach to mathematical activities of kindergarten children. *CERME8* (pp. 2248–2257).

9

MATHEMATICAL POTENTIAL, CREATIVITY AND TALENT

Demetra Pitta-Pantazi and Roza Leikin

1 Introduction

This chapter gives an overview of the main directions of the research presented and discussed in the Thematic Working Group (TWG) 'Mathematical Potential, Creativity and Talent'. The group's activities were directed at promoting discussion of research focusing on mathematical creativity and talent, encouraging in-depth empirical studies on these topics and offering a forum for presentation and discussion of mathematical activities that develop mathematical creativity and promote mathematical talent. The group was initiated in 2011 as a continuation of the work of the TWG 'Advanced Mathematical Thinking' (AMT).

The AMT group participants' research fell into two major categories (in accordance with the distinction made by Tall (1991)). The first was associated with AMT as it relates to learning and understanding of university mathematics (AMT at an absolute level in Tall's terms). For example, such studies as 'Secondary-tertiary transition and students' difficulties: the example of duality' by De Vleeschouwer (2010) or 'Conceptual change and connections in analysis' by Juter (2010). The second focused on AMT as it relates to high achievements in school mathematics, creative mathematical performance, and solving non-standard or Olympiad problems (at the relevant level, in Tall's terms). These studies explored creativity, potential and talent in school mathematics. The study by Voica and Pelczer (2010), who compared problem-posing by novices and experts, is an example of a study from the second category that was presented at the AMT group.

The issues of creativity, potential and talent were less relevant to participants from the first sub-group, and researchers from the second sub-group were less connected to the issues discussed with respect to mathematics studied at the university level. Thus, Roza Leikin suggested splitting the AMT group into two TWGs: 'University Mathematics' and 'Mathematical Potential, Creativity and Talent'.

Note that the terms mathematical giftedness, mathematical potential, high mathematical abilities and mathematical talent are often used interchangeably. However, mathematical giftedness implies high mathematical abilities with giftedness perceived as an inborn personal characteristic, in contrast to high abilities which is a dynamic characteristic that may be developed (Leikin, 2014). In this TWG, mathematical giftedness was connected to high-level performance in mathematics and to mathematical creativity.

In this chapter, we do not adopt a single definition of mathematical talent, mathematical giftedness, or exceptional mathematical abilities for several reasons. First, in education and research literature, these terms all have different interpretations and uses. Second, some researchers in this TWG used these terms interchangeably whereas others selected specific terms in specific contexts. This reflects the diversity of positions in the TWG. For instance, some researchers utilised the term 'promising students' instead of 'gifted' as per NCTM (1995), which defined mathematical promise as a complex function of abilities, motivation, self-esteem and learning opportunities provided to students. In addition to this, some researchers avoided any reference to 'gifted students' due to the impression that this label unfairly discriminates between children. The following sections will present specific studies in which these various perceptions become more apparent.

The TWG 'Mathematical Potential, Creativity and Talent' encouraged international participation and exchange of ideas on the topics of mathematical creativity, potential and talent. Participants varied in terms of their research paradigms, basic theories and research methodologies. The focus of the research studies also varied, and included individual students and classrooms, teachers and teacher education as well as experiences of research mathematicians. During the three meetings (2011, 2013 and 2015) of this TWG, 44 research studies and eight posters were discussed, and approximately 80 researchers from 22 countries participated. In particular the participants' discussions addressed the following overarching themes:

(a) theoretical underpinnings of the concepts of mathematical talent and creativity (definitions, origins and characteristics);
(b) methodological approaches and tools that are useful for identifying mathematically talented and creative students;
(c) the nature and features of mathematical approaches, tasks and activities that are challenging, fundamentally free of routine, inquiry-based, and rich in authentic mathematical problem solving, and their use in the development of mathematical creativity and realisation of mathematical talent;
(d) the relationship between mathematical talent, motivation, effort and mathematical creativity;
(e) teacher training associated with teaching gifted students as well as promoting mathematical creativity in all students;
(f) historical and sociological analysis of issues relevant to mathematical creativity and giftedness.

Potential, creativity and talent **117**

This chapter addresses these issues in the following three sections: (1) mathematical creativity, (2) mathematical talent/potential/giftedness, and (3) methodological approaches used in studies associated with these topics. The sections are organised in this form since most of the studies of the TWG dealt with one of the two strands in isolation, i.e. either with creativity or with mathematical talent/potential/giftedness. Very few studies presented in the TWG brought the two strands together and these are discussed later in the chapter. In the following sections, we attempt to bring together some of the findings and open issues of the research studies discussed during the CERME meetings.

2 Mathematical creativity

2.1 Theoretical foundations and empirical studies of mathematical creativity and its development

Discussions on the theoretical foundations of creativity focused on definitions and origins of creativity, while empirical studies on creativity concentrated on how to identify and measure it. During the meetings, no consensus was reached regarding a single definition of mathematical creativity. Researchers were more interested in understanding and gathering manifold perspectives on creativity. The numerous studies used different definitions for creativity (Krutetskii, 1976; Polya, 1973; Torrance, 1974), different methodologies (qualitative and quantitative), addressed different populations (preschool, primary, secondary and tertiary level as well as prospective teachers), various mathematical topics (equations, fractions, proofs in geometry, patterns) and assessed creativity based on cognitive and/or social characteristics. One thing researchers agreed on is that the definition of creativity depends on the context under examination.

Among the various theories that were used to measure mathematical creativity, Torrance's was the most dominant. Torrance (1974) suggested that to evaluate a person's creativity, four components need to be measured: fluency, flexibility, elaboration and originality. In mathematics education, the measurement of three components – fluency, flexibility and originality – has frequently been adopted. Difficulties have been found in assessing elaboration in mathematics. Leikin (2009) suggested that multiple solution tasks are a good means to assess mathematical creativity and also introduced a model for evaluating mathematical creativity that takes students' fluency, flexibility and originality into consideration. Leikin and Kloss (2011) showed that while 8th- and 10th-grade students' success in problem solving is highly correlated with fluency and flexibility, originality appeared to be a special mental quality. Interestingly, mathematics teaching did not seem to advance students' creativity. Through two comparative studies investigating prospective mathematics teachers, Leikin, Levav-Waynberg and Guberman (2011) found that although fluency and flexibility significantly increased, the same did not occur for originality. Thus, they concluded that fluency and flexibility are of a dynamic nature, whereas originality is a 'gift' and is also the strongest component out of the

118 Pitta-Pantazi & Leikin

three in determining creativity. However, not all researchers agreed that mental flexibility is related to the creative process. Prabhu and Czarnocha (2014) argued that Koestler's (1964) ideas of 'bisociation and simultaneity of attention' could serve as a theoretical framework for investigating creativity.

A number of studies focused on the relation between general creativity and mathematical creativity. Kattou, Christou and Pitta-Pantazi (2015) suggested that general creativity and mathematical creativity are distinct and one cannot predict mathematical creativity based on general creativity and vice versa.

Levenson (2011) explored the notion of collective creativity and examined the way in which fluency, flexibility and originality can be used to describe collective creativity. She also examined the relation between individual and collective mathematical creativity in respect to both process and product. She suggested that attempts to promote collective creativity could possibly also promote individual creativity.

Although empirical studies on mathematical creativity were conducted using participants from all levels of education, i.e. preschool (Münz, 2013), primary (Kattou et al., 2015; Levenson, 2011), secondary (Bureš & Nováková, 2015; Leikin & Kloss, 2011) and tertiary levels (De Geest, 2013; Karakok, Savic, Tang & El Turkey, 2015), Münz (2013) highlighted two problems: (1) research on mathematical creativity in early childhood is largely overlooked and (2) it is not clear how mathematical creativity is expressed and how it may be observed in these young ages. On the other end of the educational ladder, Karakok et al. (2015) investigated mathematicians' (who teach courses at the tertiary level) views on creativity. Originality and aesthetics appeared to encapsulate these mathematicians' views of creativity in their work.

2.2 Mathematical activities aimed at developing mathematical creativity

A number of researchers who investigated the development of students' creativity explored students' mathematical investigations and problem-solving culture via heuristic strategies (Bureš & Nováková, 2015) as well as the impact of high versus low guidance structured tasks on the development of creativity (Palha, Schuiterma, van Boxtel & Peetsma, 2015). Some researchers suggested problem-posing as a means of developing students' creativity (Singer, Pelczer & Voica, 2011, 2015) and others investigated the use of technology for the same (Sophocleous & Pitta-Pantazi, 2011). Most of the studies that dealt with activities aimed at developing mathematical creativity included some empirical data on the impact of these activities on students' mathematical creativity. However, there were very few studies such as the one by Vale and Pimentel (2011) which presented a collection of mathematical pattern tasks and discussed possible ways in which these could be explored in the classroom in order to develop students' creativity and higher order thinking.

Regarding the impact that the level of guidance in mathematical tasks has on the development of creativity, Palha et al. (2015) did not find any significant

Potential, creativity and talent **119**

differences in students' fluency and flexibility. On the other hand Safuanov, Atanasyan and Ovsyannikova (2015) claimed that open-ended exploratory learning is effective for the development of creativity. Singer et al. (2015) found that students fail to pose mathematically consistent problems when they must rely on deep structures of mathematical concepts and strategies (Singer et al., 2011). Regarding the impact of technology on the development of mathematical creativity, Sophocleous and Pitta-Pantazi (2011) found that students' engagement with an interactive 3D geometry software improved their creative abilities due to the opportunities it offered to imagine, synthesise and elaborate.

Among the studies that explored the development of teachers' creativity was Maj's (2011), which found that specific classroom techniques and tools are needed to develop creative mathematical activities. Maj argued that these techniques do not spontaneously develop, but rather need to be consciously thought through and implemented. Leikin and Elgrabli (2015) reached the conclusion – while investigating prospective teachers and an expert in problem solving – that discovery skills can be developed in people with different levels of problem-solving expertise, although the range of this development depends on the level of expertise.

2.3 Teaching for mathematical creativity

Fewer studies explored whole courses or practices, rather than activities, for the development of mathematical creativity. De Geest (2013) presented an undergraduate distance learning course for mathematics education students which aimed at developing students' thinking through the 'possibility thinking' framework. He found that this course offered opportunities for students to 'play' and explore, ask 'what if' questions, take risks and not have a 'fear of failure', be more imaginative, make connections with other areas of mathematics and be aware of their own learning.

Furthermore, Sarrazy and Novotná (2013) investigated the effects of teaching mathematical creativity in heterogeneous groups and especially the impact on highly able students. They claimed that teaching is more efficient if teachers take the zone of proximal development into consideration and that grouping pupils according to their performance level does not bring optimisation of results.

Finally, Émin, Essonnier, Filho, Mercat and Trgalova (2015) investigated the reactions of students when they used electronic books to work on mathematical creativity tasks. They found that high achievers gave the answers that they felt their teachers wanted but were not willing to do anything further, because it would exceed their obligations as students.

2.4 Teachers' perceptions of creativity-stimulating activities and teachers' own mathematical creativity

Some researchers focused on pre-service and/or in-service teachers' perception of creativity-stimulating activities (Desli & Zioga, 2015; Sinitsky, 2015) while others focused on pre-service and/or in-service teachers' own creativity (Birkeland, 2015). Desli and Zioga (2015) explored the features that pre-service and in-service teachers consider appropriate for tasks that promote creativity and how they envision creativity in school. They found that although both groups of teachers identified elements of tasks that promote creativity, such as arousing children's curiosity, problem-posing, own constructions, unusual questions and connections with other topics, when asked to present tasks, these characteristics were not present in the tasks that they chose. Furthermore, pre-service teachers connected mathematical creativity in the classroom to activities that arouse children's interest such as attractive stories and puzzles, students' own constructions, use of materials and non-algorithmic thinking, while in-service teachers connected creativity mainly to students' problem-posing activities. Similarly, Sinitsky (2015) investigated pre-service teachers' perceptions of creativity and found that they believed that elementary school mathematics is not suitable for promoting creativity. Regarding pre-service teachers' own creativity, Birkeland (2015) suggested that in some cases pre-service teachers' reasoning was neither imitative nor creative. Overall, researchers seem to agree that there is a need to educate teachers about mathematical creativity since both their perceptions as well as their own creativity need to be enhanced.

3 Mathematical talent/giftedness/potential

3.1 Theoretical foundations and empirical studies on mathematical talent/giftedness/potential and its development

A topic that had been of great interest to researchers of the TWG was the identification of mathematically talented students. Various theoretical backgrounds (Greenes, 1981; Kießwetter, 1992; Krutetskii, 1976; Renzulli, 1978) and methodological approaches (qualitative, quantitative, mixed) have been used for the identification of mathematically gifted students. Even though there is still no clear consensus on who may be named mathematically talented, gifted or potentially gifted, researchers seem to agree that multiple sources of information are needed to identify them.

The articles presented tried to empirically identify mathematically gifted students. Brandl (2011) argued that high attainment in mathematics does not necessarily imply mathematical giftedness and not all mathematically gifted students have high attainment in mathematics. He also claimed that creativity, curiosity, out-of-the-box thinking, flexibility and the tendency to be non-compliant are characteristics of mathematically gifted students.

Problem solving had a central role in the majority of studies that explored the identification of gifted students. Rott (2015) found that gifted students were significantly more successful in problem solving than regular students. However, gifted students used significantly fewer heuristics compared to equally successful novices, possibly because non-gifted students used heuristics to compensate for the lack of mental flexibility.

Another topic that had a central role was the relation between mathematical giftedness and mathematical creativity. To date no consensus has been reached about the relation between the two. Lev and Leikin (2013) intensively studied the relation between mathematical creativity, general giftedness and mathematical excellence as components of mathematical giftedness. They found that general giftedness and excellence in mathematics have a major effect on students' creativity. Furthermore, they revealed that students who are gifted and excel in mathematics are significantly more flexible than non-gifted students, and argued that originality in mathematics is linked to general giftedness. The findings of Kattou, Kontoyianni, Pitta-Pantazi and Christou (2011) and Kontoyianni, Kattou, Pitta-Pantazi and Christou (2011) are in accord with the above study and both studies argued that mathematical giftedness could be described in terms of mathematical ability (a term that corresponds to Lev and Leikin's (2013) 'mathematical excellence') and mathematical creativity. They also claimed that mathematical ability can be predicted by mathematical creativity.

Studies presented in the TWG were not limited to those of students' cognitive characteristics which could be observed through students' external behaviour, responses or interviews. A number of studies were interdisciplinary and multidimensional and applied cognitive and neurocognitive research methodologies. This research group distinguished between general giftedness and excellence in school mathematics. In one of their studies, Baruch-Paz, Leikin and Leikin (2013) investigated the differences in memory and speed of processing in generally gifted students and students who excel in mathematics. They found that mathematical giftedness can be defined as a combination of general giftedness and excellence in mathematics and that there are interrelated, but qualitatively different, phenomena in mathematically gifted students. They found memory and speed of processing to be important factors in explaining mathematical giftedness (Baruch-Paz, Leikin & Leikin, 2015). Szabo (2015) also found that memory plays a critical role in the students' choice of problem-solving method. When examining brain potentials of gifted students and students who excel in mathematics, Waisman, Leikin, Shaul and Leikin (2013) found that giftedness was expressed in more efficient brain functioning. Finally, Kontoyianni et al. (2011) claimed that fluid intelligence and self-perceptions could predict mathematical giftedness.

A number of researchers were interested in affective aspects that might allow the identification of mathematically gifted students. Benölken (2015) investigated students' interest in and attitudes toward mathematics. He found that affective factors were critical for the identification of mathematical talent and that among the students who were mathematically talented, boys had a stronger interest in mathematics. Thus identifying talented girls was more difficult than identifying talented boys.

Furthermore, researchers investigated the identification of twice exceptional children, and specifically mathematically gifted children with developmental impairments. Nordheimer and Brandl (2015) suggested the combination of written tests and process-based analyses of lessons. Nolte (2013) claimed that high abilities

122 Pitta-Pantazi & Leikin

can mask developmental disorders and impairments in regular classrooms and that developmental disorders can mask high cognitive abilities, putting these talents at risk of being lost. All researchers suggested a need for balanced programmes which will support these students' talents and offer intervention for their disabilities, so that these talents do not get lost.

3.2 Mathematical activities aimed at promoting mathematical talent/giftedness/potential

Another important topic covered was mathematical activities promoting mathematical talent. Some researchers offered models for analysing the typologies of problems (Ivanov, Ivanova & Stolbov, 2013), others investigated the activities in textbooks for the mathematically gifted (Karp, 2013), while others suggested approaches (Aizikovitsh-Udi & Amit, 2011; Pitta-Pantazi, Christou, Kattou, Sophocleous & Pittalis, 2015; Sarrazy & Novotná, 2011; Schindler & Joklitschke, 2015) or principles (Kasuba, 2013) that should be used to facilitate mathematical talent.

Ivanov et al. (2013) suggested a two-dimensional model to describe the typologies of mathematical problems based on the kinds of activities that students use in problem solving and the schemes of reasoning they apply. They suggested that their analysis makes it possible to identify similarities among problems and helps educators to develop new problems.

Karp (2013) investigated the differences between sets of problems from 'ordinary' textbooks and from specific textbooks for the mathematically gifted. He claimed that authors of textbooks for the mathematically gifted seem to believe that gifted students learn algorithms more easily and are able to transfer and apply mathematical ideas in new domains and contexts more quickly than ordinary students.

The principles that should guide the writing of mathematical problems and organisation of working sessions for mathematically gifted students also attracted researchers' attention. For instance, Kasuba (2013) analysed the practice of writing mathematical problems. He offered a list of principles that may guide this process: (a) prior knowledge, (b) clarity, (c) attractiveness, (d) intellectual freedom, (e) chance to act as mathematicians, (f) age appropriateness and (g) supportive environment.

Other researchers suggested or investigated specific approaches that might enhance mathematical talent. Sarrazy and Novotná (2011) investigated whether teaching cognitive and metacognitive strategies could facilitate students' mathematical learning. Aizikovitsh-Udi and Amit (2011) suggested that the infusion approach could promote students' critical thinking. Schindler and Joklitschke (2015) explored students' approaches and abilities, especially mathematisation and generalisation. They suggested that giving adequate cues to focus students' attention on a specific aspect of a graph, or asking them explicitly to consciously reflect about the similarities of problem situations could optimise the tasks and support students. Pitta-Pantazi et al. (2015) suggested that students, and especially talented ones, should be challenged with problems that combine a collection of mathematical competences (i.e. digital competence, communication in mother tongue, learning to learn and sense of initiative).

3.3 Teaching mathematically talented students

When discussing didactical principles of working with talented students, participants made distinctions between affective, social and cognitive domains. Most researchers agreed that when designing programmes for talented students, mathematics educators have to consider the social consequences of treating able children differently. If these children are offered the same curriculum as everyone else, there is no point in putting them in a different group. Effective ability grouping requires offering a qualitatively different curriculum. Furthermore, an interdisciplinary and multidisciplinary curriculum is an important principle in the education of talented students, since spending the majority of time on mathematics can sometimes stop them from doing well in other subjects which they may excel at. Additionally, any acceleration or enrichment programme should suit the individual differences of these students not only cognitively but in the affective field as well.

Moreover, regarding classroom or school culture for talented students, researchers in the WG stated that these students need to have well-trained teachers and an environment where ability is valued. The role and characteristics of teachers of mathematically gifted students were also of concern to researchers. Karp and Busev (2015) claimed that teaching gifted students requires teachers' creativity and exemplified their ideas with two cases of prominent Russian mathematicians. In addition, Pelczer, Singer and Voica (2013) studied the challenges and limitations faced by mathematics teachers when teaching talented students in mixed classrooms, and found mismatches between teachers' self-efficacy beliefs and their specialised content knowledge.

4 Methodologies used in these studies

The contributions to this TWG varied in terms of the research methodologies used, which included:

- theoretical investigation of the concepts of mathematical talent and creativity;
- empirical research on the issues underlined above;
- historical and sociological analysis of issues pertaining to mathematical creativity and giftedness;
- presentation of mathematically challenging tasks or tasks that might promote mathematical creativity.

In the quantitative studies, the majority of the researchers used tests, whereas most of the qualitative studies used a mixture of methods and tools (interviews, tests, questionnaires, observations or various artefacts such as textbooks, students'

124 Pitta-Pantazi & Leikin

work or notes). One of the innovations regarding the methodological approaches presented in this TWG was the inclusion of neuropsychological approaches by Waisman et al. (2013) investigating talented students and mathematical creativity.

However, there are still some methodological limitations that further research could address in the years to come. Primarily, we need to better establish the validity and reliability of instruments designed to measure mathematical creativity and giftedness. At the same time these tools also need to be more economical, reliable and valid. It is also critical to conduct more studies that integrate both quantitative and qualitative methods, in order to offset the methodological problems of each tradition and gain a wider and deeper understanding of mathematical creativity and talent.

Another important point is that there is a great need to link theory and practice in order to improve learning and teaching. Although the methodologies presented offered important advancements in the field, we still need methodologies that will allow us to focus more on the impact of research on practice. For instance, few studies investigated enrichment programmes fostering students' potentials and their effect on processes and learning gains, and there were no longitudinal studies. Such studies can potentially be of great value and impact. Last but not least, meta-analysis is another methodological route that could enable us to draw more causal inferences about mathematical creativity and giftedness.

5 Outlook to the future

The TWG encouraged discussions between research mathematicians, mathematics educators and educational researchers. Still, many questions remain open, either because they were left unmentioned or because they raised disagreement among the group participants. Some of these questions are the following:

- The studies presented at the TWG differed in terms of the theoretical models and frameworks used. The frameworks used determined research procedures and tools, and moreover, the interpretation of study results that we described in this chapter. It would be of interest to identify whether different models and frameworks are compatible and whether these combinations can result in deepening our understanding of the phenomena of mathematical potential and creativity.
- The variety of research approaches used by TWG participants led to the questions: What is a reasonable balance between qualitative and quantitative studies in the field of giftedness and creativity that can lead to a better understanding of these phenomena? What other research methodologies might facilitate and support the investigation and enhancement of mathematical creativity and talent?
- There was obvious interest among participants in identifying and nurturing mathematical potential and creativity. From reading the papers of this TWG, the readers can borrow various ideas about educational approaches, mathematical tasks and activities that might be beneficial for the identification

and promotion of mathematical talent. We suggest that further systematic research on the effectiveness of these approaches is needed. Such studies – comparative or descriptive, qualitative and quantitative – should help mathematics educators to make knowledgeable choices regarding the tools they use for the identification and promotion of students' creativity and giftedness in mathematics.

- Many researchers discussed mathematical creativity as one of the important components in teacher professional development. Still, teachers' competences relating to the development of students' mathematical potential and creativity are not clearly identified. Correspondingly, future research can search for answers to the questions: What should be the teachers' competences, skills and knowledge, and what kind of teachers' education programmes are needed to promote mathematical creativity and talent? What advancements have been made internationally so far?

- It might also be of interest to conduct empirical studies about the way in which technology can (a) improve mathematical creativity and mathematical abilities in heterogeneous classes, and (b) meet the needs of mathematically gifted students in heterogeneous classes.

- Furthermore, it might be relevant to propose more neuroscientific and eye-tracking studies in order to answer questions related to the relationship between mathematical creativity and giftedness.

References

Aizikovitsh-Udi, E., & Amit, M. (2011). Integrating theories in the promotion of critical thinking in mathematics classrooms. *CERME7* (pp. 1034–1043).

Baruch-Paz, N., Leikin, M., & Leikin, R. (2013). Memory and speed of processing in generally gifted and excelling in mathematics students. *CERME8* (pp. 1146–1155).

Baruch-Paz, N., Leikin, M., & Leikin, R. (2015). Visual processing and attention abilities of general gifted and excelling in mathematics students. *CERME9* (pp. 1046–1051).

Benölken, R. (2015). The impact of mathematics interest and attitudes as determinants in order to identify girls' mathematical talent. *CERME9* (pp. 970–976).

Birkeland, A. (2015). Pre-service teachers' mathematical reasoning. *CERME9* (pp. 977–982).

Brandl, M. (2011). High attaining versus (highly) gifted pupils in mathematics: A theoretical concept and an empirical survey. *CERME7* (pp. 1044–1055).

Bureš, J., & Nováková, H. (2015). Developing students' culture of problem solving via heuristic solving strategies. *CERME9* (pp. 983–988).

De Geest, E. (2013). Possibility thinking with undergraduate distance learning mathematics education students: How it is experienced. *CERME8* (pp. 1166–1174).

De Vleeschouwer, M. (2010). Secondary-tertiary transition and students' difficulties: The example of duality. *CERME6* (pp. 2256–2264).

Desli, D., & Zioga, M. (2015). Looking for creativity in primary school mathematical tasks. *CERME9* (pp. 989–995).

Émin, V., Essonnier, N., Filho, P. L., Mercat, C., & Trgalova, J. (2015). Assigned to creativity: Didactical contract negotiation and technology. *CERME9* (pp. 996–1002).

Greenes, C. (1981). Identifying the gifted student in mathematics. *Arithmetic Teacher, 28*, 14–18.

Ivanov, O., Ivanova, T., & Stolbov, K. (2013). Typologies of mathematical problems: From classroom experience to pedagogical conceptions. *CERME8* (pp. 1175–1184).

Juter, K. (2010). Conceptual change and connections in analysis. *CERME6* (pp. 2276–2285).

Karakok, G., Savic, M., Tang, G., & El Turkey, H. (2015). Mathematicians' views on undergraduate students' creativity. *CERME9* (pp. 1003–1009).

Karp, A. (2013). Mathematical problems for the gifted: The structure of problem sets. *CERME8* (pp. 1185–1194).

Karp, A., & Busev, V. (2015). Teachers of the mathematically gifted: Two case studies. *CERME9* (pp. 1010–1015).

Kasuba, R. (2013). Learning with pleasure: To be or not to be? *CERME8* (pp. 1195–1203).

Kattou, M., Christou, C., & Pitta-Pantazi, D. (2015). Mathematical creativity or general creativity? *CERME9* (pp. 1016–1023).

Kattou, M., Kontoyianni, K., Pitta-Pantazi, D., & Christou, C. (2011). Does mathematical creativity differentiate mathematical ability? *CERME7* (pp. 1056–1065).

Kießwetter, K. (1992). Mathematische Begabung: Über die Komplexität der Phänomene und die Unzulänglichkeiten von Punktbewertungen. *Der Mathematikunterricht, 38*(1), 5–10.

Koestler, A. (1964). *The act of creation*. London: Hutchinson.

Kontoyianni, K., Kattou, M., Pitta-Pantazi, D., & Christou, C. (2011). Unraveling mathematical giftedness. *CERME7* (pp. 1066–1074).

Krutetskii, V. A. (1976). *The psychology of mathematical abilities in schoolchildren*. Chicago, IL: The University of Chicago Press.

Leikin, R. (2009). Exploring mathematical creativity using multiple solution tasks. In R. Leikin, A. Berman, & B. Koichu (Eds.), *Creativity in mathematics and the education of gifted students* (pp. 129–145). Rotterdam: Sense Publishers.

Leikin, R. (2014). Giftedness and high ability in mathematics. In S. Lerman (Ed.), *Encyclopedia of mathematics education*. Dordrecht: Springer. Electronic Version. doi: 10.1007/s11858-017-0837-9.

Leikin, R., & Elgrabli, H. (2015). Creativity and expertise – the chicken or the egg? Discovering properties of geometry figures in DGE. *CERME9* (pp. 1024–1031).

Leikin, R., & Kloss, Y. (2011). Mathematical creativity of 8th and 10th grade students. *CERME7* (pp. 1084–1094).

Leikin, R., Levav-Waynberg, A., & Guberman, R. (2011). Employing multiple solution tasks for the development of mathematical creativity: Two comparative studies. *CERME7* (pp. 1094–1103).

Lev, M., & Leikin, R. (2013). The connection between mathematical creativity and high ability in mathematics. *CERME8* (pp. 1204–1213). Ankara, Turkey: Middle East Technical University.

Levenson, E. (2011). Mathematical creativity in elementary school: Is it individual or collective? *CERME7* (pp. 1104–1114).

Maj, B. (2011). Developing creative mathematical activities: Method transfer and hypotheses' formulation. *CERME7* (pp. 1115–1124).

Münz, M. (2013). Mathematical creative solution processes of children with different attachment patterns. *CERME8* (pp. 1214–1224).

NCTM (National Council of Teachers of Mathematics) (1995). *Report of the NCTM taskforce on the mathematically promising*. NCTM News Bulletin 32 (December): Special Insert, NCTM Inc., Reston, Virginia.

Nolte, M. (2013). Twice exceptional children: Mathematically gifted children in primary schools with special needs. *CERME8* (pp. 1225–1234).

Nordheimer, S., & Brandl, M. (2015). Students with hearing impairment: Challenges facing the identification of mathematical giftedness. *CERME9* (pp. 1032–1038).

Palha, S., Schuiterma, J., van Boxtel, C., & Peetsma, T. (2015). The effect of high versus low guidance structured tasks on mathematical creativity. *CERME9* (pp. 1039–1045).

Pelczer, I., Singer, M., & Voica, C. (2013). Teaching highly able students in a common class: Challenges and limits of a case-study. *CERME8* (pp. 1235–1244).

Pitta-Pantazi, D., Christou, C., Kattou, M., Sophocleous, P., & Pittalis, M. (2015). Assessing mathematically challenging problems. *CERME9* (pp. 1052–1058).

Polya, G. (1973). *How to solve it.* Princeton, NJ: Princeton University.

Prabhu, V., & Czarnocha, B. (2014). Democratizing mathematical creativity through Koestler Bisociation Theory. In C. Nicol, S. Oesterle, P. Liljedahl, & D. Allan (Eds.), *Proceedings of the Joint Meeting of PME 38 and PME-NA 36, Vol. 5* (pp. 1–8). Vancouver, Canada: PME.

Renzulli, J. S. (1978). What makes giftedness? Reexamining a definition. *Phi Delta Kappan, 60*(3), 180–184.

Rott, B. (2015). Heuristics and mental flexibility in the problem solving processes of regular and gifted fifth and sixth graders. *CERME9* (pp. 1059–1065).

Safuanov, I., Atanasyan, S., & Ovsyannikova, I. (2015). Exploratory learning in the mathematical classroom (open-ended approach). *CERME9* (pp. 1097–1098).

Sarrazy, B., & Novotná, J. (2011). Didactical vs. mathematical modelling of the notion competence in mathematics education: Case of 9–10-year-old pupils' problem solving. *CERME7* (pp. 1125–1132).

Sarrazy, B., & Novotná, J. (2013). Mathematical creativity and highly able students: What can teachers do? *CERME8* (pp. 1245–1253).

Schindler, M., & Joklitschke, J. (2015). Designing tasks for mathematically talented students – according to their abilities. *CERME9* (pp. 1066–1072).

Singer, F. M., Pelczer, I., & Voica, C. (2011). Problem posing and modification as a criterion of mathematical creativity. *CERME7* (pp. 1133–1142). Rzeszów, Poland: University of Rzeszów.

Singer, F. M., Pelczer, I., & Voica, C. (2015). Problem posing: Students between driven creativity and mathematical failure. *CERME9* (pp. 1073–1079).

Sinitsky, I. (2015). What can we learn from pre-service teachers' beliefs on – and dealing with – creativity stimulating activities? *CERME9* (pp. 1080–1086).

Sophocleous, P., & Pitta-Pantazi, D. (2011). Creativity in three-dimensional geometry: How can an interactive 3D-geometry software environment enhance it? *CERME7* (pp. 1143–1153).

Szabo, A. (2015). Mathematical problem-solving by high achieving students: Interaction of mathematical abilities and the role of the mathematical memory. *CERME9* (pp. 1087–1093).

Tall, D. (1991). *Advanced mathematical thinking.* Dordrecht: Kluwer.

Torrance, E. P. (1974). *Torrance Tests of Creative Thinking.* Bensenville, IL: Scholastic Testing Service.

Vale, I., & Pimentel, T. (2011). Mathematical challenging tasks in elementary grades. *CERME7* (pp. 1154–1164).

Voica, C., & Pelczer, I. (2010). Problem posing by novice and experts: Comparison between students and teachers. *CERME6* (pp. 2656–2665).

Waisman, I., Leikin, M., Shaul, S., & Leikin, R. (2013). Brain potentials during solving area-related problems: Effects of giftedness and excellence in mathematics. *CERME8* (pp. 1254–1263).

10

AFFECT AND MATHEMATICAL THINKING

Exploring developments, trends, and future directions

Markku S. Hannula, Marilena Pantziara and Pietro Di Martino

1 Affect and Mathematical Thinking Group: looking back

The initial caution of the mathematics education community in considering affective issues is shown by the fact that CERME has hosted a Thematic Working Group (TWG) on 'Affect' only since its third meeting. Yet the first contribution on affect (Zan & Poli, 1999) had already appeared in CERME 1. Now the relevance of affect is recognised in the community of mathematics educators. In particular in ERME Conferences, the Affect and Mathematical Thinking TWG has been consistently present, and increasingly well attended.

1.1 Development of theories in the field

The affect group in CERME has discussed extensively matters of conceptual framework and terminology, leading to more extensive theorisation of the area:

> [The discussions] increased our awareness of being specific about the concepts that we use. We have realized that it is not sufficient to give definitions of the concepts that are being used in a particular study, but we have to explicate their relations to the other dimensions of affect research as well.
>
> *(Hannula, 2011, p. 41)*

There are three theoretical frameworks that have been influential in CERME for structuring the area of affect. The first is McLeod's (1992) framework that identified three main topics of research in mathematics related to affect: emotions, attitudes, and beliefs (Figure 10.1). Moreover, the framework suggested that emotions are the most intensive, the least stable and the least cognitive of the three, while beliefs are at the other end of the continuum, and attitudes are in the middle.

Beliefs	Attitudes	Emotions
←――――――――――――――――――――――→		
Most cognitive		Least cognitive
Most stable		Least stable
Less affective		Most affective

FIGURE 10.1 McLeod's (1992) framework on affect

A significant step forward was the graphic representation of the conceptual field that Peter Op 't Eynde composed during CERME 5 (Figure 10.2). This model captured new ideas discussed in the CERME affect group: it recognised motivation as an important concept and identified different levels of social context.

These ideas were further elaborated by Hannula in his CERME plenary (2011) and in an article for a CERME special issue of RME (2012). Hannula described three dimensions that can be used to identify and define affective theoretical concepts (Figure 10.3). The first dimension recognises that the concepts may be either cognitive (what one believes), affective (what one feels), or motivational (what one desires). The second, temporal dimension, separates state-type constructs that aim to describe dynamical processes from trait-type constructs that aim to describe rather stable dispositions. The third dimension recognised the social theories in mathematics education research, and also the embodied nature of affect, identifying three ontologically different traditions for affect-related research: psychological, social, and embodied theories.

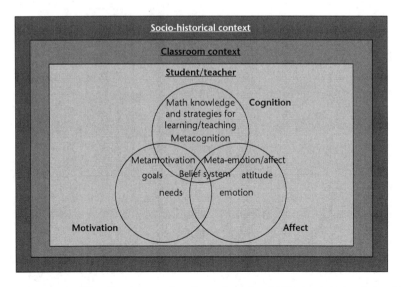

FIGURE 10.2 The dimensions of mathematics-related affect and their relationships, presented at CERME 5 (Hannula, Op 't Eynde, Schlöglmann & Wedege, 2007, p. 204)

130 Hannula, Pantziara & Di Martino

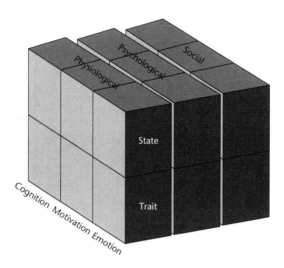

FIGURE 10.3 Hannula's (2011, p. 46; 2012) model of the dimensions for affective constructs

1.2 Shifting focus of attention

In order to empirically explore the relevance of different constructs and dimensions suggested above, we analysed the terminology for affect appearing in 134 CERME affect papers (Figure 10.4). We first identified altogether 51 different affect terms appearing in the titles and then searched for and counted all appearances of these words in the texts. After removing false positives, the number of occurrences of affect terms was 17,368. The most frequently appearing terms

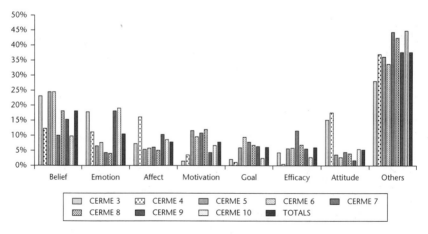

FIGURE 10.4 Frequency of affect term appearance in CERME Affect TWG research reports as percentage of all affect terms used

were Belief, Emotion, Affect(ive), Motivation, Goal, Efficacy, and Attitude. The frequencies of the term Attitude fell after CERME 4 while the frequencies of the terms Motivation and Goal increased. The term Emotion was popular in CERME 3 and then again in CERMEs 9 and 10. New concepts have also been introduced to the group over the years; for example, the concept of personal meaning (Vollstedt, 2010) emerged in CERME 6.

Next, we present Hannula and Moreno-Esteva's (2017) results of a network analysis of affect papers in CERMEs 4–10. The analysis used graph theory to identify groups of papers citing the same authors. The results suggest ten groups: Foundation (30 papers), Self-Efficacy (11 papers), Motivation (11 papers), Teacher Development (8 papers), Resilience (5 papers), Academic Emotions (4 papers), Metacognition (4 papers), Meaning (4 papers), Identity (4 papers), and Teacher Beliefs (3 papers). As many of the groups are quite small a more detailed examination is useful.

The largest group, Foundation, was united by citing some key researchers in the field of mathematics-related affect (e.g. McLeod, Schoenfeld, and Goldin) and some active CERME affect group participants (Hannula, Zan, Pehkonen, and Di Martino). Most of these papers used theoretical frameworks where affect (7 papers), attitude (8 papers), belief (7 papers), or emotion (10 papers) were among the key concepts. Rather than identifying a separate research tradition, this group represents the common ground largely shared by CERME affect papers. Looking at the other groups identified, we see that belief research appears in three groups: Foundation, Self-Efficacy, and Teacher Beliefs. Likewise, emotion appears in Foundation and Academic Emotions.

Taken together, the results confirm those of the previous analysis of affect terms in articles, identifying affect, belief, attitude, motivation, and emotion as the key concepts. Moreover, both analyses suggest that research on self-efficacy beliefs is somewhat separated from other belief research. Results support the distinction between cognition (beliefs), motivation, and emotions as proposed by Op 't Eynde's (Figure 10.2) and Hannula's (Figure 10.3) models. In the highly influential McLeod (1992) framework, motivation was addressed as a subset of beliefs (motivational beliefs) and may explain why motivation was a rare term until CERME 5. However, since then motivation has become an established key concept for studying mathematics affect, Philippou and Pantziara being most frequent contributors. Among the several theoretical approaches that have been developed in the realm of educational psychology the most influential in CERME has been the achievement goal theory (Elliot, 1999).

The separate group on Identity advocates the relevance of the distinction between psychological and sociological theories. The first papers that discussed identity (Kaasila, Hannula, Laine & Pehkonen, 2006; Gómez-Chacón, 2006) were classified in the Foundation group, suggesting their need to relate to the dominating frameworks of that time, which has no longer been necessary in more recent CERME papers.

Finally, the analysis suggests dynamics of change as a possible additional characterising feature for research, exemplified by the groups Teacher Development and

2 Key findings of the CERME affect research

In order to discuss further some key findings of the studies presented in affect meetings, we chose to concentrate on studies that involve students, since Chapters 12 and 13 of this book specifically consider teachers' knowledge, beliefs, identity, and development. The main focus of such studies over the years concerned (1) the structure of affect and the relation between affective variables and achievement, (2) the role of affect in mathematical problem solving and problem posing, and (3) change in students' affect. Some results in these areas are presented below.

2.1 The structure of affect and the relation between the different affective variables and achievement

In this focus area, the results of the different studies concentrate on the most frequently appearing terms described earlier (beliefs, attitudes, emotions, and motivation).

The results of many studies in the TWG with various age-groups of students supported the belief structure suggested by McLeod in 1992 (which refers to beliefs about mathematics, about self, about mathematics teaching, and about the social context), and also revealed important patterns regarding the relation between beliefs and other affective variables. Studies in the group also confirmed the role of social variables such as the school context and gender in the formation of different beliefs. Particularly, in different CERME meetings, researchers (Gagatsis, Panaoura, Deliyianni & Elia, 2010; Kapetanas & Zachariades, 2007; Rösken, Hannula, Pehkonen, Kaasila, & Laine, 2007) reported the development and properties of scales meant to investigate students' beliefs in mathematics and mathematics learning/teaching. Some scales were more extended than others, including various dimensions. A closer look at these scales revealed factors such as beliefs about mathematics and mathematics learning, beliefs related to the personal aspect (self-efficacy, emotional expression, competence, effort) as well as the social aspect, the role of the teacher and family encouragement.

The results show that positive beliefs and self-efficacy beliefs are correlated with high performance in mathematics (Gagatsis et al., 2010; Kapetanas & Zachariades, 2007; Nicolaidou & Philippou, 2004). The direction of this relationship, and specifically the reciprocal relationship between 8th- and 9th-grade students' mathematics self-efficacy beliefs and their performance, was investigated by Sørlie Street, Malmberg, and Stylianides (2017) in a longitudinal study.

Studies also showed that the social context was an influential factor in students' beliefs. The study by Kapetanas and Zachariades (2007) traced differences in 10th–12th-grade students' beliefs related to the type of school (public, private, and technical) they attended. The findings by Rösken et al. (2007) included significant differences in the beliefs subscales related to 11th-grade students of general or advanced courses, with students in advanced courses having more positive beliefs. Hannula (2010) revealed that 11th graders of the same class tended to have similar effort, enjoyment of mathematics, and evaluation of the teacher, while their mathematical confidence was influenced by gender. Panaoura, Deliyianni, Gagatsis, and Elia (2011) showed differences in students' beliefs in respect to school grade (5th–8th grades).

The structure of attitudes has received considerable attention in the discussions of the TWG. Di Martino (2010) investigated the structure of attitudes which emerged from the essays of 1,600 students across grades 1–13. Three dimensions of the attitude construct emerged. These are the emotional disposition (concisely expressed by "I like/do not like maths"), an affective one (expressed by "to like" and "to adore"), and one correlated with the idea of success in mathematics (expressed by "to understand" and "clever"). Nicolaidou and Philippou (2004) reported positive correlation between primary students' attitudes and mathematics achievement.

Research on mathematics-related emotions has become more intense in more recent CERME meetings (Figure 10.4). A main source for the studies was Pekrun's framework of academic emotions (Pekrun, Frenzel, Goetz & Perry, 2007). The framework provides a three-dimensional taxonomy of academic emotions referring to activity vs. outcome emotions, to their valence (positive vs. negative emotions), and to the degree of activation (activated vs. deactivated). The framework assumes that control and value appraisals relating to learning, teaching, and achievement are of great importance for students' and teachers' emotions. Studies about emotions focused on investigating different emotions at the same time (Martínez-Sierra, 2015), in investigating the intensity of emotions, factors that develop certain emotions such as teaching and types of tasks (Schukajlow, 2015), performance (Pantziara, Pitta-Pantazi & Philippou, 2007), and emotion regulation strategies (Op 't Eynde, De Corte & Mercken, 2007).

Regarding motivation, some main directions were observed in the research papers. Pantziara and Philippou (2010, 2011), using achievement goal theory, investigated primary students' different motivational goals (mastery, performance, and performance-avoidance) and found that mastery goals are related to positive affective variables (self-efficacy, interest) and behaviour (achievement). Another direction referred to intrinsic and extrinsic motivation (Wæge, 2007). Motivation in terms of needs and goals was investigated qualitatively by Wæge (2010). Tuohilampi (2011) combined self-beliefs and motivational theories (achievement goals) to investigate the discrepancy between real and ideal self. In the realm of motivation, the absence of studies related to values was stressed in the CERME 9 and 10 meetings.

2.2 The role of affect in mathematical problem solving and problem posing

One important area of research on mathematics-related affect is the role of different affective constructs in the process of mathematical thinking with problem solving. Some quantitative research reports in the TWG investigated the relationship between multiple affective variables and problem solving and posing. Nicolaou and Philippou (2007) found in their study with students of grade 5 and 6 that students' perceived efficacy was a stronger predictor of their problem-posing ability and their general mathematics achievement than their attitudes.

Studies using qualitative methods described students' emotions during problem solving. The analysis of emotional states revealed their significant role in students' difficulties in problem solving, emphasising the relationship between affective states and cognitive aspects. Antognazza, Di Martino, Pellandini, and Sbaragli (2015) discussed that the distinction between students' positive or negative emotions in a specific activity derived from an assessment of the difficulty of the activity proposed (intrinsic aspects), or from more general aspects (e.g. "I do not like mathematics"). Liljedahl (2017) studied students' problem-solving behaviour when faced with an imbalance between their skills and the challenge of the task. Results indicate that most students have perseverance in the face of challenge and tolerance in the face of the mundane and they use these as a buffer in order to autonomously correct the imbalance between skill and challenge that they experience.

Assuming the existence of a mutual influence of affective and cognitive factors, a case study by Furinghetti and Morselli (2006) revealed that among the elements that shaped the behaviour of a good problem solver were aesthetic values and feelings of freedom in facing the problem. In the same vein, Viitala (2015) described a grade 9 high achiever's mathematical thinking through problem solving and mathematics-related affect. The results revealed a successful, though quite unsure, problem solver whose affective state (connected to problem solving) seemed to tell the same story as her affective trait (view of mathematics). The differences between results on affective state and trait seemed to be connected mostly to emotions.

2.3 Change in students' affect

The analysis of the papers suggested dynamics of change as a possible additional characterising feature for research. In CERME 6, the affect group identified and reported four different aspects of stability (Hannula, Pantziara, Wæge, & Schlöglmann, 2010): (1) The state and trait aspects of affect; (2) Resistance to change; (3) Robustness of constructs; and (4) Relative stability in relation to other persons.

Studies investigating changes in affect involving students, referred to changes in students' affect as they move to upper school grades, or the importance of the classroom microculture on students' affect. Change in students' affect was also observed through their engagement in structured interventions such as problem solving and modelling cases or through specific instructional practices.

Past research evidently indicates that students' mathematics-related affect develops detrimentally during school years. A decline in students' positive affect was documented by the studies of Athanasiou and Philippou (2010) during the transition to secondary school. Tuohilampi, Näveri, and Laine (2015), trying to prevent this decline in students' affect, applied a three-year intervention designed to improve primary school students' problem-solving skills, and their mathematics-related affect. The impact was restricted but crucial: girls' affect regarding mathematics decreased less in the intervention group.

Dropping out of mathematics, and especially advanced mathematics, has become a major concern for society (Moscucci, Piccione, Rinaldi & Simoni, 2006). Moreover, several reports indicate a low rate of tertiary students around the world that are enrolled in science, technology, engineering, and mathematics (STEM) related careers with an even lower rate for women (Sánchez Aguilar, Romo Vázquez, Rosas Mendoza, Molina Zavaleta & Castañeda Alonso, 2013). Moscucci et al. (2006), highlighting the role of affect in this situation, showed that students who dropped out of school had failed in mathematics the same school year. Other studies have investigated factors that influence students' enrolment in advanced mathematical courses. Factors referred to students' perceived competence in mathematics, their future expectations but also to their teachers and relatives (Kleanthous & Williams, 2011; Sánchez Aguilar et al., 2013).

Several studies reported positive change in students' affect after structured interventions through problem solving (Marcou & Lerman, 2007; Stylianides & Stylianides, 2011) and modelling (Schukajlow & Krug, 2013). Barnes (2015) reported on an intervention that explored perseverance in mathematical reasoning in children aged 10–11. The findings suggest improved perseverance because of the effect the intervention seemed to have on the bidirectional interplay between affect and cognition.

Students' affect is influenced by the learning context and the teacher (Liljedahl & Hannula, 2016). However, the experiences of students in one class may differ and the development of their affect might follow very different paths. Some studies (e.g.Vankúš, 2007) showed increase in students' affect after implementing some new practices such as didactical games and humour in the mathematics classroom. Other studies showed that students' motivation for learning mathematics can be influenced and altered by changes in teaching (Pantziara & Philippou, 2010; Wæge, 2010).

3 The development of methodology in the field of affect

Methodological issues have always been central in the discussions of the Affect TWG in CERME. Since the first meeting of the TWG there have been contributions focused on methodological issues: not only concerning the development of new methods, but also related to theoretical aspects (Evans, 2004). In particular, it has been underlined that researchers' choices about methodology are related to their affective traits (Zan & Di Martino, 2004) and the choices can condition and constrain the findings (Pantziara, Wæge, Di Martino & Rösken-Winter, 2013).

The reflections about methodology developed over the years in CERME have had a great impact on the development of several critical issues in the field of affect.

136 Hannula, Pantziara & Di Martino

In the first meetings of the TWG, attention was focused on the definition and delimitation of affective traits. The literature showed the lack of a generally accepted conceptualisation of the principal affective constructs (attitudes, beliefs, emotions, motivation): the different constructs tended to be defined implicitly and a posteriori through the instruments used to measure them (Furinghetti & Pehkonen, 2002; Di Martino & Zan, 2001) and there was no clear distinction between these constructs. Schlöglmann (2004) underlined that the research methods used in the field could not establish a distinction between the categories above, and he used this argument to encourage the development of new research methods and approaches. In particular, he suggested considering methods developed in other domains, such as neuroscience.

A second aspect concerns the nature of affective constructs: that it is difficult to infer them. There are essentially two schools of thought: one sees affective constructs as an inner awareness or process of interpretation of events rather than an overt behaviour, and consequently they are not directly observable and, moreover, individuals themselves are often not conscious of these processes (Panaoura & Philippou, 2006). Another school of thought sees affective constructs (such as attitudes and beliefs) not as a quality of an individual but rather as a researcher's model for describing and understanding some mathematical behaviour (Zan & Di Martino, 2004). In both views, the problems connected with the methodology are evident.

Another important issue affecting the discussion about methodology is the distinction between rapidly changing states and relatively stable traits (Hannula, 2011). Observing affective states seems particularly complicated and there is an imbalance in favour of studies that focus on traits over studies that focus on states. Only in the more recent CERMEs have some studies focused on affective states, and some specific theories (e.g. reversal theory) been introduced as interpretative frameworks (Lewis, 2015; Antognazza et al., 2015). Schlöglmann (2004) argued that quantitative methods reveal stable and less intense categories, while qualitative methods are able to grasp quickly changing and very intense reactions. Actually, the first qualitative studies presented in our CERME group were typically case studies (e.g. Furinghetti & Morselli, 2006) or small sample studies (Liljedahl, Rolka & Rösken, 2007) developed to observe changes in action.

New issues and new goals in the research on affect have been identified in the last 20 years. We can highlight two main directions: the first follows the traditional approach in the field of affect, searching for causal relationship between affective variables and mathematical performance or behaviour. In this frame, the crucial action is to measure, privileging quantitative methods. It demands isolating, clearly identifying, and measuring variables in order to interpret statistical results. A necessary part of the studies conducted within this frame is the development of means for the efficient measurement of affective constructs, but also of mathematical performance. The second direction – following the interpretive paradigm in the social sciences – abandons the goals of explaining behaviour through measurements and of determining general rules based on a cause-effect model to describe the interaction

Affect and mathematical thinking **137**

between affective and cognitive constructs in mathematics education, and focuses on trying to make sense of the observed phenomena from the perspective of participants. This implies a significant shift in focus and, in particular, a movement toward the use of qualitative approaches (Evans, Hannula, Philippou & Zan, 2004).

An evident consequence of these considerations is the trend toward a balancing between the use of quantitative and qualitative methods in the research on affect, overcoming the initial preponderance of quantitative methods (Table 10.1). Since the early TWGs, mixed methods or a hybrid approach (i.e. collecting qualitative data and analysing them with systematic categorisation and basic statistical analysis, for example, cross-tabulations) have become more popular. Moreover, the shift in focus from the description of a phenomenon to the interpretation of the same phenomenon intensifies the attention on how the collected data are interpreted (Di Martino et al., 2015).

Concerning quantitative methods, the development of measurement tools is particularly critical. The majority of the quantitative papers have used classic and consolidated questionnaires and scales, but two interesting trends emerge: the first, coherent with the consideration of the complexity of the affective factors, is the trend to modify and combine two or more scales for the same studies (Pantziara & Philippou, 2011); the second is the trend to adopt the more complex computational tools that have become available to analyse the data (Mosvold, Fauskanger, Bjuland & Jakobsen, 2011).

Concerning qualitative approaches, CERME papers have often introduced new methods to collect data. An exemplary case is offered by Kaasila et al.'s paper (2006): it discusses the potential of autobiographical narratives to reconstruct the mathematical identity of a person. The key assumption is that humans are storytelling beings who, individually and socially, lead storied lives. The study of narrative, therefore, is the study of the ways humans experience the world. Such new approaches and observational or other tools were developed for specific goals and specific contexts. Perhaps, precisely their specificity is a point of weakness, because it is very difficult to use these instruments when some conditions change.

TABLE 10.1 The proportions of different methodological approaches in the CERME Affect TWG

Edition	Qualitative	Quantitative	Theoretical	Mixed/hybrid
CERME3	20%	30%	50%	0%
CERME4	36%	36%	18%	9%
CERME5	20%	58%	11%	11%
CERME6	36%	50%	14%	0%
CERME7	54%	46%	0%	0%
CERME8	35%	35%	6%	24%
CERME9	41%	21%	18%	20%
CERME10	47%	22%	9%	22%

4 Affect and Mathematical Thinking Group: looking ahead

The research on mathematics-related affect has repeatedly raised the terminological issues as a problem (e.g. Furinghetti & Pehkonen, 2002; Hannula, 2011, 2012). This problem is related to the cumulative nature of the research, and therefore to the need that new research builds on a critical analysis of previous research. The CERME group has been, and will be, an important place to highlight this fundamental issue and to tackle it. Yet, it is important to keep a way open for new concepts to emerge. It seems clear, in retrospect, that motivation and identity were terms that were necessary for the research field. It also seems reasonable that we need specific terms, for example, for 'Perseverance' (Barnes, 2015) and 'Resilience' (Lee & Johnston-Wilder, 2011).

Despite the improved understanding of mathematics-related affect, the general trend still is that enjoyment of mathematics decreases over the school years. We need to develop teaching approaches that promote a positive relationship with mathematics without compromising understanding of concepts. Such approaches should be tested through systematic longitudinal intervention studies.

There are three specific methodological possibilities that can open yet newer understandings of the dynamics of mathematics-related affect. The first possibility would be to analyse the dynamics of group-level processes: How does the teacher initiate and maintain excitement and good working climate in the class? What kind of processes lead to the 'energy' of the class being lost? Concepts such as classroom climate would be useful for this kind of analysis. The other new methodological possibility is to implement physiological measures (e.g. heart rate monitoring) to gain a continuous indication of participants' affective states. These methods have been used for a long time in laboratories, but have only recently become more affordable and usable in actual classrooms. The third methodological possibility is strongly associated with the CERME spirit of collaboration. Comparative research on mathematics-related affect has confirmed that while some research findings about affect are universal, some other findings are contextual. Therefore, we should examine which results about affect are transferable to different sociocultural contexts, and Europe with its diverse educational systems and linguistic groups is a wonderful test bed for such comparative studies.

Acknowledgements

We wish to thank Enrique Garcia Moreno-Esteva for analysing the appearances of affect terms in CERME papers, and Gareth Lewis for his help with polishing the language of this chapter.

References

Antognazza, D., Di Martino, P., Pellandini, A., & Sbaragli, S. (2015). The flow of emotions in primary school problem solving. *CERME9* (pp. 1116–1122).

Athanasiou, C., & Philippou, G. (2010). The effects of changes in the perceived classroom social culture on motivation in mathematics across transitions. *CERME6* (pp. 114–123).

Barnes, A. (2015). Improving children's perseverance in mathematical reasoning: Creating conditions for productive interplay between cognition and affect. *CERME9* (pp. 1131–1138).

Di Martino, P. (2010). "Maths and me": software analysis of narrative data about attitude towards math. *CERME6* (pp. 54–63).

Di Martino, P., & Zan, R. (2001). Attitude toward mathematics: Some theoretical issues. In M. van den Heuvel-Panhuizen (Ed.), *Proceedings of the 25th IGPME Conference* (Vol. 3, pp. 351–358). Utrecht, the Netherlands.

Di Martino, P., Gómez-Chacón, I., Liljedahl, P., Morselli, F., Pantziara, M., & Schukajlow, S. (2015). Introduction to the papers of TWG08: Affect and mathematical thinking. *CERME9* (pp. 1104–1107).

Elliot, A. (1999). Approach and avoidance motivation and achievement goals. *Educational Psychologist, 34*, 169–189.

Evans, J. (2004). Methods and findings in research on affect and emotion in mathematics education. *CERME3*: www.mathematik.uni-dortmund.de/~erme/CERME3/Groups/TG2/TG2_evans_cerme3.pdf.

Evans, J., Hannula, M., Philippou, G., & Zan, R. (2004). Introduction to the papers of the thematic working group Affect and Mathematical Thinking. *CERME3*: www.mathematik.uni-dortmund.de/~erme/CERME3/Groups/TG2/TG2_introduction_cerme3.pdf.

Furinghetti, F., & Morselli, F. (2006). Reflections on creativity: The case of a good problem solver. *CERME4* (pp. 184–193).

Furinghetti, F., & Pehkonen, E. (2002). Rethinking characterizations of beliefs. In G. C. Leder, E. Pehkonen, & G. Törner (Eds.), *Beliefs: A hidden variable in mathematics education?* (pp. 39–57). Dordrecht: Kluwer.

Gagatsis, A., Panaoura, A., Deliyianni, E., & Elia, I. (2010). Students' beliefs about the use of representations in the learning of fractions. *CERME6* (pp. 64–73).

Gómez-Chacón, I. M. (2006). Affect, mathematical thinking and intercultural learning: A study on educational practice. *CERME4* (pp. 194–204).

Hannula, M. S. (2010). The effect of achievement, gender and classroom context on upper secondary students' mathematical beliefs. *CERME6* (pp. 34–41).

Hannula, M. S. (2011). The structure and dynamics of affect in mathematical thinking and learning. *CERME7* (pp. 34–60).

Hannula, M. S. (2012). Exploring new dimensions of mathematics related affect: Embodied and social theories. *Research in Mathematics Education, 14*(2), 137–161.

Hannula, M. S., & Moreno-Esteva, G. (2017). Identifying subgroups of CERME affect research papers. *CERME10* (pp. 1098–1105).

Hannula, M. S., Op 't Eynde, P., Schlöglmann, W., & Wedege, T. (2007). Affect and mathematical thinking. *CERME5* (pp. 202–208).

Hannula, M. S., Pantziara, M., Wæge, K., & Schlöglmann, W. (2010). Introduction: Multimethod approaches to the multidimensional affect in mathematics education. *CERME6* (pp. 28–33).

Kaasila, R., Hannula, M. S., Laine, A., & Pehkonen, E. (2006). Autobiographical narratives, identity and view of mathematics. *CERME4* (pp. 215–224).

Kapetanas, E., & Zachariades, T. (2007). Students' beliefs and attitudes concerning mathematics and their effect on mathematical ability. *CERME5* (pp. 258–267).

Kleanthous, E., & Williams, J. (2011). Students' dispositions to study further mathematics in higher education: The effect of students' mathematics self-efficacy. *CERME7* (pp. 1229–1238).

Lee, C., & Johnston-Wilder, S. (2011). The pupils' voice in creating a mathematically resilient community of learners. *CERME7* (pp. 1189–1198).

Lewis, G. (2015). Patterns of motivation and emotion in mathematics classrooms. *CERME9* (pp. 1216–1222).

Liljedahl, P. (2017). On the edges of flow: Student engagement in problem solving. *CERME10* (pp. 1146–1153).

Liljedahl, P., & Hannula, M. S. (2016). Research on mathematics-related affect in PME 2005–2015. In A. Gutierrez, G. C. Leder, & P. Boero (Eds.), *The second handbook of research on the psychology of mathematics education* (pp. 417–446). Rotterdam: Sense Publisher.

Liljedahl, P., Rolka, K., & Rösken, B. (2007). Belief change as conceptual change. *CERME5* (pp. 278–287).

Marcou, A., & Lerman, S. (2007). Changes in students' motivational beliefs and performance in a self-regulated mathematical problem-solving environment. *CERME5* (pp. 288–297).

Martínez-Sierra, G. (2015). Students' emotional experiences in high school mathematics classroom. *CERME9* (pp. 1181–1187).

McLeod, D. B. (1992). Research on affect in mathematics education: A reconceptualization. In D. A. Grouws (Ed.), *Handbook of research on mathematics learning and teaching* (pp. 575–596). New York: MacMillan.

Moscucci, M., Piccione, M., Rinaldi, M. G., & Simoni, S. (2006). Mathematical discomfort and school drop-out in Italy. *CERME4* (pp. 245–254).

Mosvold, R., Fauskanger, J., Bjuland, R., & Jakobsen, A. (2011). Using content analysis to investigate student teachers' beliefs about pupils. *CERME7* (pp. 1389–1398).

Nicolaidou, M., & Philippou, G. (2004). Attitudes towards mathematics, self-efficacy and achievement in problem-solving. *CERME3*: www.mathematik.uni-dortmund. de/~erme/CERME3/Groups/TG2/TG2_nicolaidou_cerme3.pdf.

Nicolaou, A. A., & Philippou, G. N. (2007). Efficacy beliefs, problem posing, and mathematics achievement. *CERME5* (pp. 308–317).

Op 't Eynde, P., De Corte, E., Mercken, I. (2007). Students' self-regulation of emotions in mathematics learning. *CERME5*, (pp. 318–3280).

Panaoura, A., & Philippou, G. (2006). The measurement of young pupils' metacognitive ability in mathematics: The case of self-representation and self-evaluation. *CERME4* (pp. 255–264).

Panaoura, A., Deliyianni, E., Gagatsis, A., & Elia, I. (2011). Self-beliefs about using representations while solving geometrical problems. *CERME7* (pp. 1167–1178).

Pantziara, M., & Philippou, G. (2010). Endorsing motivation: Identification of instructional practices. *CERME6* (pp. 106–113).

Pantziara, M., & Philippou, G. (2011). Fear of failure in mathematics: What are the sources? *CERME7* (pp. 1269–1278).

Pantziara, M., Pitta-Pantazi, D., & Philippou, G. (2007). Is motivation analogous to cognition? *CERME5* (pp. 339–348).

Pantziara, M., Wæge, K., Di Martino, P., & Rösken-Winter, B. (2013). Introduction to the papers of the thematic working group Affect and Mathematical Thinking. *CERME8* (pp. 1272–1278).

Pekrun, R., Frenzel, A., Goetz, T., & Perry, R. P. (2007). The control-value theory of achievement emotions: An integrative approach to emotions in education. In P. A. Schutz & R. Pekrun (Eds.), *Emotion in education* (pp. 13–36), San Diego, CA: Academic Press.

Rösken, B., Hannula, M. S., Pehkonen, E., Kaasila, R., & Laine, A. (2007). Identifying dimensions of students' view of mathematics. *CERME5* (pp. 349–358).

Sánchez Aguilar, M., Romo Vázquez, A., Rosas Mendoza, A., Molina Zavaleta, J. G., & Castañeda Alonso, A. (2013). Factors motivating the choice of mathematics as a career among Mexican female students. *CERME8* (pp. 1409–1418).

Schlöglmann, W. (2004). Can neuroscience help us better understand affective reactions in mathematics learning? *CERME3*: http://www.mathematik.uni-dortmund.de/~erme/CERME3/Groups/TG2/TG2_schloeglmann_cerme3.pdf.

Schukajlow, S. (2015). Is boredom important for students' performance? *CERME9* (pp. 1273–1279).

Schukajlow, S., & Krug, A. (2013). Uncertainty orientation, preference for solving task with multiple solutions and modelling. *CERME8* (pp. 1428–1437).

Sørlie Street, K., Malmberg, L., & Stylianides, G. (2017). Self-efficacy and mathematics performance: Reciprocal relationships. *CERME10* (pp. 1186–1193).

Stylianides, G., & Stylianides, A. (2011). An intervention on students' problem-solving beliefs. *CERME7* (pp. 1209–1218).

Tuohilampi, L. (2011). An examination of the connections between self-discrepancies and effort, enjoyment and grades in mathematics. *CERME7* (pp. 1239–1248).

Tuohilampi, L., Näveri, L., & Laine, A. (2015). The restricted yet crucial impact of an intervention on pupils' mathematics-related affect. *CERME9* (pp. 1287–1293).

Vankúš, P. (2007). Influence of didactical games on pupils' attitudes towards mathematics and process of its teaching. *CERME5* (pp. 369–378).

Viitala, H. (2015). Emma's mathematical thinking, problem solving and affect. *CERME9* (pp. 1294–1300).

Vollstedt, M. (2010). "After I do more exercise, I won't feel scared anymore": Examples of personal meaning from Hong Kong. *CERME6* (pp. 124–135).

Wæge, K. (2007). Intrinsic and extrinsic motivation versus social and instrumental rationale for learning mathematics. *CERME5* (pp. 379–388).

Wæge, K. (2010). Students' motivation for learning mathematics in terms of needs and goals. *CERME6* (pp. 84–93).

Zan, R., & Di Martino, P. (2004). The role *of* affect in the research *on* affect: The case of 'attitude'. *CERME3*: www.mathematik.uni-dortmund.de/~erme/CERME3/Groups/TG2/TG2_zan_cerme3.pdf.

Zan, R., & Poli, P. (1999). Winning beliefs in mathematical problem solving. *CERME1* (pp. 97–104).

11

TECHNOLOGY AND RESOURCES IN MATHEMATICS EDUCATION

Jana Trgalová, Alison Clark-Wilson and Hans-Georg Weigand

1 The Technology Thematic Working Group at CERME: history, background, aims and scope

The Thematic Working Group (TWG) on 'Technology' at the Congress of European Research in Mathematics Education (CERME) was established at the very first congress in 1999, as one of only seven themes, which highlights the importance that the mathematics education community had placed on research on technology in mathematics teaching and learning at that time. Over the subsequent congresses, the TWG continued to grow, from nine contributions at CERME 1 to more than 30 (papers and posters) at CERMEs 7 and 8, which led to a sub-division into two TWGs addressing the theme from the perspectives of students and teachers, respectively.

From the outset, the TWG has considered mostly digital tools and technology encompassing mathematical software and applications, programming languages, communication platforms and mobile devices. Recently, more elaborated concepts of resources have led to tools and technology being considered more systematically as a component of the full range of resources available for students, teachers or teacher educators. Thus, since CERME 6, the TWG welcomes contributions not only on digital tools, but also on more traditional mathematical tools, textbooks and other resources.

Since that very first congress, the work of the Technology TWG has been framed by the following "three embedded levels" that can be considered "when analysing the use of tools in mathematics education" (Gutiérrez, Laborde, Noss & Rakov, 1999, pp. 183–184):

- the level of the interactions between tool and knowledge,
- the level of interactions between knowledge, tool and the learner, and
- the level of integration of a tool in a mathematics curriculum and in the classroom.

These three levels highlight the four components that can be distinguished in a didactic system involving any technological tool, namely the tool, some knowledge, student(s) and a teacher, and the inevitable relationships between these poles. Such a system can be represented by a *didactic tetrahedron*, as shown in Figure 11.1, inspired by Tall (1986, p. 25). The didactic tetrahedron introduces a fourth component (vertex), a technology or a resource, into the traditional representation of a didactic system as a triangle: teacher–learner–knowledge; and allows for considerations of the impact of this introduction on the other three vertices.

The chapter aims to capture the work of the Technology TWG over time with respect to three sub-themes that have permeated this collaborative work: deep articulation of the nature of technological tools and resources and their related interactions (Section 2); explanations of the principles and theories relating specifically to task design in technology-mediated environments (Section 3); and an expansion of our knowledge of theories and approaches that underpin and/or explain research in this field (Section 4).

We conclude the chapter with a summary of the significant research by the TWG over the past 20 years, its potential and limitations; and the impact of this work within and beyond the CERME community. We close by offering a prospective vision for the possible trajectories for future research. This is set within the context of a world where the rapid growth in both access to, and design of, new technologies within and beyond mathematics education is increasingly hard to understand.

2 Tools and resources

A *resource* or *medium* (lat.: medium = middle, midpoint) is something that is positioned between two domains. Language, gestures, paper, pencil, books, videos, ruler and compass, computer or interactive whiteboards are media and, in mathematics, they are positioned between mathematical objects (concepts, statements, algorithms) and human thinking; they mediate between mathematics and understanding. *Tools* are a special form of media. Monaghan, Trouche and Borwein (2016) give a quite general and "somewhat crude" definition of a "tool" as "something you use to do

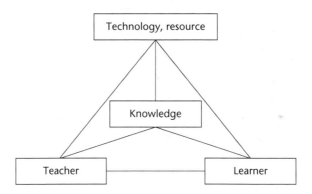

FIGURE 11.1 Didactic tetrahedron

144 Trgalová, Clark-Wilson & Weigand

something" (p. 5). This already shows the fundamental and global aspects of a tool in human activities, which makes it difficult to give a meaningful and satisfactory definition. Mathematical tools enable us to create, to operate with and to transform mathematical objects. Computers, and especially programs such as spreadsheets, dynamic geometry systems (DGS) and computer algebra systems (CAS), are digital mathematical tools. Mathematics education questions the meaning of media and tools for the teaching and learning of mathematics, which is the primary interest of the CERME Technology TWG. In what follows, we aim to not only reflect these discussions from within the CERME Technology TWG(s), but also connect the TWG's work with the global discourse on digital technologies in mathematics and mathematics education.

Some common threads include:

- the recurring discussions on the relationships between the theoretical and practical aspects of tool use;
- the construction and development of theories that have been modified for (digital) tools and resources;
- the interrelationships between digital tools and other (non-digital) resources.

2.1 From suggestions of classroom use to more general reflections on technology-enhanced teaching and learning

Prior to CERME 1 in 1999, an intensive discussion about new – nowadays digital – technologies in mathematics education had already begun. At that time, the main goal was to develop and evaluate strategies for the integration of digital tools in mathematics curricula and classrooms. For example, as one of the first DGS, Cabri-Géomètre had appeared in 1988, by CERME 1 the TWG was able to draw upon experiences of over 10 years of its use. Consequently, the most important features of this class of tools, such as variation of objects by dragging, visualisations of loci, and authoring macro-constructions, had begun to be widely discussed. In addition, the didactical implications of these features created opportunities for new problem-solving strategies, the discovery of geometrical theorems as invariant features, developing conjectures and making or discovering mathematical proofs.

CERME 1 built on this earlier work by adopting a research-oriented view with questions such as: What are students' views and interpretations of dynamic tools? What is the relationship between drawings and symbols? How can technologies support the learning of the concepts of variables and functions? The contributions to this congress concentrated mainly on particular uses of tools in the classroom with many examples of the dynamic affordances of these tools.

The following two congresses, CERMEs 2 and 3, led to a shift from the fascination of the new technological possibilities and novel examples of tool use in particular content-oriented environments, to theoretical reflections concerning comparisons, relationships and connections between tools. Moreover, other considerations of technology design and use emerged, such as the use of technology

Technology and resources **145**

for distance education and the potential impact of technology on the nature of examinations. In this period, increasing attention was paid to teachers' roles in technology-mediated mathematical activities and their associated knowledge base.

Simultaneously, the focus of the scientific discussion concerning digital technologies outside CERME was CAS. In 1995 the first calculator with CAS became available, the Texas Instruments' TI-92, followed in 1999 by the Casio FX 2.0. These calculators generated a high level of expectation within the community. The possibility for students to have a readily available tool, which would do (nearly) all symbolic transformations of high school mathematics at the press of a button, was predicted to lead to deep, far-reaching changes for the content of the curriculum and its examination. These particular tools promoted the research community to rethink existing theories and develop new interpretations (see Section 4).

By the time of CERME 4, the questions concerning tool classification and design were more specified. It was evident that the word *tool* meant a variety of objects with different characteristics. There were special pedagogy-free environments such as CAS, DGS, graphing and programming tools. Other tools, often called applets, microworlds or special learning programs could be considered as local dedicated environments. This raised the question concerning the characteristics of each of these tools: What kind of technological tools do we need in our teaching? Moreover, it was also apparent that the reflective use of tools in the learning process needed theoretical frameworks specific to the tool and mathematical content, for which the Instrumental Approach emerged as a central theoretical framework. CERME 5 drew on the concepts introduced by the Instrumental Approach and opened questions about the design and appearance of tools to support successful appropriation, integration and institutionalisation for both students and teachers. These ideas are described further in Section 4.

2.2 The move to technologies as tools within a resource system

In the beginning of the new century, there were two demands concerning the use of technology in the teaching and learning process. On the one hand, there was a request for more sustained and longitudinal projects to obtain significant and convincing research findings in 'real' classroom situations. A number of empirical studies had taken place over longer periods that had revealed results that were common to many studies. For example, e-CoLab[1] project in France, RITEMATHS[2] project in Australia, or the '**M**[3]-**M**odel Project New **M**edia in **M**athematics Education' (Weigand, 2008).

These increasingly robust findings established that using technology provides opportunities to:

- work within dynamically linked multiple mathematical representations;
- construct new problem-solving environments;
- design more personalised learning;
- integrate more realistic modelling problems into the mathematics classes;

146 Trgalová, Clark-Wilson & Weigand

while highlighting that technology use demanded:

- new types of learners' knowledge, e.g. to move between representational forms with understanding;
- new types of teachers' knowledge in relation to design, implementation and assessment;
- some rethinking of the content and hierarchies of the mathematics curriculum and its assessment.

On the other hand, questions around connectivity emerged, e.g. how to connect students and mathematics through technology, students and teachers, and technology to other resources for teaching and learning (Monaghan et al., 2016, p. 433). On the tool level, CERME 6 reacted to this aspect by adding the word *resources* in the name of the TWG, which was previously called 'Tools and Technologies in Mathematical Didactics'. This expressed the need for considering technologies within the full range of resources available for students, teachers and teacher educators. Resources might be software, computers, interactive white-boards, online resources, but also traditional geometry tools and textbooks. This demanded a deeper understanding of the relationship between technologies and the traditional tools and resources. How can these old and new resources interact with each other? For example, how can digital features be incorporated in new forms of textbooks? However, in the subsequent period there have been only a few contributions to the CERME Technology TWG that have addressed this demand, mainly with reference to textbooks, e-textbooks and some online courses.

2.3 From the students' uses of tools to that of teachers

The introduction of the term *resources* was also in line with the development of a Documentational Approach to Didactics (elaborated in Section 4). This places a greater emphasis on the roles and actions of teachers, a theme first introduced at CERME 3 through the following questions:

- How can we understand how mathematics teachers integrate technology in their teaching?
- How might we encourage more mathematics teachers to use technology?
- How does using technology change the ways mathematics teachers think about teaching and learning?

(Jones & Lagrange, 2004)

These questions, supplemented by others, have remained an important focus for the TWG, and have sowed the seeds for the more recent sub-division of the TWG (see Section 1). However, with respect to tools, this has meant that, in many cases the user is no longer the student alone. The teacher perspective is now not only considered alongside, but specific functionalities and environments have been developed for the primary purpose of supporting the teacher.

Adopting a more holistic perspective of tools that includes teachers, we should consider all of the processes inherent in the design of teaching: looking for resources, integrating these in a personal resource system, implementing resources in practice, sharing resources with colleagues, revising resources to take account of feedback, etc. This wider discussion concerning teachers' integration of technology continued at CERME 8 through particular examples from research: for example, using interactive whiteboards in geometry, creating tests and examinations in a CAS-environment and exploring the potential of technology for the teaching and learning of functions.

In addition, the TWG again emphasised the need to focus research more intensively toward longer-term studies involving practising teachers within 'real' classroom settings. Another request was to concentrate more on emerging research themes present within the general technology literature, which at that time had been underrepresented at ERME congresses. Examples included the design and use of innovative technologies such as Web 2.0, mobile technologies, the development of e-textbooks or the design and use of technologies and resources for learners with special educational needs. CERME 9 and CERME 10 reflected the great variety of digital books or e-books for classroom use. Participants also explored the meaning and the impact on mathematical learning of: free and widely available online courses, such as the Khan Academy (n.d.), that offer free tools to allow teachers to monitor students' activity and provide them with feedback and guidance; tablets that emphasise the meaning of gestures, e.g. zooming with finger movements, drawing graphs by using the finger; and particular digital learning environments, which also included different kinds of computer games.

2.4 From local empirical studies to scaling up good practices with digital resources

In 2010, the 17th ICMI study *Mathematics Education and Technology: Rethinking the Terrain* (Hoyles & Lagrange, 2010) was published, revisiting the theme of the very first ICMI study *The Influence of Computers and Informatics on Mathematics and its Teaching* (Churchhouse, 1986). Given the great enthusiasm 20 years earlier for the new possibilities that computers and technology might open to mathematics and mathematics education, the 2010 study gave a disappointing account of the current situation concerning the dissemination of technology. Despite a high number of research studies and accounts of classroom practices, the use of technologies in mathematics education and the impact on curriculum and assessment change was still limited.

The TWG at CERME 5 had also first highlighted the predominance of small empirical studies, calling for longer-term, larger-scale research. More recent ERME congresses have begun to include contributions on this theme (Clark-Wilson, Hoyles & Noss, 2015; Lavicza et al., 2015), which explore ways to implement or transfer ideas or consequences of empirical investigations concentrated in the word *scaling* or *scaling-up*, i.e. researching how to realise the results of research into the reality of daily teaching.

Since an effective use of technological tools requires the thoughtful development of relevant tasks, we move now to consider in more detail the nature of mathematical digital tasks and aspects of their design and use for mathematical learning.

3 The design and implementation of digital mathematical tasks

Mathematical tasks are an integral element of mathematics education and its associated research agenda. The design of digital mathematical tasks has recently received particular attention, as a sub-theme of the 22nd ICMI Study *Task Design in Mathematics Education* (Watson & Ohtani, 2015) and within a dedicated volume of the *Mathematics in the Digital Era* book series (Leung & Baccaglini-Frank, 2016).

This section focuses on technology-mediated tasks with an emphasis on the explicit design decisions that influence how tasks are subsequently used in and for mathematical learning. It considers how such tasks are combined or developed to produce learning sequences or courses and finishes by addressing an important emerging theme – the design of tasks for prospective and practising teachers/ lecturers for their professional development to introduce and use technology in mathematics classrooms.

3.1 Tasks and task designers

At the very first meeting of the TWG, it was highlighted that:

> One of the key issues for teachers is how to design tasks based on tools or technologies in which real questions for the learner emerge from the use of the tool, in which the tool is relevant and gives a new dimension to the task.
>
> *(Gutiérrez et al., 1999, p. 187)*

Indeed, elements of task design have featured within many contributions to the early congresses, varying from the individual design decisions for tasks within classroom or research laboratory settings to those concerning whole courses within large-scale university courses. However, more usually the tasks were offered as a *given*, often subsumed within the notion of the tool or activity. Consequently, unless the research was specifically reporting aspects of task design, the constraints of the length and format of a CERME research paper/presentation often limited the opportunity for the task design to be explicitly described or theorised about. Early collaborations within the Technology TWG at CERME highlighted the possible gap between a task designer's intended learning goals for a digital task and the mathematical meaning that learners ultimately construct. While it has always

Technology and resources **149**

proved challenging to separate aspects of the design of the tool from that of the task itself, in this section we try to distil the contribution of ERME research to the community knowledge concerning task design.

The term *task designer* has always held a broad definition to include teachers, researchers, teacher educators and technology developers – with many of the CERME participants representing one or more of these roles – and, as a result, offering enriched perspectives. Task designers appear to have been motivated by two broad approaches:

1. Development of innovative technological tasks that provide access to traditional mathematical knowledge and activity – often attempting to create a technology-mediated version of the equivalent paper and pencil task. For example, an early paper at CERME 1 by Gélis and Lenne (1999) described the design of tasks using CAS-based technology to support upper secondary French students to learn about arithmetic sequences.
2. Development of innovative technological tasks that might lead to new forms of mathematical knowledge and activity. For example, the advent of dynamic geometry software and the dragging affordance led to a re-examination of the role of empirical measurement and 'checking by dragging' within the processes of justification and proof (e.g. Olivero & Robutti, 2002). By contrast, researchers, who have been involved in the design of digital environments that have aimed to disrupt mathematics education norms by offering new ways of mediating existing mathematical knowledge or suggesting new epistemologies, have concluded that resulting tasks lack educational legitimacy at a system level. From the perspective of teachers and schools, tasks might not align with institutional constraints such as the prevailing classroom norms or assessment regimes, which might not have kept pace.

However, it has been common to integrate technology into research, although it started with motivations that were more pragmatic. This happened in order to improve the learning experiences in relation to the traditional curriculum to report findings in relation to new epistemologies and learning hierarchies. However, it has been common for research that began with more pragmatic motivations to integrate technology to improve the learning experiences in relation to the traditional curriculum, to report findings in relation to new epistemologies and learning hierarchies.

3.2 Contexts and theories for task design

Reviewing the earlier contributions to the Technology TWG, the vast majority of papers included examples of tasks that were used in the context of the research or study. However, it was rare for authors to describe explicitly their motivations for the design of the task, choice of representational forms and intended mathematical progression of potential pathways through the task. An exception to this is the example by Jones (1999), who described an empirical study in which pairs of students worked through a sequence of "specially designed tasks" involving the construction of quadrilaterals using Cabri-Géomètre over a nine-month period (see Figure 11.2).

In addition, although by reading the task carefully it might be possible to assume the researcher's mathematical objectives inherent in its design, Jones still holds much 'tacit knowledge' (Polanyi, 1966) related to the design, sequencing and classroom implementation of the sequence of tasks that featured in his study. This highlights a challenge for researchers, which is to articulate clearly the epistemological, psychological and sociocultural perspectives that underpin their task design decisions.

Interestingly, the CERME Technology TWG seems to have arrived at some common understandings of what is inherent in such 'specially designed tasks' in that they are often constructivist in nature – allowing students to explore and create mathematical knowledge, often working in pairs or small groups. It is only since CERME 8 that a number of researchers have begun to share their analyses of aspects of task design using theories such as Brousseau's Theory of Didactic Situations (Lagrange & Psycharis, 2013; Mackrell, Maschietto & Soury-Lavergne, 2013) and Variation Theory (Attorps, Björk, Radic & Viirman, 2013).

3.3 From the design of individual tasks to the design of courses

The sequencing of individual digitally mediated tasks to form a distinct course involves complex design decisions. For example, Belousova and Byelyavtseva (1999)

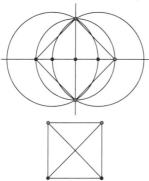

The Square
Construct these figures so that they are not "messed up".

What do you know about this shape from the way in which you constructed it?
Think about . . . . . sides
    diagonals
Explain why the shape is a square.

FIGURE 11.2 A task asking pupils to construct a square that is invariant under drag (Jones, 1999, p. 258)

at CERME 1 described the development of a university-based numerical methods course, which raised important questions about the balance between students' empirical and theoretical work; individual work and group work; and a range of mediational roles for the technology. These themes have recurred in many subsequent contributions to the Technology TWG as researchers have responded to the pragmatic decisions by institutions and policy-makers to design digitally mediated courses, while simultaneously seeking to theorise about course designs, implementations and subsequent impacts on mathematical learning.

Task design within digital environments incorporates an articulation of the way in which the task is initiated/mediated. Given that a potential role of technology is to support the communication of both mathematical and meta-cognitive knowledge between the student and 'teacher', there have been many CERME contributions that describe research in which aspects of the teacher's role is outsourced to the technology in the form of scaffolding and feedback to the learner. The notion of *e-learning* can often imply a learning pathway mediated in the absence of an obvious teacher. For example, at CERME 8, Fredriksen (2013) reported on design decisions concerning prospective teachers' uses of video lectures within an online course in Norway, while at CERME 9, Jančařík and Novotná (2015) researched how teachers scaffolded student learning in an online course for talented children in the Czech Republic, concluding the importance of offline discussions.

There have been many research contributions to CERME that have focused on the design and use of digitally mediated courses that have been developed over decades. Such examples are the Digital Mathematics Environment (Freudenthal Institute, Netherlands) and the Pepite resources in France. In both cases, an important feature is the provision of formative assessment data for the teacher on the students' mathematical responses to tasks.

More recently, the concept of a mathematics course is blurred with the emergence of the electronic book (e-book), which offers both sequences of tasks and embedded dynamic digital objects. The 'creative book' developed by Geraniou and Mavrikis (2017) is such an example. However, there remain many unanswered questions related to their best design, implementation and impact.

3.4 Implementing digital mathematical tasks in research and classroom settings

While it is easy to conceive that the task is an artefact that is offered to learners, a crucial component of task design relates to the many decisions about when, how and with whom a task is implemented in a research or classroom setting. While this detail is a fundamental component of any research methodology, it is often an under-reported aspect of task design. An exception to this is the research reported by Gallopin and Zuccheri (2002), which included detailed description of the phases of the 'didactical path' adopted for their study, which used two contrasting dynamic geometry softwares with Italian secondary school students.

152 Trgalová, Clark-Wilson & Weigand

3.5 Designing digital tasks for prospective and practising teachers

More recently, attention in the Technology TWG is shifting to research that concerns the design, implementation and impact of tasks that are intended for prospective and practising teachers with the particular aim to develop their professional learning concerning technology use in classrooms. The TWG has always acknowledged that teachers need to undertake specific professional learning to achieve this aim, however the nature and complexities of this learning have often been under-defined. Since CERME 7, a number of theories have been developed that articulate aspects of teachers' technological knowledge and practices, see for example Ruthven (2009), Drijvers (2011), Haspekian (2011) and Rocha (2015). However, research on the application of these theories within the design of tasks intended for teachers' professional learning initiatives is still in its infancy.

4 Theories and approaches concerning technology and resources

Theoretical frameworks became an explicit theme discussed within the Technology TWG since CERME 3. This was a clear milestone in the progress of the TWG. In what follows, we outline the variety of theories and approaches that have permeated the TWG and trace the evolution of theories used in research on technologies in mathematics education over the past 20 years.

In charting the progress of the TWG since this time, the considerations of theories presented below are organised with reference to the didactic tetrahedron, attempting to isolate faces or edges according to the research focus to gain a deeper insight into the strengths and limitations of these frames, although we are aware that in technology-supported teaching/learning situations, all four vertices are intertwined and interact with each other.

4.1 Technology and knowledge

Prior to CERME 3 most of the contributions investigate "a new epistemology of mathematics created by the use of the technology" (Gutiérrez et al., 1999, pp. 185–186). Referring to the didactic tetrahedron (Figure 11.1), these issues are related to the *technology–knowledge* edge and address mostly the epistemological dimension of the use of technology in mathematics education. They can be classified into two categories:

1. Exploring how technology mediates knowledge and the consequences of this mediation on the knowledge itself. A paradigmatic example is provided by the new behaviour of objects in a dynamic geometry environment, which actually gives birth to a new kind of objects (ibid., 1999, p. 185). A consequence of interacting with such new objects on the students' conceptualisation of linear algebra notions is explored by Dreyfus, Hillel and Sierpinska (1999).

Knowledge mediation by technology can be addressed in terms of the 'computerised transposition' (Balacheff, 1993) bringing to the fore tool-designed constraints (internal, command, interface) introduced by the use of a computer likely to impact upon the mathematical knowledge at stake. Alternatively, knowledge mediation can be interpreted through the notion of the 'epistemological triangle' (Figure 11.3), which represents

> the connection between the mathematical signs, the reference contexts and the mediation between signs and reference contexts which is influenced by the epistemological conditions of mathematical knowledge.
>
> *(Steinbring, 2006, p. 135)*

Both theoretical constructs are usually combined with an epistemological content analysis, which defines the essence of the mathematical knowledge at stake.

2. Investigating better ways to learn mathematical concepts with technology. This category of research focuses on software or task design and on *a priori* analysis of the potential of the task as instrumented by a chosen technology in order to achieve a known learning goal. Among the frameworks mobilised by the researchers are:

- Theory of Semiotic Mediation (Bartolini Bussi & Mariotti, 2008) assuming that, in social contexts, mathematical meaning can be created from specific uses of a tool (e.g. in the L'Algebrista microworld (Cerulli, 2002), expressions and commands may be thought as external signs of the algebraic theory and transforming an expression into another using available buttons as proving a theorem);
- Situated Abstraction (Noss & Hoyles, 1996) used to "describe how learners construct mathematical ideas by drawing on the webbing of a particular setting which, in turn, shapes the way the ideas are expressed". The 'situatedness' emphasises the "specificities of the situation, and in particular [. . .] the linguistic and conceptual resources available for expressing mathematically within them" (p. 122).

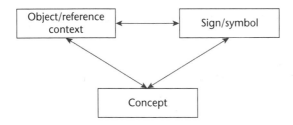

FIGURE 11.3 Epistemological triangle

154 Trgalová, Clark-Wilson & Weigand

4.2 Technology, knowledge and student(s)

From the first ERME congresses, researchers investigated "the complex inter-play between the work in a technological environment and the development of mathematical understanding and skills" (Barzel, Drijvers, Maschietto & Trouche, 2006, p. 928), addressing the cognitive and instrumental dimensions of the use of technology.

Several theoretical constructs are used to explore such interplay:

- The concepts of 'embodied cognition' and 'metaphors' (Lakoff & Nuñes, 2000) viewing mathematical meaning as rooted in our experience of common phenomena such as movement and working as a metaphor. This framework is used for instance to analyse students' cognitive processes while doing activities involving artefacts such as movement sensors and the corresponding develop-ment of the function concept.
- The Theory of Didactic Situations (Brousseau, 1997) considering digital tech-nology as a component of the 'milieu' with which a learner interacts and allowing for the analysis of the possible learner – milieu interactions alongside the related learning outcomes.
- The notion of 'instrumentation' appears at CERME 3 as the approach that "distinguishes the instrument (a psychological construction) from the artefact (the material object involved in an instrumented action)" (Jones & Lagrange, 2004). The Instrumental Approach (Rabardel, 2002), which emerged as the most central theoretical framework at CERME 4 (and subsequently), pinpoints that "given a tool, the genesis of a fruitful instrument is far from self-evident, but is the result of a social process, guided by a set of tasks in a given institution" (Barzel et al., 2006, p. 929). This construct has proved particularly helpful when studying the evolution of technology use.

4.3 Technology, knowledge and teacher

As previously alluded to, the awareness of the importance of the teacher dimension in research on technology in mathematics education (the *technology–teacher–knowledge* face of the didactic tetrahedron) emerged slowly. It was first considered explicitly during CERME 3 and also featured in a small number of contributions at CERME 4, as pointed out by Barzel et al. (2006) who commented:

> On the issue of the second theme, the role of the teacher in technology-rich mathematics education, we observe that in spite of the relevance that is attributed to this theme, little research was reported in this working group.
> *(p. 937)*

Subsequently, four aspects of the teacher dimension have been addressed by the Technology TWG:

Technology and resources **155**

- investigating the role of the teacher in technology-based settings;
- analysing teachers' practices involving technology;
- characterising the new knowledge and skills required for efficient use of technology, and its evolution;
- designing and assessing teacher education/teacher training programmes.

The interest of researchers in the Instrumental Approach, used so far for studying issues related to the *knowledge–technology–learner* face of the didactic tetrahedron, raised the question of its adaptation for teachers. Soon, new concepts and approaches were developed and shared within the mathematics education community through, in particular, the CERME Technology TWG:

- 'Instrumental orchestration' introduced by Trouche (2004) and further developed by Drijvers, Doorman, Boon and van Gisbergen (2010) who define it as "the intentional and systematic organisation and use of the various artefacts available in [. . .] computerised-learning environment by the teacher in a given mathematical task situation, in order to guide students' instrumental genesis" (p. 1350).
- 'Double instrumental genesis' (Haspekian, 2011) highlighting that, in order to efficiently use a digital tool in her teaching, a teacher has to develop not one, but two instruments: a 'personal instrument' for mathematical work and a 'professional instrument' for teaching mathematics.
- Documentational Approach to Didactics (Gueudet & Trouche, 2009), which, drawing on the Instrumental Approach, substitutes 'artefact' by 'resource' and 'instrument' by 'document' and takes into account a great variety of resources intervening in teachers' work: textbooks, students' worksheets, Internet resources, discussions with colleagues, etc. The development of this framework has led to the re-conceptualisation of tools and technology within the TWG by considering them within a wider range of resources (see Section 2.2).
- The role of 'hiccups' in technology-mediated lessons, the perturbations experienced by the teachers triggered by the use of the technology that illuminate discontinuities in their knowledge and offer opportunities for the teachers' epistemological development (Clark-Wilson, 2013).

Several studies aim to analyse teachers' practices with the use of technology in order to get a deeper insight into the complexity of technology integration. Assude (2007) elaborated a theoretical tool that characterises the degree of teachers' integration of technology, taking into account both the *instrumental* (i.e. how instrumental integration is taken into account by the teacher, focusing in particular on orchestrations she uses and types of tasks she proposes) and the *praxeological* (i.e. how the pupil's mathematical work is organised by the teacher, focusing in particular on the relationship between pencil-and-paper tasks and techniques, and tasks and techniques within the tool) dimensions.

Ruthven's Structuring Features of Classroom Practice framework (2009) introduces five key components that structure teachers' classroom practices with technology: working environment, resource system, activity format, curriculum script, and time economy. These features "shape patterns of technology integration into classroom practice and require teachers to develop their craft knowledge accordingly" (p. 52). Further developments of this approach were envisaged recently at CERME 10 by suggesting introducing a sixth component "relating to teacher craft knowledge for managing different types of student behaviours or attitudes" (Gustafsson, 2017, p. 2380).

Abboud-Blanchard and Vandebrouck (2013) developed a model for studying evolutions of teachers' practices in terms of technology uses, combining the Activity Theory (Engeström, 1999), the Instrumental Approach and the Double Approach to Teachers' Practices (Robert & Rogalski, 2005). The latter considers a teacher's activity through five components (personal, mediative, cognitive, institutional and social), a frame also used by Emprin (2007) to analyse training courses aiming at the use of technology.

The Technology, Pedagogy and Content Knowledge (TPACK) framework (Mishra & Koehler, 2006) is a dominant frame used to address teachers' professional knowledge and skills, suggesting seven categories of this knowledge: mathematical content knowledge (CK), pedagogical knowledge (PK), technological knowledge (TK) and all possible intersections of these (PCK, TCK, TPK and TPCK). However, to date, only a few studies have been reported to the TWG on the application of this frame to mathematics education.

At CERME 10, Abboud and Rogalski (2017) introduced new theoretical concepts of 'tensions' and 'disturbances', developed within a model of instrumented activities of teacher and students. 'Tensions' in a teacher's activity are "manifestations of 'struggles' between maintaining the intended cognitive route and adapting to phenomena linked to the dynamics of the class situation", whereas 'disturbances' are "consequences of non-managed or ill-managed tensions that lead to an exit out of the intended cognitive route".

The notion of *community* is another concept widely used, mainly in relation with teacher professional development. Communities of practice (Wenger, 1998), or communities of inquiry (Jaworski, 2005) are either established purposefully by the researchers or teacher educators to accompany teachers' efforts with integrating technology in their everyday practice (Fuglestad, 2011), or they develop spontaneously around Web2.0 tools enabling sharing resources and practices (Trgalová, Jahn & Soury-Lavergne, 2010).

This overview of theories used in CERME contributions reveals a wide variety of frames, which can be seen as a wealth of the research field, but there is a risk of "the framework compartmentalisation that could hinder the capitalisation of knowledge and its practical exploitation" (Artigue, Bosch, & Gascón, 2011, p. 2381). The awareness of this risk appears quite early within the Technology TWG, at CERME 4, which concluded:

a more ecological and systematic approach is needed rather than a unifying theory, which takes into account the existing subsystems, and which combines various theories focusing on each of these subsystems (didactics, instrumental approach, situated and distributed cognition, community of practice).

(Barzel et al., 2006, p. 929)

The subsequent congresses call for further development of theories toward a comprehensive and more coherent landscape of articulated frames.

5 Conclusion

In concluding this chapter, a key question to ask is, what do we know now about technology integration in mathematics teaching and learning that we did not know before the Technology TWG was established? The collective knowledge and experience of the community would suggest the following conclusion. It takes a significant amount of time for learners and teachers to become fully instrumental-ised, that is to learn to use and apply the technology for their relevant mathematical purpose, which for teachers includes important didactic considerations and the development of their resource systems. The ongoing innovation of technologi-cal tools, within and outside of mathematics education, requires us to continue to address the *learner–knowledge–technology* face of the didactic tetrahedron. This might mean that we appear to reinvent the wheel again and again, as new research-ers, teachers and technology designers encounter for the first time known issues and challenges concerning technology integration for mathematics education. However, although it is inevitable that educational goals are continually re-evaluated as a result of new developments in society, science and education, it might be that knowledge of related theory *alone* is insufficient to short-cut this process of rethinking. A key role for the community of the CERME Technology TWG is to support important connections between educational goals, theory and practice to be made. In the beginning, the focus of the research was on the effects of using technology on students' learning and teachers' practices. Now, as we know more about these effects, our attention has shifted to be concerned with researching how we can scale 'successful' innovations in mainstream education systems. However, at the most recent CERME 10, within the submitted papers, little attention was paid to digital assessment of and for the learning of mathematics, and large-scale experimental studies were not presented.

It is notable that technology is now a visible element across all CERME TWGs. This is an indication of how it now permeates and has wide legitimacy across the mathematics education research landscape. This leads us to question our special role and to justify the important questions or topics to which we can make a distinct contribution. In some ways, this justification comes from the continued disappointment in the lack of widespread uptake of technology, first highlighted at CERME 5:

The use of technologies has simply not scaled up and the changes promised by the case study experiences have not really been noticed beyond the empirical evidence given by the studies themselves.

(Kynigos, Bardini, Barzel & Maschietto, 2007, p. 1541)

It is the Technology TWG that perseveres to confront this issue and, although new, possibly more exciting technologies arrive on the scene with great promise to revolutionise classrooms, we find ourselves expanding our experiential and theoretical knowledge in order to be able to better inform future attempts to scale.

One challenge faced by the TWG is that it has grown from only nine accepted paper submissions at CERME 1 to 27 accepted paper submissions (and eight posters) at CERME 8. The TWG leaders responded by dividing the TWG to differentiate research that foregrounded students from that which foregrounded teachers. However, given the frame of the didactic tetrahedron, this is somewhat dissatisfying as most studies feature both perspectives. An alternative might be to divide the TWG by educational phase, as in the ICME congress series. However, we question whether it is possible to partition in a wholly satisfying way and this will be an ongoing topic for discussion for the TWG.

There is still a need to develop more comprehensive theoretical frameworks to address 'old' but still topical themes, such as task design and methods for large-scale dissemination of research-informed practices with digital technologies. The role of technologies within processes of formative assessment, networked classroom technologies and e-learning (particularly Massive Open Online Courses) appear among the emergent issues that require further theoretical and methodological development. It is worth noticing that this issue becomes shared with the TWG on theoretical perspectives and approaches as its call for contributions at CERME 10 includes "Theories for research in technology use in mathematics education".[3]

Finally, with a focus on emergent technologies, we expect that future ERME congresses will feature research on touch screens and human–computer interaction, which is already a frequently discussed topic in present classrooms, questioning how gestures can help visualising and, hopefully, understanding mathematical concepts; 3D technology, including the use of 3D printers within mathematics education; virtual and augmented reality in mathematics education; artificial intelligence features to include intelligent tutoring and support systems that take account of large data sets; ICT-support and special needs students, particularly students with physical disabilities; and digital technologies that support individuality, for example, the creation of portfolios and personalised e-textbooks.

Notes

1 e-CoLab = Expérimentation Collaborative de Laboratoires mathématiques, see http://educmath.ens-lyon.fr/Educmath/ressources/lecture/dossier_mutualisation/ecolab.pdf.
2 RITEMATHS = The project is about the use of real problems (R) and information technology (IT) to enhance (E) students' commitment to, and achievement in, mathematics (MATHS). http://extranet.edfac.unimelb.edu.au/DSME/RITEMATHS.
3 See http://cerme10.org/scientific-activities/twg-teams.

References

Abboud, M., & Rogalski, J. (2017). Real uses of ICT in classrooms: Tensions and disturbances in the mathematics teacher's activity. *CERME10* (pp. 2334–2341).

Abboud-Blanchard, M., & Vandebrouck, F. (2013). Geneses of technology uses: A theoretical model to study the development of teachers' practices in technology environments. *CERME8* (pp. 2504–2513).

Artigue, M., Bosch, M., & Gascón, J. (2011). Research praxeologies and networking theories. *CERME7* (pp. 2381–2390).

Assude, T. (2007). Teacher's practices and degree of ICT integration. *CERME5* (pp. 1339–1348).

Attorps, I., Björk, K., Radic, M., & Viirman, O. (2013). Teaching inverse functions at tertiary level. *CERME8* (pp. 2524–2533).

Balacheff, N. (1993). La transposition informatique, un nouveau problème pour la didactique. In M. Artigue et al. (Eds.), *Vingt ans de didactique des mathématiques en France* (pp. 364–370). Grenoble: La Pensée Sauvage éditions.

Bartolini Bussi, M. G., & Mariotti, M. A. (2008). Semiotic mediation in the mathematics classroom: Artifacts and signs after a Vygotskian perspective. In L. English, M. Bartolini Bussi, G. Jones, R. Lesh, & D. Tirosh (Eds.), *Handbook of international research in mathematics education* (2nd ed.) (pp. 746–783). Mahwah, NJ: Lawrence Erlbaum.

Barzel, B., Drijvers, P., Maschietto, M., & Trouche, L. (2006). Tools and technologies in mathematical didactics. *CERME4* (pp. 927–938).

Belousova, L., & Byelyavtseva, T. (1999). Training explorations on numerical methods course using technology. *CERME1* (pp. 201–208).

Brousseau, G. (1997). *Theory of didactical situations in mathematics*. Dordrecht: Kluwer Academic Publishers.

Cerulli, M. (2002). Introducing pupils to theoretical reasoning: The case of algebra. *CERME2* (pp. 139–151).

Churchhouse, R. F. (Ed.) (1986). *The influence of computers and informatics on mathematics and its teaching*. ICMI Study Series. Cambridge: Cambridge University Press.

Clark-Wilson, A. (2013). How teachers learn to use complex new technologies in secondary mathematics classrooms: The notion of the hiccup. *CERME8* (pp. 2544–2553).

Clark-Wilson, A., Hoyles, C., & Noss, R. (2015). Scaling mathematics teachers' professional development in relation to technology: Probing the fidelity of implementation through landmark activities. *CERME9* (pp. 2333–2339).

Dreyfus, T., Hillel, J., & Sierpinska, A. (1999). Cabri based linear algebra: Transformations. *CERME1* (pp. 209–221).

Drijvers, P. (2011). Teachers transforming resources into orchestrations. *CERME7* (pp. 2178–2187).

Drijvers, P., Doorman, M., Boon, P., & van Gisbergen, S. (2010). Instrumental orchestration: Theory and practice. *CERME6* (pp. 1349–1358).

Emprin, F. (2007). Analysis of teacher education in mathematics and ICT. *CERME5* (pp. 1399–1408).

Engeström, Y. (1999) Activity theory and individual social transformation. In Y. Engeström et al. (Eds.), *Perspectives on activity theory* (pp. 19–38). Cambridge: Cambridge University Press.

Fredriksen, H. (2013). Mathematics teaching on the web for student teachers: Action research in practice. *CERME8* (pp. 2554–2563).

Fuglestad, A. B. (2011). Challenges teachers face with integrating ICT with an inquiry approach in mathematics. *CERME7* (pp. 2328–2338).

Gallopin, P., & Zuccheri, L. (2002). A didactical experience carried out using, at the same time, two different tools: A conceptual one and a technological one. *CERME2* (pp. 152–162).

Gélis, J.-M., & Lenne, D. (1999). Integration of learning capabilities into a CAS: The SUITES environment as example. *CERME1* (pp. 222–232).

Geraniou, I., & Mavrikis, M. (2017). Investigating the integration of a digital resource in the mathematics classroom: The case of a creative electronic book on reflection. *CERME10* (pp. 2555–2562).

Gueudet, G., & Trouche, L. (2009). Towards new documentation systems for mathematics teachers? *Educational Studies in Mathematics, 71*(3), 199–218.

Gustafsson, P. (2017). Exploring a framework for technology integration in the mathematics classroom. *CERME10* (pp. 2374–2381).

Gutiérrez, A., Laborde, C., Noss, R., & Rakov, S. (1999). Tools and technologies. *CERME1* (pp. 183–188).

Haspekian, M. (2011). The co-construction of a mathematical and a didactical instrument. *CERME7* (pp. 2298–2307).

Hoyles, C., & Lagrange, J.-B. (Eds.) (2010). Mathematics education and technology: Rethinking the terrain. The 17th ICMI Study. New York: Springer.

Jančařík, A., & Novotná, J. (2015). Scaffolding in e-learning course for gifted children. *CERME9* (pp. 2354–2360).

Jaworski, B. (2005). Learning communities in mathematics: Creating an inquiry community between teachers and didacticians. *Research in Mathematics Education, 7*(1), 101–119.

Jones, K. (1999). Student interpretations of a dynamic geometry environment. *CERME1* (pp. 245–258).

Jones, K., & Lagrange, J.-B. (2004). Tools and technologies in mathematical didactics: Research findings and future directions. *CERME3*: www.mathematik.uni-dortmund. de/~erme/CERME3/Groups/TG9/TG9_introduction_cerme3.pdf.

Khan Academy. (n.d.) www.khanacademy.org [accessed 3 October 2017].

Kynigos, C., Bardini, C., Barzel, B., & Maschietto, M. (2007). Tools and technologies in mathematical didactics. *CERME5* (pp. 1499–1508).

Lagrange, J.-B., & Psycharis, G. (2013). Exploring the potential of computer environments for the teaching and learning of functions: A double analysis from two traditions of research. *CERME8* (pp. 2624–2633).

Lakoff, G., & Nuñes, R. (2000). *Where mathematics comes from: How the embodied mind brings mathematics into being.* New York: Basic Books.

Lavicza, Z., Juhos, I., Koren, B., Fenyvesi, K., Csapodi, C., et al. (2015). Integrating technology into primary and secondary school teaching to enhance mathematics education in Hungary. *CERME9* (pp. 2430–2431).

Leung, A., & Baccaglini-Frank, A. (2016). *Digital technologies in designing mathematics education tasks: Potential and pitfalls (Mathematics Education in the Digital Era, Vol. 8).* Dordrecht: Springer.

Mackrell, K., Maschietto, M., & Soury-Lavergne, S. (2013). Theory of didactical situations and instrumental genesis for the design of a Cabri elem book. *CERME8* (pp. 2654–2663).

Mishra, P., & Koehler, M. J. (2006). Technological pedagogical content knowledge: A framework for integrating technology in teacher knowledge. *Teacher College Records, 108*(6), 1017–1054.

Monaghan, J., Trouche, L., & Borwein, J. (2016). *Tools and mathematics.* Dordrecht: Springer.

Noss, R., & Hoyles, C. (1996). *Windows on mathematical meaning: Learning cultures and computers.* Dordrecht: Kluwer Academic Publisher.

Olivero, F., & Robutti, O. (2002). An exploratory study of students' measurement activity in a dynamic geometry environment. *CERME2* (pp. 215–226).

Polanyi, M. (1966). *The tacit dimension.* Chicago, IL: The University of Chicago Press.

Rabardel, P. (2002). *People and technology: A cognitive approach to contemporary instruments.* Université Paris 8. https://halshs.archives-ouvertes.fr/file/index/docid/ 1020705/filename/ people_and_technology.pdf [accessed 18 December 2016].

Robert, A., & Rogalski, J. (2005). A cross-analysis of the mathematics teacher's activity: An example in a French 10th-grade class. *Educational Studies in Mathematics, 59*(1–3), 269–298.

Rocha, H. (2015). Knowledge for teaching mathematics with technology and the search for a suitable viewing window to represent function. *CERME9* (pp. 2403–2409).

Ruthven, K. (2009). Towards a naturalistic conceptualisation of technology integration in classroom practice: The example of school mathematics. *Education et Didactique, 3*(1), 131–149.

Steinbring, H. (2006). What makes a sign a mathematical sign? An epistemological perspective on mathematical interaction. *Educational Studies in Mathematics, 61*(1–2), 133–162.

Tall, D. (1986). Using the computer as an environment for building and testing mathematical concepts: A tribute to Richard Skemp. In *Papers in Honour of Richard Skemp* (pp. 21–36). Warwick. http://homepages.warwick.ac.uk/staff/David.Tall/pdfs/dot1986h-computer-skemp.pdf [accessed 15 September 2016].

Trgalová, J., Jahn, A. P., & Soury-Lavergne, S. (2010). Quality process for dynamic geometry resources: The Intergeo project. *CERME6* (pp. 1161–1170).

Trouche, L. (2004). Managing the complexity of human/machine interactions in computerized learning environments: Guiding students' command process through instrumental orchestrations. *International Journal of Computers for Mathematics Learning, 9*(3), 281–307.

Watson, A., & Ohtani, M. (2015). *Task design in mathematics education: An ICMI Study 22.* Cham: Springer International Publishing.

Weigand, H.-G. (2008). Teaching with a symbolic calculator in 10th grade: Evaluation of a one year project. *International Journal for Technology in Mathematics Education, 15*(1), 19–32.

Wenger, E. (1998). *Communities of practice: Learning, meaning, identity.* New York: Cambridge University Press.

12

CLASSROOM PRACTICE AND TEACHERS' KNOWLEDGE, BELIEFS AND IDENTITY

Jeppe Skott, Reidar Mosvold and Charalampos Sakonidis

1 Introduction

Issues pertaining to teachers and teaching have gained ever-increasing attention in mathematics education over the last few decades. While teachers and teaching were conspicuously absent in mathematics education research until the mid-1980s, Sfard (2005) reported on behalf of a research team for ICME X that "the last few years have been the era of the teacher as the almost uncontested focus of researchers' attention" (p. 409).

These issues seem to have gained even further momentum since Sfard and the survey team conducted their study, not least in Europe. At the CERME conferences, they have been dealt with under several headlines. Initially there was one relevant Thematic Working Group (TWG), 'From a Study of Teaching Practices to Issues in Teacher Education'. This TWG has now been split into several others, and issues related to teachers, teaching and teacher education were, at the last two CERMEs, dealt with in three TWGs, named 'Mathematics Teacher Education and Professional Development', 'Mathematics Teachers and Classroom Practices', and 'Mathematics Teachers' Knowledge, Beliefs, and Identity'. In this chapter, we deal with research on the last two of these themes presented at the first 10 CERMEs. We leave issues pertaining to the first theme to Chapter 13.

1.1 Change and continuity in research on and with teachers

It has been a recurrent theme in papers on teachers and teaching presented at CERMEs what a professionalisation of teaching entails. In particular, there has been a persistent effort to develop understandings of how teachers contribute to classroom practice and of the types of knowledge they need in instruction. Issues related to teachers' beliefs and the impact their beliefs might have on instruction have also played a recurrent but smaller role.

Teachers and classroom practice **163**

In Sections 3 to 5 below, we outline developments in research on the thematic issues of teachers' knowledge, teachers' beliefs, and classroom practice. At CERMEs, issues related to beliefs are usually dealt with in the TWG on 'Affect' (cf. Chapter 10 in this volume); in this chapter, we limit our discussion to papers on beliefs presented in the TWGs on teachers and teaching.

Unsurprisingly, there are elements of both continuity and change in the papers presented. Beyond the fact that the three themes mentioned above are, to some extent, addressed at all CERMEs, continuity is also found at least to some extent in the frameworks and the methods used to study them. One important reference that is relevant at all CERMEs is Shulman's general categorisation of teachers' knowledge (Shulman, 1986). There are still frequent references to Shulman's original work, and the most significant subject-specific frameworks for understanding teachers' knowledge explicitly draw and elaborate on Shulman's initial categories. These frameworks are used to describe what teachers know, how this knowledge impacts practice, and how it might be developed.

There is no core reference in belief research with an influence that parallels Shulman's in the field of teachers' knowledge. However, the underlying construct of beliefs seems to be understood in a somewhat similar sense over the years, although the expectation of belief impact on practice might have changed. Also, it is a recurrent theme if, what and how teachers might learn from their engagement in practice.

Changes appear within each of the thematic issues, in the relative significance attached to them, and in the relationship among them. In our concluding discussion, we argue that lately there seems to have been a move toward less causal and more dynamic interpretations of the relationships between knowledge and beliefs on the one hand and classroom practice on the other. We consider it likely that this trend continues, and we suggest that greater attention to some of the social theories of human functioning that have so far played a relatively minor role in these TWGs could be helpful for the further development of such dynamic perspectives.

Before engaging in the main discussion, however, we introduce what appears to be a dominant vision of school mathematics underlying most papers presented at these working groups.

2 The vision of mathematics teaching and learning

The vision of school mathematics that orients the larger part of studies on teachers and teaching at CERMEs is to some extent informed by reform efforts that have also been promoted on the other side of the Atlantic (e.g. NCTM, 2000). Studies presented at CERMEs often seek to understand or support instruction and classroom practice that emphasise mathematical processes, and, explicitly or implicitly, they ask what it takes for teachers to engage students in, for instance, mathematical investigations and problem solving.

It seems, then, that contributions to CERMEs on teachers and teaching are often informed by what is sometimes referred to as the reform in mathematics education.

164 Skott, Mosvold & Sakonidis

This vague and ambiguous term refers to developments that draw on constructivist as well as sociocultural approaches to learning and on a view of mathematics as an ever-evolving human construction. Consequently, the reform promotes a vision of school mathematics that focuses on students' creative engagement in exploratory and problem-solving activities as they develop their understandings of significant mathematical concepts and procedures. In a draft to the Principles and Standards for School Mathematics (NCTM, 2000), the role of the teacher is outlined as follows:

> Curricular frameworks and guides, instructional materials, and lesson plans are only the first elements needed to help students learn important mathematics well. Teachers must balance purposeful, planned classroom teaching with the ongoing decision-making that can lead the teacher and the class into unanticipated territory from an effective mathematical and pedagogical knowledge base.
>
> *(NCTM, 1998, p. 33)*

According to such principles, the teacher is expected to play a decisive role that goes beyond a tradition of presenting ready-made concepts and procedures for the students to follow and copy.

References to Standards 2000 and other documents in the same tradition are recurrent in CERME proceedings, and, even when no such reference is made, inspiration is often apparent from what Ellis and Berry (2005) call the cognitive-cultural paradigm in mathematics education. A question underlying many of the papers presented at CERMEs is, therefore, what it takes in terms of teachers' knowledge and beliefs for them to support classroom practices in line with those promoted by the reform.

3 Studies focusing on teachers' knowledge

Teachers' knowledge has been an important theme in educational research for decades. The dominant approaches in mathematics education build on the parts of the work of Shulman and his colleagues from the 1980s that concern aspects of content knowledge.

The focus on mathematics teachers' knowledge has been prevalent at every CERME conference. At CERME 1, Carrillo, Coriat and Oliveira (1999) emphasise how teachers' knowledge develops from practice, and they call for a way of talking about knowledge that is gained from experience. Drawing heavily upon the legacy of Shulman (1986), Carrillo et al. describe different attempts to characterise types and components of mathematics teacher knowledge, and they discuss the challenging connections between teachers' knowledge and their classroom performance. Ever since CERME 1, there have been ongoing discussions on how to characterise mathematics teacher knowledge, and different approaches have been suggested. In the following, we highlight three lines of work that have been prevailing in discussions of mathematics teacher knowledge throughout the history of CERME.

3.1 Mathematical Knowledge for Teaching

A first major approach to conceptualising mathematics teacher knowledge evolved directly from Shulman's (1986) ideas about professional teacher knowledge. Ball and her colleagues developed their practice-based theory of Mathematical Knowledge for Teaching (MKT) from careful analyses of video recordings of mathematics teaching in the USA. They decomposed the work of teaching mathematics into mathematical tasks of teaching and defined MKT as the knowledge needed to carry out these tasks (Ball, Thames & Phelps, 2008). They extended Shulman's categories of subject matter knowledge, pedagogical content knowledge, and curriculum knowledge into six subcategories of MKT, two of which have attracted particular attention at CERMEs. These are the ones of horizon content knowledge and specialised content knowledge. Horizon content knowledge concerns "an awareness of how mathematical topics are related over the span of mathematics included in the curriculum" (Ball et al., 2008, p. 403). Specialised content knowledge, on the other hand, is mathematical knowledge that does not require knowledge of children, learning or instruction, but that is specific to the profession.

With Ball and colleagues' (2008) seminal article on MKT still in publication, Ribeiro, Monteiro and Carrillo (2010) presented and discussed the main aspects of the framework in a paper at CERME 6. At the same conference, Stylianides and Stylianides (2010) also presented a paper that described some early versions of the MKT framework. From CERME 7 and onwards, Ball et al.'s MKT framework attracted much attention in the European research community, and a growing number of papers refer to studies that build on it. Several studies investigate and discuss subcategories of MKT. For instance, Fernández, Figueiras, Deulofeu and Martínez (2011) discuss horizon content knowledge and suggest re-defining this category of MKT to connect it more closely with the work of teaching. At CERME 8, Jakobsen, Thames and Ribeiro (2013) continued this discussion, highlighting the importance and usefulness of advanced mathematical knowledge in their working definition of this subcategory of MKT. Carreño, Ribeiro and Climent (2013) carry on the discussions of specialised content knowledge and horizon content knowledge, point at some limitations of the MKT framework, and call for a reconceptualisation of mathematics teacher knowledge. In more recent years, there has tended to be a shift in focus from categorising knowledge to investigating how MKT is used in teaching, and, at CERME 10, Mosvold and Hoover (2017) presented a literature review of studies on the influence of MKT on teaching. They suggest a stronger focus on the mathematical entailments of the work of teaching, and they also call for more shared conceptualisations of the mathematical work of teaching.

3.2 Mathematics Teachers' Specialised Knowledge

A second line of study proposes a reconceptualisation of mathematics teacher knowledge, with a particular focus on specialised content knowledge. Investigations of this aspect of mathematics teacher knowledge led to the development of Mathematics Teachers' Specialised Knowledge (MTSK) framework in Spain. Like the MKT framework, it draws upon Shulman's distinction between mathematical knowledge and pedagogical content knowledge, and it describes three subcategories of each. For mathematical knowledge, the MTSK framework distinguishes between knowledge of topics, knowledge of the structure of mathematics, and knowledge of practices in mathematics. The subcategories of pedagogical content knowledge are knowledge of mathematics teaching, knowledge of features of learning mathematics, and knowledge of mathematics learning standards. Unlike the MKT framework, the MTSK framework also includes beliefs about mathematics and beliefs about mathematics teaching and learning (Montes & Carrillo, 2015).

At CERME 8, several papers referred to this framework. For instance, Flores, Escudero and Carrillo (2013) argue that specialised content knowledge is inextricably connected with teaching. They thereby challenge the definition of Ball et al. (2008) that presents specialised content knowledge as a purely mathematical knowledge that does not require any knowledge or experience of teaching or students. Based on some examples of tasks that aim at illustrating specialised content knowledge, they argue that "there is no purely mathematical knowledge here that can be seen as exclusive to mathematics teachers, or to which the pupil cannot have access" (Flores et al., 2013, p. 3061).

An emphasis in the MTSK framework is thus not on the attributes of teachers' mathematical knowledge as such, but on how this knowledge is used in teaching. With this as a starting point, Montes and Carrillo (2015) use the framework to investigate three teachers' mathematical knowledge of infinity emerging when teaching addition of successive powers. They emphasise various aspects of the teachers' mathematical knowledge of infinity. In another study, Vasco, Climent, Escudero-Ávila and Flores-Medrano (2015) use the MTSK framework to investigate the specialised knowledge revealed by a university lecturer when teaching linear algebra. They suggest that several sub-domains of specialised knowledge are prevalent – among them knowledge of topic. More recently, Espinoza-Vasques, Zakaryan and Carrillo Yañez (2017) investigate the relationships between specialised knowledge of functions and the teaching of this topic.

3.3 The knowledge quartet

A third line of work on conceptualising mathematical knowledge specific to teaching originates from England. This framework also builds upon the legacy of Shulman, but with a different emphasis from the MKT and MTSK frameworks. At CERME 3, Rowland, Thwaites and Huckstep (2004) presented a paper that reports on an investigation of prospective teachers' use of knowledge in classroom teaching. Based on grounded analyses of classroom videos, the knowledge quartet (KQ) is a framework consisting of the four components of foundation, transformation, connection and contingency.

The KQ was initially developed to qualify communication among prospective teachers, teacher educators, and school-based mentors during school-based teacher education. Although it was considered a tool for teacher education and teacher development, the framework has also been used as an analytical tool for interpreting prospective and practising teachers' engagement with classroom practice. The fourth element of the quartet, that of contingency, emphasises knowledge in interaction and can be regarded as an important contribution to understanding the dynamic character of classroom processes (Huckstep, Rowland & Thwaites, 2006). Also focusing on dynamic classroom processes, Ainley and Luntley (2006) refer to attention-dependent knowledge as both highly contextual and to some extent generalisable, but not as an expression of the application of general rules in a particular situation. Rowland, Jared and Thwaites (2011) show that the KQ, originally developed for use in the education of elementary teachers, is also useful for the analysis of teaching in secondary mathematics classrooms. Also at CERME 7, Turner (2011) returns to an investigation of one of Shulman's original forms of teacher knowledge: propositional knowledge. She uses the KQ as a framework for investigating differences between elementary teachers' propositional knowledge and knowledge in practice, and suggests that mathematics teachers' propositional knowledge is activated by reflections on practice.

In recent CERME conferences, the KQ framework has been applied in a steady stream of studies from different European countries such as Ireland (Corcoran, 2007), Cyprus (Petrou, 2010), Norway (Kleve & Solem, 2015) and Turkey (Koklu & Aslan-Tutak, 2015).

3.4 Other frameworks

Although the above are the most frequently used frameworks for studying mathematics teacher knowledge in the history of ERME, not all studies in this field employ these frameworks. Some studies use French frameworks such as the Anthropological Theory of Didactic or Brousseau's theory of didactical situations in their analysis of teachers' mathematical knowledge. Other studies develop their own theoretical frameworks. On a side note, the German COACTIV project has also investigated mathematics teacher knowledge and received attention internationally (Kunter et al., 2013). For some reason, however, this framework has not been widely used in studies of teacher knowledge at CERME.

Although authors frequently refer to the frameworks of Shulman and Ball, they tend to supplement or modify them. Such adjustments generally emphasise the dynamic character of classroom interaction and the complexities involved as contingencies arise. In the former case, reference is often made to some form of reflective practice or teacher inquiry. This is so because reflection in or on practice is considered an important prerequisite for supporting students' learning in line with reform recommendations as well as a significant vehicle for professional learning. Also in line with reform recommendations, the notion of inquiry generally refers to the activity of both students and teachers. In other cases, different types

168 Skott, Mosvold & Sakonidis

of knowledge or knowing are pointed to as necessary for teachers, if they are to flexibly orchestrate classroom interaction in line with the visions of teaching outlined earlier. More recently, there has been a shift of focus from investigating and discussing the content and definition of subcategories of teacher knowledge to this knowledge as a whole. Few of the more recent studies investigate the influence of knowledge on student learning, or how teacher knowledge develops in and from practice, but an increased focus on examining how mathematical knowledge is embedded in teaching practice seems to be emerging.

4 Studies focusing on teachers' beliefs

The character, development and possible impact of teachers' beliefs have been recurrent themes in research on and with teachers for decades, and they are also important in studies presented at CERMEs. In many papers, however, the notion of belief is used merely in passing and as an unproblematic reference to comments made by the teacher in question. In what follows, we restrict the discussion of beliefs to papers with more explicit focus on beliefs or on one or more of the terms that are considered close proxies to one of beliefs, that is, the ones of conceptions, perceptions, values or orientations.

The original rationale of belief research in mathematics education is the expectation that beliefs serve as an explanatory principle for how mathematics is taught and learnt. Indeed, without the expectation of belief impact it is unlikely that the field would have attracted more than minimal attention. This expectation is also apparent in papers presented at CERMEs. In his introduction to the section on teacher education and teachers' beliefs at CERME 1, da Ponte (1999) contended that "to have some insight into the way teachers understand and carry out their job one needs to know their conceptions and beliefs about curriculum, learning, and teaching" (p. 46). At later CERMEs similar expectations have been phrased differently, from using "practice" synonymously with "beliefs in action" (Furinghetti, Grevholm & Krainer, 2002, p. 268), over talking about teachers' beliefs as "reflected in their classroom practices" (Ferreira, 2007, p. 1996) to claiming that beliefs "play a key role for [teachers'] decision-making" (Schueler, Roesken-Winter, Weißenrieder, Lambert & Römer, 2015, p. 3255). Some of these wordings indicate that it is helpful to distinguish between espoused and enacted beliefs, and/or that contextual factors need to be considered to account for the role of teachers' beliefs in practice.

4.1 Conceptual issues in relation to beliefs

One of the challenges for belief research is that there is no agreement about an explicit definition of the key construct of beliefs. In spite of that, it has been argued that there is a core to how the concept is generally used (Skott, 2015). Beyond the expectation of impact the term is used about mental constructs that are (1) subjectively true, (2) value-laden, and (3) relatively stable. Despite some agreement on this, conceptual difficulties remain. This is apparent also in studies presented at CERMEs and is reflected in the multitude of terms used almost synonymously with beliefs, for instance conceptions, perceptions, views, worldviews and priorities.

One particular aspect of the difficulties with the construct of beliefs is how to distinguish it from knowledge. Philipp (2007) suggested that one way to go about the quandary is to use the term belief about some conviction if one can accept an alternative or opposite position as reasonable and to call it knowledge if such alternatives are considered unreasonable. This suggestion does away with the absolutist connotations of knowledge, as the truth-value and justifications of knowledge now reside in the social or individual domain. We have not come across similar wordings in CERME proceedings, but the intention of doing away with the absolute difference between knowledge and beliefs is sometimes apparent. An example is the KQ, where the foundation dimension encompasses elements of both knowledge and beliefs (e.g. Rowland, Turner & Thwaites, 2013; Koklu & Aslan-Tutak, 2015; Samková & Hošpesová, 2015). Also, Kuntze et al. (2011) work with an overarching concept of teachers' professional knowledge that "integrates the spectrum between knowledge on the one hand and convictions/beliefs on the other" (p. 2620), because a clear-cut distinction between the two is unobtainable.

It is inherent in the notion of belief stability (cf. the core of the concept mentioned above) that beliefs take a long time to develop and that it takes substantial new experiences for them to change. Teachers' mathematics-related beliefs, then, are considered robust, and because they are an outcome of the teachers' own schooling, they are at odds with current reform efforts. This seems to be the dominant understanding also at CERMEs, irrespective of whether the beliefs in question are about mathematics in general, about the teaching and learning of specific mathematical topics (algebra, data handling, number, etc.), or about particular aspects of the reform (e.g. problem solving or the use of ICT). It follows that belief change is considered important in many PD-programmes, and there seem to be three approaches to this. Sometimes the aim is to change teachers' beliefs by engaging them in reflection on their own or their students' mathematical activity (Hodgen, 2004), on their modes of classroom communication (Ferreira, 2007), or on the character of their own beliefs (Helmerich, 2013). Others base development initiatives on the expectation that although beliefs have some impact, there is no linear and direct causality between beliefs and classroom practice. These studies draw on the distinction between espoused and enacted beliefs and try to make the latter (traditional) version align with the former (reformist) one (Malara & Navarra, 2015). Still others adopt a comprehensive approach to teacher development and seek to change teachers' beliefs, knowledge, and practice in the same project, without basing the initiative on a specific premise concerning the character of the connections among the three (Zehetmeier & Krainer, 2011).

4.2 Methodological issues in relation to beliefs

It is generally acknowledged that beliefs are elusive and that there is no easy access to whatever teachers believe. The methodological problems in the field are also acknowledged at CERMEs. In most of the studies, teachers' beliefs are inferred from – or attributed to – teachers on the basis of surveys, interviews, or observations from classrooms or seminars. A combination of these methods is sometimes used for purposes of triangulation. A few studies directly address the methodological challenges of belief research and develop approaches other than the dominant ones mentioned above. For instance, Reid, Savard, Manuel and Wan Jung Lin (2015) ask teachers to select and discuss video recorded examples of typical and exceptional teaching from the teachers' own and from their colleagues' classrooms. Positioning teachers as observers and analysing what they say about their observations, Reid and his colleagues seek to gain insight "into the implicit criteria that guide [teachers'] observations" (p. 3116). It is indicative of the general methodological problems of belief research that in Reid et al.'s study the teachers focused on other issues when asked to select typical and exceptional examples of teaching than when they discussed these examples with their colleagues. As Reid et al. observe, this points to a methodological problem, if one relies on only one type of data. However, the observation might also indicate that 'implicit criteria' or beliefs are in some sense situated. If this is the case, the traditional argument for methodological triangulation may be challenged, as different methods do not necessarily provide access to the same mental constructs, beliefs.

4.3 From beliefs to dynamic affect systems

We argued above that the expectation of belief impact is a fundamental premise for research on teachers' beliefs. However, the thesis of congruity between the two has been refuted as much as confirmed in empirical studies (Fives & Buehl, 2012). Further, and as indicated above, belief research is almost notorious for its conceptual and methodological problems. Few studies in and beyond CERME (e.g. Larsen, Østergaard & Skott, 2013) react to these problems by challenging the key assumptions of belief research. Over the last decade, however, one trend in the field of beliefs has been to rely less on expectations of linear and causal relationships between beliefs and practice and point to other types of connections between the two. This development is hinted at in the title of two volumes that at different points in time intended to present the state of the art in this field. The first one, edited by Leder, Pehkonen and Törner in 2002, is called *Beliefs: A hidden variable in mathematics education?* It seems that the variable they were looking for was, at least to some extent, considered an independent variable in relation to practice. The second book, edited by Pepin and Roesken-Winter in 2015, is called *From beliefs to dynamic affect systems in mathematics education*. This title suggests both that the relation between beliefs and practice could be viewed as less causal and more reciprocal than before and that dynamic interconnections among different mental constructs, including beliefs and knowledge, should be considered.

This dynamic trend is also, to some extent, apparent in papers presented at CERMEs. Some focus on the belief–practice interface and point to the contextual framing of belief enactment (e.g. Turner, 2007). This challenges the expectation of belief impact. Others discuss beliefs as part of a broader set of teaching resources or competences (Bosse & Törner, 2015; Schueler et al., 2015; Wittmann, Schuler & Levin, 2015) and by doing so question the isolated significance of the ones that are traditionally referred to as beliefs. In this sense developments at CERMEs reflect the more general trend toward dynamic perspectives on the role of teachers in mathematics classrooms.

5 Understanding teacher–practice relationships

While it is a continuous effort in the CERME community to understand relationships between or among teachers' knowledge and beliefs on the one hand and the practices that evolve in their classrooms on the other, understandings of that relationship have changed over the last 20 years. As indicated previously in the case of beliefs, research on teachers and teaching has increasingly moved toward a concern for the complex, dynamic and emerging character of classroom practice. This implies that a different set of frameworks might be useful, frameworks that acknowledge the significance of the multiple micro and macro factors that may influence how learning and lives in classrooms unfold.

Before embarking on our discussion of CERME studies focusing on mathematics classroom practices, a short note on the meaning of the term 'practice' and its development over time is needed. Da Ponte and Chapman (2006) noted that the early uses of the term 'practice' in mathematics education research are mostly related to individuals' 'actions', 'acts' or 'behaviours'. Gradually its meaning evolved to include what a teacher does, knows, believes and intends. Nowadays, two main considerations of the term can be identified: the cognitive and the sociocultural. The former pays special attention to teachers' action plans and decisions taking into account their knowledge, beliefs and goals (e.g. Schoenfeld, 2000). The sociocultural perspective, on the other hand, sees practice as a social phenomenon, that is, as "the recurrent activities and norms that develop in classrooms over time, in which teachers and students engage" (Boaler, 2003, p. 3). Sometimes based on similar social interpretations, the notion of practice has been extended so as to include teacher engagement in professional activities beyond the classroom, such as planning (e.g. Grundén, 2017) and collaboration with colleagues (e.g. Kempe, Lövström & Hellqvist, 2017). The use of the term 'practice' in ERME conferences appears to reflect the above evolution of its meaning.

5.1 Focusing on teachers' actions

There are shifting trends in the prevailing research interest in papers presented in CERME TWGs devoted to teachers and teaching. In the limited number of studies appearing in the early CERMEs, the focus is on aspects of teachers' actions typically exhibited within mathematics classrooms. These studies focus, for instance, on classroom management of pupils' errors (Tzekaki, Kaldrimidou & Sakonidis, 2002); on selecting and implementing tasks/examples (Rowland et al.,

172 Skott, Mosvold & Sakonidis

2004); and on reacting to students' unexpected answers (Coulange, 2006). Such papers offer some insights into how the acts of teaching affect opportunities for students' mathematical learning, most often in elementary school classrooms. The frameworks used are often (explicitly or implicitly) constructivist/acquisitionist, and the methods are mainly qualitative and small scale, although they are sometimes combined with quantitative approaches.

This line of research does not cease to appear at later CERMEs. Examples are the effort by Gunnarsdóttir & Pálsdóttir (2015) to shed light on how teachers structure everyday mathematics lessons, or Drageset's (2015) study of teachers' response to students' answers which do not explain how a result was obtained. However, three new research trends become evident. The first is similar to the one above, concentrating though on the study of 'desirable' instructional practices such as practices compatible with (reform) initiatives. These initiatives may be related to problem solving (Asami-Johansson, 2011; Georget, 2007) or integrating ICT in mathematics teaching (Abboud-Blanchard, Cazes & Vandebrouck, 2007); to an effective teacher's actions in the classroom (Taylan, 2015) or the use of research findings in instruction (Nowińska, 2011). Overall, these studies employ similar theoretical and methodological approaches to the ones of the first era.

5.2 Focusing on classroom interaction

The second research trend identified in later CERMEs includes studies examining the dynamic nature of classroom interaction between teachers and students (Berg, 2011; Nunes & da Ponte, 2011; Wester, Wernberg & Meaney, 2015). This research recognises the complexity of teaching and aspires to provide analytical tools to further conceptualise it (Badillo, Figueiras, Font & Martinez, 2013; Felmer, Perdomo-Díaz, Giaconi & Espinoza, 2015). Sometimes one of three more specific foci is employed. The first concerns communication in the mathematics classroom and emphasises teachers' orchestration of mathematical discourse, dealing with students' unpredictable contributions, mathematical argumentation, or effective questioning (Drageset, 2013; Kwon, 2015; da Ponte & Quaresma, 2015; Baldry, 2017). The second focus is on teaching and classroom dynamics in specific mathematical contexts such as algebra or statistical graphs (Kaldrimidou, Sakonidis & Tzekaki, 2011; Cusi & Malara, 2013; Velez & da Ponte, 2015). Finally, the third focus is related to teachers' reflection on their own teaching with emphasis on classroom interaction (Berg, 2011; Nunes & da Ponte, 2011; Wester et al., 2015). Several different theoretical frameworks are used to address these issues, few of them cognitive, the majority interactionist or participatory in other ways. From a methodological point of view, all but a few of the papers in this group follow an interpretative paradigm, conducting research through case-studies based on classroom observations and/or interviews with practising or, more rarely, prospective teachers.

5.3 Practice as a context for teacher learning

A third research trend was already emerging in early CERMEs (e.g. Krainer, 2002), but has become more prominent in the later ones. The relevant studies do not only view instruction as habits or norms for actions that facilitate student learning, but also as situated/participatory activities that offer contexts for teachers' own learning. The aim of this line of research is to explore how teachers' reflection on and inquiry into classroom practices may contribute to their knowledge for teaching mathematics and to their sense of 'being' or 'becoming' teachers (Azcárate, Cardeñoso & Serradó, 2006; Baş et al., 2008; Caseiro, da Ponte & Monteiro, 2015; Gade, 2015; Mhuirí, 2017). This is so, for example, when teachers are involved in discussions of real classroom episodes (Clivaz, 2013) or seek to unpack students' ideas (Aslan-Tutak & Ertas, 2013).

As mentioned previously, the notion of practice has come to include different forms of teacher collaboration and relevant studies presented at CERMEs follow this development. Teacher learning has been connected to such collaborative efforts among the teachers themselves and between teachers and researchers since the first CERME, and the call for collaboration has been a recurrent theme. However, new formats for collaboration have emerged. Most notably a number of papers have been presented at recent conferences on the use of lesson study. While certainly not new in Japan, lesson study is relatively new in the West, including at CERMEs. It was first mentioned at CERME 5, both in a paper (Corcoran, 2007) and in the introduction to the activities of TWG 12 on teachers and classroom practice, in which the organisers point to lesson study as one of the emerging issues, one that has gained more attention since then.

6 Trends and possible future developments

Research on teachers and teaching reported at CERMEs over the last 20 years is characterised by both continuity and change. One recurrent theme is how teachers contribute (or not) to the types of classroom practice envisaged by current reform efforts and, in particular, how they deal with contingencies that might arise. Phrased in the terminology of the draft of the NCTM standards (cf. Section 1), this amounts to if and how the teacher manages to supplement planned classroom activities with leading the class into unanticipated territory and – taking the metaphor a little further – getting the students home safe again. Another theme that has been dealt with continuously is what it takes for teachers in terms of their professional qualifications to do so.

It is apparent from the previous sections that the latter question has often been addressed in studies of teachers' knowledge and beliefs. Sometimes knowledge and beliefs have been studied as characteristics of the individual teacher and as if they function almost independently of one another. Also, it has often been a premise that they are the basis of teachers' acts and meaning-making and that each of them significantly impacts practice. If taken to extremes, this means that 'practice' is viewed as a function of two variables: teachers' knowledge and beliefs. For obvious reasons, this has fuelled significant amounts of research on and with teachers.

6.1 Open beginnings – focused follow-up

Already at CERME 1 there was considerable research interest in teachers' knowledge based on the expectation that it influences classroom practice in significant ways. The expectation, however, was not one of immediate causality. Technical rationality was challenged, practice as a source of knowledge was emphasised, and the notion of situatedness was discussed as a supplementary perspective on teaching and teacher learning. Beliefs were also discussed at CERME 1 with explicit concern for their practical significance. But in line with the discussion of knowledge, readers of the proceedings are cautioned that connections between beliefs and practice might not be as unidirectional and causal as one might expect and that mere descriptions of what teachers believe might not be helpful for understanding of belief change and impact (da Ponte, 1999). A number of very different approaches and frameworks were suggested to shed more light on the character and role of teachers' beliefs, including phenomenology, anthropology and socio-cultural studies.

Other theoretical approaches were also discussed, and much in line with the dominant trend at the time, constructivism seems to have been the most prominent among them. In fact, it might be argued that CERME 1 presented an eclectic and somewhat incoherent set of perspectives on teachers and teaching, apparently in an attempt to remain open to new approaches to address complex issues.

Significant efforts were made in the following years to develop new categorisations of teachers' knowledge. The content-related categories of Shulman's general framework were discussed, further developed, and applied in studies presented at CERMEs. They became the basis for the subject-specific, home-grown frameworks presented on teachers' knowledge included in Section 2.

There is no theoretical import to belief research in mathematics education with a significance that parallels Shulman's categorisation of teachers' knowledge. Also, there are no theories developed in mathematics education that have significant support in the field and that address the conceptual and methodological problems of belief research in the same comprehensive manner as the frameworks employed for studying teachers' knowledge.

Somewhat in line with studies of teachers' knowledge, papers on teachers' beliefs presented at CERME have aimed for subject and even topic specificity. A major concern, then, has been what teachers believe not only about mathematics and its teaching and learning, but what they believe about particular content areas (e.g. algebra) and reform issues (e.g. problem solving). In both fields a stronger focus on approaches to describing and analysing their respective key constructs has emerged.

6.2 Current trends and ways ahead – a social turn?

Hand in hand with the increased tendency to content specificity, there has been an apparently contradictory trend to broaden the perspective in CERME papers on teachers' knowledge and beliefs. In belief research this is evident in attempts to view beliefs as dynamically interdependent with other mental constructs and with classroom contingencies. A similar tendency can be discerned in research on teachers' knowledge. In this sense the focus on teachers' beliefs and knowledge is linked more immediately to classroom practice, a shift of emphasis that may be phrased as one from research on *teachers* to research on *teaching*.

To some extent this challenges the idea that what teachers think and do in their classroom is determined by previously reified mental constructs in terms of their knowledge and beliefs. This more dynamic perspective is apparent especially in a small, but visibly growing, field of research on teacher identity.

Identity is sometimes mentioned in passing in CERME papers and as an unproblematic reference to prospective or practising teachers, who need to think differently about themselves as teachers or teachers-to-be. There are, however, also studies that focus explicitly on identity. Sometimes identity development is linked to teachers' knowledge and interpretations of particular issues such as the use of ICT (da Ponte & Oliveira, 2002) or of curricular documents (Nunes & da Ponte, 2011; Sayers, 2013). Other studies focus on changes in how teachers engage in the practices of profession (Bosse & Törner, 2015) or in long-term teacher education or development programmes (Hodgen, 2004; Ebbelind, 2013, 2015).

These studies of identity share an interest in how teachers participate in, or come to view themselves as part of, a professional community. Doing so, they draw for instance on social practice theory (e.g. Wenger, 1998; Holland, Skinner, Lachicotte & Cain, 1998) or discourse analysis (e.g. Gee, 2000). This suggests that identity studies are part of the trend that Lerman (2000) refers to as *the social turn* in mathematics education research, that is, of the tendency to "see meaning, thinking, and reasoning as products of social activity" (p. 23). One might wonder if the general shift of emphasis from teachers to teaching at CERMEs is part of the same trend.

Reading across the papers from all CERMEs, however, there is no evidence of an increased interest in a *social turn*, if this were to be reflected in a growing number of references to frameworks that see learning as participation or expansion, that is, frameworks such as sociocultural theory, later developments in activity theory, social practice theory, or symbolic interactionism. At all CERMEs there are references to Vygotsksy, Wertsch, Leont'ev, Engeström, Lave, Wenger, Holland or Mead in less than 20 per cent of the papers, and the proportion of such references has been fairly stable. Sfard's (2008) theory of commognition has received some attention lately, and there are rare references to people like Bernstein, Van Oers, Forman and Cole. In spite of that, it would be an exaggeration to suggest that a social turn plays prominently in these working groups.

This means that the dynamic trend in research on teaching at CERMEs is primarily interested in the multiple relationships among different teacher characteristics, most notably their knowledge and beliefs, and how they interact with the contingencies that arise as classroom processes unfold. This approach might still be in line with the original acquisitionist orientation, but elaborates on it by working with relationships among more and more complex mental constructs, as well as with the interface between these constructs and social setting in which teaching unfolds.

There are still many studies conducted on *teachers* in mathematics education, but the trend at present is moving toward studying *teaching* and adopting either a dynamic acquisitionist or, in fewer cases, a more participatory stance. Both these approaches link the acts of teaching to the social practices of which they are part. One might wonder what the potentials are of each of them and whether the tensions between them should be seen as an opportunity for dynamic growth in the overall field of research on teaching or as an obstacle to the accumulation of knowledge – or possibly as both.

The moral of the story is that there appears to be some consensus that a more dynamic and less causal approach is needed, but little consensus on what this might mean in terms of choice of framework. We suggest that a more elaborate theoretical discussion is needed on the advantages and disadvantages of each of the options mentioned above and possibly others. This means that there seems to be a need for more profound discussions also of participatory frameworks, that have played an important part in mathematics education more generally and that were introduced already at CERME 1, but that have always played a relatively minor role.

References

Abboud-Blanchard, M., Cazes, C., & Vandebrouck, F. (2007). Teachers' activity in exercises-based lessons: Some case studies. *CERME5* (pp. 1827–1836).

Ainley, J., & Luntley, M. (2006). What teachers know: The knowledge bases of classroom practice. *CERME4* (pp. 1410–1419).

Asami-Johansson, Y. (2011). A study of a problem solving oriented lesson structure in mathematics in Japan. *CERME7* (pp. 2549–2558).

Aslan-Tutak, F., & Ertas, F. G. (2013). Practices to enhance preservice secondary teachers' specialized content knowledge. *CERME8* (pp. 2917–2926).

Azcárate, P., Cardeñoso, J. M., & Serradó, A. (2006). The learning portfolio as an assessment strategy in teacher education. *CERME4* (pp. 1430–1439).

Badillo, E., Figueiras, L., Font, V., & Martinez, M. (2013). Visualizing and comparing teachers' mathematical practices. *CERME8* (pp. 2927–2935).

Baldry, F. (2017). Analyzing a teacher's orchestration of mathematics in a 'typical' classroom. *CERME10* (pp. 3041–3048).

Ball, D. L., Thames, M. H., & Phelps, G. (2008). Content knowledge for teaching: What makes it special? *Journal of Teacher Education, 59*(5), 389–407.

Baş, S., Didis, M. G., Erbas, A. K., Cetinkaya, B., Cakıroglu, E., & Alacacı, C. (2008). Teachers as investigators of students' written work: Does this approach provide an opportunity for professional development? *CERME8* (pp. 2936–2945).

Berg, C. V. (2011). Adopting an inquiry approach to teaching practice: The case of a primary school teacher. *CERME7* (pp. 2580–2589).

Boaler, J. (2003). Studying and capturing the complexity of practice: The case of the dance of agency. In N. Pateman, B. J. Dougherty, & J. T. Zilliox (Eds.), *Proceedings of the 27th PME International Conference, 1,* 3–16.

Bosse, M., & Törner, G. (2015). Teachers' professional development in terms of identity development: A shift in perspective on mathematics teachers' learning. *CERME9* (pp. 2769–2775).

Carreño, E., Ribeiro, C. M., & Climent, N. (2013). Specialized and horizon content knowledge: Discussing prospective teachers' knowledge on polygons. *CERME8* (pp. 2966–2975).

Carrillo, J., Coriat, M., & Oliveira, H. (1999). Teacher education and investigations into teachers' knowledge. *CERME 1* (III) (pp. 99–145).

Caseiro, A., da Ponte, J. P., & Monteiro, C. (2015). Elementary teacher practice in project work involving statistics. *CERME9* (pp. 2995–3001).

Clivaz, S. (2013). Teaching multidigit multiplication: Combining multiple frameworks to analyze a class episode. *CERME8* (pp. 2995–3005).

Corcoran, D. (2007). "You don't need a tables book when you have butter beans!" Is there a need for mathematics pedagogy here? *CERME5* (pp. 1856–1865).

Coulange, L. (2006). Teacher's activity and knowledge: The teaching of setting up equations. *CERME4* (pp. 1462–1472).

Cusi, A., & Malara, N. A. (2013). A theoretical construct to analyze the teacher's role during introductory activities to algebraic modelling. *CERME8* (pp. 3015–3024).

da Ponte, J. P. (1999). Introduction: Teachers' beliefs and conceptions as a fundamental topic in teacher education. *CERME1* (I) (pp. 43–49).

da Ponte, J. P., & Chapman, O. (2006). Mathematics teachers' knowledge and practices. In A. Gutierrez & P. Boero (Eds.), *Handbook of research on the psychology of mathematics education: Past, present and future* (pp. 461–494). Rotterdam: Sense.

da Ponte, J. P., & Oliveira, H. (2002). Information technologies and the development of professional knowledge and identity in teacher education. *CERME2* (pp. 310–321).

da Ponte, J. P., & Quaresma, M. (2015). Conducting mathematical discussions as a feature of teachers' professional practice. *CERME9* (pp. 3100–3106).

Drageset, O. G. (2013). Using redirecting, progressing and focusing actions to characterize teachers' practice. *CERME8* (pp. 3025–3034).

Drageset, O. G. (2015). Teachers' response to unexplained answers. *CERME9* (pp. 3009–3014).

Ebbelind, A. (2013). Disentangling prospect teacher's participation during teacher education. *CERME8* (pp. 3045–3054).

Ebbelind, A. (2015). Systemic Functional Linguistics as a methodological tool when researching Patterns of Participation. *CERME9* (pp. 3185–3191).

Ellis, M. W., & Berry III, R. Q. (2005). The paradigm shift in mathematics education: Explanations and implications of reforming conceptions of teaching and learning. *The Mathematics Educator, 15*(1), 7–17.

Espinoza-Vasquez, G., Zakaryan, D., & Carrillo Yañez, J. (2017). Use of analogies in teaching the concept of function: Relation between Knowledge of Topics and Knowledge of Mathematics Teaching. *CERME10* (pp. 3288–3295).

Felmer, P., Perdomo-Díaz, J., Giaconi, V., & Espinoza, C. G. (2015). Problem solving teaching practices: Observer and teacher's view. *CERME9* (pp. 3022–3028).

Fernández, S., Figueiras, L., Deulofeu, J., & Martínez, M. (2011). Re-defining HCK to approach transition. *CERME7* (pp. 2640–2649).

Ferreira, R. A. T. (2007). The teaching modes: A conceptual framework for teacher education. *CERME5* (pp. 1994–2003).

Fives, H., & Buehl, M. M. (2012). Spring cleaning for the messy construct of teachers' beliefs: What are they? Which have been examined? What can they tell us? In K. R. Harris, S. Graham, & T. Urdan (Eds.), *APA educational psychology handbook. Vol. 2: Individual differences and cultural and contextual factors* (pp. 471–499). Washington, DC: APA.

Flores, E., Escudero, D. I., & Carrillo, J. (2013). A theoretical review of specialised content knowledge. *CERME8* (pp. 3054–3064).

Furinghetti, F., Grevholm, B., & Krainer, K. (2002). Introduction to WG3. Teacher education between theoretical issues and practical realization. *CERME2* (pp. 265–268).

Gade, S. (2015). Teacher-researcher collaboration as formative intervention and expansive learning activity. *CERME9* (pp. 3029–3035).

Gee, J. P. (2000). Identity as an analytic lens for research in education. *Review of Research in Education, 25*(1), 99–125.

Georget, J.-P. (2007). Facilitate research activities at the primary level: Intentional communities of practice, teaching practices, exchanges about these practices. *CERME 5* (pp. 1866–1875).

Grundén, H. (2017). Practice of planning for teaching in mathematics: Meaning and relations. *CERME10* (pp. 3065–3072).

Gunnarsdóttir, G. H., & Pálsdóttir, G. (2015). Instructional practices in mathematics classrooms. *CERME9* (pp. 3036–3042).

Helmerich, M. A. (2013). Competence in reflecting: An answer to uncertainty in areas of tension in teaching and learning processes and teachers profession. *CERME8* (pp. 3095–3104).

Hodgen, J. (2004). Reflection, identity, and belief change in primary mathematics. *CERME 3*: www.mathematik.uni-dortmund.de/~erme/CERME3/Groups/TG12/TG12_Hodgen_cerme3.pdf.

Holland, D., Skinner, D., Lachicotte Jr, W., & Cain, C. (1998). *Identity and agency in cultural worlds*. Cambridge, MA: Harvard University Press.

Huckstep, P., Rowland, T., & Thwaites, A. (2006). The knowledge quartet: Considering Chloe. *CERME4* (pp. 1568–1578).

Jakobsen, A., Thames, M. H., & Ribeiro, C. M. (2013). Delineating issues related to horizon content knowledge for mathematics teaching. *CERME8* (pp. 3125–3134).

Kaldrimidou, M., Sakonidis, H., & Tzekaki, M. (2011). Readings of the mathematical meaning shaped in the mathematics classroom: Exploring different lenses. *CERME7* (pp. 2680–2689).

Kempe, U. R., Lövström, A., & Hellqvist, B. (2017). Making distinctions: Critical for the learning of the 'existence' of negative numbers? Exploring how the 'instructional products' from a theory informed lesson study can be shared and enhance student learning. *CERME10* (pp. 3153–3160).

Kleve, B., & Solem, I. H. (2015). A contingent opportunity taken investigating in-between fractions. *CERME9* (pp. 3051–3057).

Koklu, O., & Aslan-Tutak, F. (2015). 'Responding to student ideas' as an indicator of a teacher's mathematical knowledge in teaching. *CERME9* (pp. 3206–3212).

Krainer, K. (2002). Investigation into practice as a powerful means of promoting (student) teachers' professional growth. *CERME2* (pp. 281–291).

Kunter, M., Baumert, J., Blum, W., Klusman, U., Krauss, S., & Neubrand, M. (Eds.). (2013). *Cognitive activiation in the mathematics classroom and professional competence of teachers: Results from the COACTIV project*. New York: Springer.

Kuntze, S., Lerman, S., Murphy, B., Kurz-Milcke, E., Siller, H.-S., & Winbourne, P. (2011). Professional knowledge related to Big Ideas in mathematics: An empirical study with pre-service teachers. *CERME7* (pp. 2717–2726).

Kwon, M. (2015). Supporting students' development of mathematical explanation: A case of explaining a definition of fraction. *CERME9* (pp. 3058–3064).

Larsen, D. M.,Østergaard, C. H., & Skott, J. (2013). Patterns of participation: A framework for understanding the role of the teacher for classroom practice. *CERME8* (pp. 3165–3174).

Lave, J. (1997). The culture of acquisition and the practice of learning. In D. Kirshner & J. A. Whitson (Eds.), *Situated cognition: Social, semiotic, and psychological perspectives* (pp. 17–35). Mahwah, NJ: Lawrence Erlbaum.

Leder, G. C., Pehkonen, E., & Törner, G. (Eds.). (2002). *Beliefs: A hidden variable in mathematics education?* Dordrecht: Kluwer.

Lerman, S. (2000). The social turn in mathematics education research: Multiple perspectives on mathematics teaching and learning. In J. Boaler (Ed.), *Multiple perspectives on mathematics teaching & learning* (pp. 19–44). Westport, CT: Greenwood Publishing Group.

Malara, N. A., & Navarra, G. (2015). Principles and tools for teachers' education and the assessment of their professional growth. *CERME9* (pp. 2854–2860).

Mhuirí, S. N. (2017). Considering research frameworks as a tool for reflection on practices: Grain-size and levels of change. *CERME10* (pp. 3113–3120).

Montes, M., & Carrillo, J. (2015). What does it mean as a teacher to 'know infinity'? The case of convergence series. *CERME9* (pp. 3220–3226).

Mosvold, R., & Hoover, M. (2017). Mathematical knowledge for teaching and the teaching of mathematics. *CERME10* (pp. 3105–3112).

NCTM (1998). *Principles and standards for school mathematics: Discussion draft*. Reston, VA: NCTM.

NCTM (2000). *Principles and standards for school mathematics*. Reston, VA: NCTM.

Nowińska, E. (2011). A study of the differences between the surface and the deep structures of math lessons. *CERME7* (pp. 2777–2786).

Nunes, C. C., & da Ponte, J. P. (2011). Teachers managing the curriculum in the context of the mathematics' subject group. *CERME7* (pp. 2787–2797).

Pepin, B., & Roesken-Winter, B. (Eds.). (2015). *From beliefs to dynamic affect systems: Exploring a mosaic of relationships and interactions*. Cham, Switzerland: Springer.

Petrou, M. (2010). Adapting the knowledge quartet in the Cypriot mathematics classroom. *CERME6* (pp. 2020–2029).

Philipp, R. A. (2007). Mathematics teachers' beliefs and affect. In F. K. Lester (Ed.), *Second handbook of research on mathematics teaching and learning* (pp. 257–315). Charlotte, NC: NCTM & IAP.

Reid, D. A., Savard, A., Manuel, D., & Wan Jung Lin, T. (2015). Québec anglophone teachers' pedagogies: Observations from an auto-ethnography. *CERME9* (pp. 3115–3121).

Ribeiro, C. M., Monteiro, R., & Carrillo, J. (2010). Professional knowledge in an improvisation episode: The importance of a cognitive model. *CERME6* (pp. 2030–2029).

Rowland, T., Jared, L., & Thwaites, A. (2011). Secondary mathematics teachers' content knowledge: The case of Heidi. *CERME7* (pp. 2827–2837).

Rowland, T., Thwaites, A., & Huckstep, P. (2004). Elementary teachers' mathematics content knowledge and choice of examples. *CERME3*: www.mathematik.uni-dortmund. de/~erme/CERME3/Groups/TG12/TG12_Rowland_cerme3.pdf.

Rowland, T., Turner, F., & Thwaites, A. (2013). Developing mathematics teacher education practice as a consequence of research. *CERME8* (pp. 3227–3236).

Samková, L., & Hošpesová, A. (2015). Using Concept Cartoons to investigate future teachers' knowledge. *CERME9* (pp. 2141–3247).

Sayers, J. (2013). The influence of early childhood mathematical experiences on teachers' beliefs and practice. *CERME8* (pp. 3247–3256).

Schoenfeld, A. (2000). Models of the teaching process. *Journal of Mathematical Behavior, 18*(3), 243–261.

Schueler, S., Roesken-Winter, B., Weißenrieder, J., Lambert, A., & Römer, M. (2015). Characteristics of out-of-field teaching: Teacher beliefs and competencies. *CERME9* (pp. 3254–3261).

Sfard, A. (2005). What could be more practical than good research? *Educational Studies in Mathematics, 58*(3), 393–413.

Sfard, A. (2008). *Thinking as communicating: Human development, the growth of discourses, and mathematizing.* Cambridge, UK: Cambridge University Press.

Shulman, L. S. (1986). Those who understand: Knowledge growth in teaching. *Educational Researcher, 15*(2), 4–14.

Skott, J. (2015). The promises, problems, and prospects of research on teachers' beliefs. In H. Fives & M. G. Gill (Eds.), *International handbook of research on teachers' beliefs* (pp. 13–30). New York: Routledge.

Stylianides, G. J., & Stylianides, A. J. (2010). The mathematical preparation of teachers: A focus on tasks. *CERME6* (pp. 1931–1940).

Taylan, R. D. (2015). Characterizing a highly-accomplished teacher's instructional actions in response to students' mathematical thinking. *CERME9* (pp. 3136–3143).

Turner, F. (2007). The mathematics content knowledge of beginning teachers: The case of Amy. *CERME5* (pp. 2004–2013).

Turner, F. (2011). Differences in the propositional knowledge and the knowledge in practice of beginning primary school teachers. *CERME7* (pp. 2898–2907).

Tzekaki, M., Kaldrimidou, M., & Sakonidis, X. (2002). Reflections on teachers' practices in dealing with pupils' mathematical errors. *CERME2* (pp. 322–332).

Vasco, D., Climent, N., Escudero-Ávila, & Flores-Medrano, E. (2015). The characterisation of the specialised knowledge of a university lecturer in linear algebra. *CERME9* (pp. 3283–3288).

Velez, I., & da Ponte, J. P. (2015). Promoting the understanding of graph representations by grade 3 students. *CERME9* (pp. 3143–3149).

Wenger, E. (1998). *Communities of practice: Learning, meaning, and identity.* Cambridge, UK: Cambridge University Press.

Wester, R., Wernberg, A., & Meaney, T. (2015). Students' perceptions of Norms in a reformed classroom. *CERME9* (pp. 3150–3156).

Wittmann, G., Schuler, S., & Levin, A. (2015). To what extent can kindergarten teachers and primary school teachers initiate and foster learning mathematics in typical situations? *CERME9* (pp. 3289–3296).

Zehetmeier, S., & Krainer, K. (2011). How to promote sustainable professional development. *CERME7* (pp. 2868–2877).

13

MATHEMATICS TEACHER EDUCATION AND PROFESSIONAL DEVELOPMENT

Alena Hošpesová, José Carrillo and Leonor Santos

1 Introduction

Mathematics teacher education (MTE)

> consists of processes and practices through which teachers or student teachers learn to teach mathematics. It involves as participants, primarily, student teachers, teachers, and teacher educators . . .
>
> *(Jaworski, 2014, p. 76)*

Professional development (PD) has been sometimes understood as the process by which to arrive at a previously fixed model of beliefs and practices, sometimes conceptualised as a change or increase in knowledge, at other times envisaged as a process of flexibility, autonomy and adaptation to the teaching context, or as a reflective process of better understanding teaching practice (Climent & Carrillo, 2002). Research on MTE and PD has been a strong theme of ERME conferences from its beginning; its importance being expressed by the fact that a Thematic Working Group (TWG) has always been organised dealing with this issue. The participants of these TWGs have focused on a wide range of issues connected with the different activities of teachers, teacher educators and researchers; their influence on the participants' professional competence and/or teaching; and impact of contextual variables connected with them.

The practice of MTE and PD is complex, being institutionally and culturally specific. Career structures for teachers affect their motivation to take part in PD programmes (differently in different countries) and relationships between university and school need to be handled carefully. Small-scale PD work with schools necessarily creates studies of special cases, whereas large-scale work runs the risk of producing surface or naive interpretations of 'reform' messages.

To identify the main ideas discussed during CERMEs, we used an 'evolutionary' approach. We started from the list of summaries of the activities of the TWGs published in conference proceedings and searched for the main ideas that were common across CERMEs. We identified three interconnected themes.

From the beginning, there has been a focus on the link between *theory and practice*. In CERME 1, possible bridges between theory and practice in teacher education were discussed, with different studies on the relation between the researchers and practitioners being directed to answer the question: What kinds of relationships between theory and practice are beneficial for both researchers and teachers? In CERMEs 2 and 3 there was a working group directly related to the issue (called 'Theory and practice of teaching from pre-service to in-service teacher education' in 2001, and 'Inter-relating theory and practice' in 2003). Since 2005 the issue was mostly discussed in the working group named 'From a study of teaching practices to issues in teacher education'.

One of the answers to the question above is associated with *collaborative environments* in MTE and PD (teachers, teachers as researchers, teacher educators, academic researchers). Jaworski (1999, p. 213) stressed that "such collaboration can be extremely fruitful in the enhancement of mathematics teaching and the development of related knowledge in the public domain". Later, Krainer presented a theoretical framework (see Figure 13.1) that highlighted cooperative work as the crucial component of the networking dimension included in a four-dimensional model (action, autonomy, reflection and networking) of teachers' professional practice (Krainer, 1996, p. 310). According to the author: "The (student) teachers share their experiences with other participants or with the teacher educators, they use electronic means and research literature, and thus enriched their personal views (and those of others)" (Krainer, 2002, p. 289).

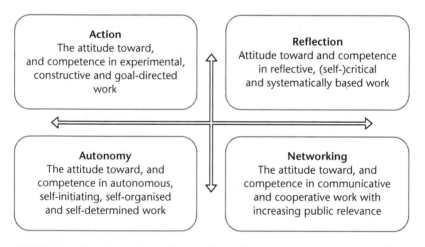

FIGURE 13.1 Four dimensions of teacher's professional practice (schematised in accordance with Krainer (1996))

Mathematics teacher education and PD **183**

Reflection was discussed as being an important ingredient in the process of MTE and PD. Krainer's (1996) idea about the unbalanced relationship between four dimensions of teacher's professional practice (Figure 13.1), "there is a lot of action and autonomy but less reflection and networking" (p. 310), can be taken as a starting point for the discussion on reflection and its role in teachers' professional lives.

In what follows, we will discuss, in depth, each of the three themes mentioned above. Within each theme, we searched for the main emerging ideas, aiming for as wide coverage as possible in respect of topics and researchers (participants in CERMEs).

All issues were studied in different environments (in-service, pre-service), by different groups of participants. Most papers dealt with pre- or in-service teachers in grades from 1 to 12, preschool and university teachers being almost absent. Different mathematical contexts (often algebra and word problems) have been taken as the kind of environment in which MTE and PD take place.

2 Linking theory and practice in mathematics teacher education and professional development

2.1 Theoretical perspectives

The tension between theory and practice was expressed by Seeger and Steinbring (1992) as the relationship "between the teacher's immediate involvedness and critical distance" (cited in Scherer & Steinbring, 2004, p. 1). In the process of MTE and PD, we have to cope with the gap between theory and practice constantly. When dealing with theory and practice, the former is frequently associated with research or an academic context (the content to be learnt in the pre-service institutions), and the second is associated with teachers (e.g. Schön (1983, p. 26) distinguishes between "practitioners, educators and researchers").

Different constructs were used to explain this interplay between theory and practice. Skott (2006) introduced the notion of *theoretical loop* to explain the process of starting with teachers' practice, going on to theorising it and coming back to practice. This loop pays special attention to the usefulness of theory in practice:

> It is reasonable to assume that theoretical constructs grounded in the sites of practice are of greater potential use to practitioners (e.g. teachers) than constructs developed without such grounding and without recognition of the contextual complexities of teaching.
>
> *(pp. 1599–1600)*

In Carrillo, Climent and Muñoz (2006), the case under study serves as a means to reflect on the authors' former conceptualisation of PD. The process of (1) conceptualisation, (2) the practice of a novice teacher and (3) reconceptualisation of PD, shows a singular relationship between theory and practice, where the analysis of the latter introduces new considerations in the previously built theory.

The attitude of teachers toward theorising their practice and questioning theories is considered by Zehetmeier (2010) to be one of the factors that promotes the impact of PD projects. In this case, the mode of relationship between theory and practice serves as a means for developing professionally.

2.2 Questions and settings

The shift from pre- to in-service education was considered to be a crucial point. Carrillo, Coriat and Oliveira (1999) said that "the subjects are usually theoretically biased and one expects that the prospective teachers shall make the integration . . . when they become teachers. And this perspective goes on in the in-service education" (Carrillo et al., 1999, p. 135).

Verhoef and Terlouw (2007) dealt with the same issue directly via the relationship between trainee teachers' experiences and "the teachers' role in the classroom setting with reference to discussed theory and trainers' didactical remarks" (p. 2017). Cusi and Malara (2011), being aware of the lack of relationship between theory and practice among undergraduates who have not yet worked in the schools, concentrated on "giving them (the students) theoretical and methodological tools to learn how to interpret their future actions in the classes" (p. 2625).

The transition between theory and practice takes place in different milieus. The substantial learning environments (in the sense used by Wittmann (2001)) are considered by Tichá and Hošpesová (2011) as "one of the fields where researchers' and teachers' objectives interlink, . . . mathematics and didactics of mathematics mingle and which is open to natural long-term, systematic cooperation of researchers and teachers" (Tichá & Hošpesová, 2011, p. 2890).

2.3 Methodological approaches and tools

Many authors see the analysis of classroom episodes as a means to link theory and practice: in in-service education, for example, Goffree and Oonk (1999) and Martignone (2015). Brown (2006), adopting an enactivist position, showed how theory became the result of a progressive awareness of what happens in practice. Ribeiro, Monteiro and Carrillo (2010), differentiating actions and cognition (including beliefs and knowledge[1]), named classroom episodes as a good "starting point for an approximation between theory and practice . . . researchers and teachers 'speak the same language', using the same codifications; in doing so, a great degree of collaboration is needed" (p. 2037). The authors underline the role of collaboration in the process of linking theory and practice. Berg (2011) approached the link between theory and practice within the context of developmental research and using the notion of inquiry cycle (Jaworski, 2007).

The pre- and in-service courses are, in some investigations, analysed or used from the perspective of the relationship between theory and practice. Malara and Navarra (2007) approached the promotion of this relationship through the implementation of early algebra tasks in in-service courses, the aim being to test the effectiveness of such task as an element of mediation between theory and practice and, more specifically, as a tool to set trainees "inside" a virtual class, so that they can get aware of what it means to "act in the moment".

(p. 1930)

Joubert, Back, De Geest, Hirst and Sutherland (2010) present an in-service course driven by an advisor and a teacher educator where the participants were asked to read some research papers and reflect on their application in their own teaching contexts. The role of the course leaders (and also the benefit of the relationship between in-service and pre-service training institutions) is emphasised by one of the participant teachers: "they [the course leaders] provide a link between the theory and practice in both [my] own classroom and what other schools are doing" (p. 1767).

Aizikovitsh-Udi, Clarke and Star (2013) contributed the idea that connecting theories and practice in in-service programmes is essential, but not enough to provoke teachers' change in the classroom. The lack of influence of methods courses on teachers is commented on by Somayajulu (2013):

> [T]he methods courses were designed without much sensitivity towards helping the teachers' [*sic*] bridge theory and practice . . . I recommend that pre-service teacher education programs should take into consideration the prior experiences of teachers as learners and build on them while designing activities that are more effective.
>
> *(p. 3275)*

Finally, coming from the investigations into teacher knowledge, Ribeiro and Carrillo (2011) propose moving beyond the analysis of a teacher's knowledge to contribute to the improvement of his/her practice, that way building a "bridge between theory and practice" (p. 2824), helpful in initial as well as in in-service education. Ribeiro, Aslan-Tutak, Charalambous and Meinke (2015) write that focusing on what teachers know and how they know it can

> lead to the development of practices that enrich the levels of awareness and connections that contribute to improved education and, ultimately, practice. Bridging theory and practice is essential for such improvement, and core to building such bridges is defining the nature and goals of proposed tasks . . . for enhancing teacher acquisition of ideal knowledge that would allow them to foster fruitful mathematical understanding in their students.
>
> *(p. 3179)*

3 Collaborative environments in mathematics teacher education and professional development

Collaborative environments, in which mathematics researchers, teachers and prospective teachers work together, reflecting on and designing classroom situations,

186 Hošpesová, Carrillo & Santos

emerged as an appropriated context to build a bridge between theory and practice. Since the beginning of the creation of ERME, collaboration has been a central concept, being one element of the CERME spirit, the three Cs of communication, cooperation and collaboration.

3.1 Theoretical perspectives

Although collaboration is a familiar concept between ERME members, it has to be noticed that the meaning of collaboration is, in general, not defined or clarified. We can remark, reading the different CERME proceedings, that it is rare that the authors are explicit about which meaning they are using for collaboration. A common idea is that collaboration implies working jointly. For instance, Witterholt and Goedhart (2010) studied the development of practical knowledge of two mathematics teachers working together during a period of time. Although these teachers maintained their different ideas about what it is to teach, they were also committed to the success of the peer work and "experienced the interest of combining each other's ideas and constructing an educational design to which everybody could commit" (p. 1998). Nevertheless, according to several authors, "collaborating is not just sitting or working together" (Santos, Carrillo, Hošpesová & Blanchard, 2010, p. 1680).

Once in a while, we may find clarification of the term 'collaboration'. For example, Pesci (2010), focusing on the content of the collaboration process, wrote: the term collaboration has a more general meaning: a positive inter-relationship amongst the people involved, not necessarily connected to a specific modality of acting in groups" (p. 1987). Berg (2011), highlighting the type of group constitution, defined collaboration as "teachers and researchers work together as co-learners" (p. 2588).

Looking at the proceedings in sequential order of publication, we may remark that contexts of others are progressively included instead of the collaboration context. When these new concepts were introduced, they were defined and clarified, compared to what we have now observed. This is the case with Wenger's (1998) community of practice (for example, in Reinup, 2010 or García, Sánchez, Escudero & Llinares, 2004) that envisages collaboration under the notion of community of practice (of teaching) in which prospective teachers, teachers, researchers and educators took part. A second concept used instead of collaboration is Jaworski's (2008) inquiry community (for example, in Oliveira & Henriques, 2015).

3.2 Questions and settings

There is a great diversity of settings of collaborative work, naturally associated to particular issues. The studies that involved mathematics teachers and researchers, in pre-service or in-service educational environments, were the ones that appeared at CERME more often. However, there are other contexts of collaborative work, such as the ones that involved only mathematics teachers in the same school. This is the case for Nunes and da Ponte (2010) who aimed to understand the curriculum development of a mathematics teacher and the relationship between such curriculum development and the collaborative work undertaken by the mathematics teachers' subject group (14 mathematics teachers of a secondary school). The results pointed out that the curriculum development has a collective level (annual planning and the construction of units and tasks) and an individual one (decisions about what is important to develop with their own students, such as autonomy in mathematics learning; responsibility for their own actions; and the capacity to think independently). Some conflicts appear in the group when different perspectives collide, but the mathematics teacher participating in the study, "the natural leader of the group, nurtures his relationship with his colleagues using curriculum management as a focal activity" (p. 1721). Moreover, the dynamics from the collaborative context "support the PD of the mathematics teachers and their capacity to accept new challenges" (Nunes & da Ponte, 2011, p. 2795).

There are differences, in the Nunes and da Ponte collaborative context, between the members of the group. According to Bräuning and Nührenbörger (2010), teachers react more spontaneously than reflected and "construct knowledge by observation, experience, transfer and interrelation" (p. 1744). Moreover, although teachers and researchers seem, in general, to have identical objectives, that is not always really true: "Researchers look for answers to theoretical questions, while teachers deal with practical problems" (Hošpesová, Tichá & Macháčková, 2007, p. 1914).

Whatever the focus of the collaborative work, the problem of sustainability of the learning that occurs in this context in pre-service environments is an important issue (Gunnarsdóttir & Pálsdóttir, 2013). According to these authors, from the perspective of the newly graduated teachers, the collaborative practices they experienced in their pre-service teacher programme were important not only during the programme but also in their practice as teachers. Nevertheless, they still felt the need of support from outside the school.

3.3 Methodological approaches and tools

If working collaboratively is an important capacity for teachers, then pre-service teachers' education programmes need to pay attention to this aspect and create favourable contexts to develop such a capacity. One context may be created through certain types of activities such as the development of project work, which requires "predisposition and the teacher's ability for self-exposure" (Santos & Bento, 2007, p. 1974) or planning and discussing collaboratively with the mathematics teacher educator, following the design of a lesson study (Gunnarsdóttir & Pálsdóttir, 2011).

In in-service educational environments, there are, similarly, studies that focused, for example, on: (1) planning teacher mathematical activities, namely in

188 Hošpesová, Carrillo & Santos

task design using an experience-based approach to identify the need for supporting theories (Nilsson, Sollervall & Milrad, 2010); (2) the process of interaction between educational professionals (Muñoz-Catalán, Carrillo & Climent, 2010); (3) teacher discourse in the mathematics classroom (Martinho & da Ponte, 2010); or (4) strategies to support changes in mathematics assessment practices (Dias & Santos, 2015). In all of these studies, the collaborative context has been considered to be an important support to the continuing development of teachers' professional knowledge.

4 Approaching reflection in mathematics teacher education and professional development

4.1 Theoretical approaches

Reflection was considered to be an essential feature of human thought in Dewey's early works, where it was defined as "active, persistent, and careful consideration of any beliefs or supposed form of knowledge in the light of the grounds that support it, and the further conclusions to which it tends" (Dewey, 1910, p. 6). MTE and PD of mathematics teachers is often considered to be the result of participation in specific activities and diverse forms of reflection on them.

Although the concept of reflection has different meanings in different papers, it mostly includes (a) pedagogical considerations such as observation and contemplation about decisions taken and strategies used; (b) mathematical (didactical) analysis of key content elements, such as, possible ways of explaining a concept. Hodgen (2004) stressed the relation to teacher's practice and spoke about "reflection as the reconstruction of experience and knowledge" (p. 1). Other authors agree that reflection creates space for the transition from an intuitive to a conscious and justified action: "It is possible to treat reflection as connected with interpretation of teaching/learning situations to be the best way to develop the teachers' professional way of thinking and to present practical didactical theory" (Slavík, 2004, p. 1, own translation).

Hošpesová and Tichá (2006) emphasised the analysis of the teacher's own thinking and dealing with pupils in a way suitable for their ability to plan their own lifelong education and consider qualified pedagogical (self-)reflection as one of the teacher's competences.

4.2 Questions and settings

The process of reflection was discussed as being among the promising ways in PD for supporting teacher development. Often it was created as a particular phase of the activities. Climent & Carrillo (2002, pp. 270–271) said that "the most important thing is to make available opportunities for reflection on one's own knowledge and beliefs, this being a possible point of departure for change chosen by the teacher". Jaworski (2003) suggested investigating the reflexive pair "inquiry and reflection".

Scherer and Steinbring (2004, p. 2) and others claimed that it is necessary to create opportunities for "common reflection on everyday instructional activities" and emphasised that a mathematics teacher's work is so demanding that it is necessary to come to a conception of collective reflection on everyday teaching activities to enable teachers to develop a deeper view of their activity. The authors discussed the role of researchers and/or teacher educators in reflecting groups and stressed the specific nature of mathematics education leading to one focus being on pupils' understanding of mathematical structures.

The most frequent issues discussed during CERMEs were the changes observed in research situated within the co-learning inquiry paradigm (Jaworski, 2004). It was shown that being involved in action and reflection collaboratively enables the participants, teachers and teacher educators, to achieve a deeper understanding of both their own world and the world of the other participants in the community (Pesci, 2007; Berg 2007). Georget (2007) concluded that even a simple opportunity to discuss or to have a reaction to their own teaching helps teachers to access existing curricular documents and integrate them into their own practice.

It was stated by Hodgen (2004) that to replicate such intense experiences for teachers "would be an extremely difficult and costly task" and he suggested a model of teacher education in which

> teachers not only engage critically with the mathematics curriculum as teachers and as learners of mathematics, but also places them in situations where, as teacher tutors and curriculum makers, they encourage other teachers to engage critically in similar ways.
>
> *(p. 8)*

4.3 Methodological approaches and tools

Supporting (collective and/or self-) reflection via video episodes and the creation of a friendly environment for collaborative discussion were discussed quite often. This collaborative context included primary and lower secondary school teachers, student teachers, doctoral students in didactics of mathematics, teacher educators and researchers. In this scenario, "the collaborative discussions support teachers' attempts to unpack the practice observed on the screen, through the implementation of mechanisms such as ascribing goals and weighing alternatives" (Karsenty, Arcavi & Nurickp, 2015, p. 2831). Moreover, it is possible for teachers to get external points of view, contributing to a better understanding of their practices (Muñoz-Catalán, Carrillo & Climent, 2007; da Ponte, Serrazina & Fonseca, 2004), and to pay more attention to the importance of the role of error in mathematics learning (Guerreiro & Serrazina, 2010).

190 Hošpesová, Carrillo & Santos

Other modes of intervention with specific tasks (Pesci, 2010; Helmerich, 2013); online environments (Sánchez, 2011); portfolios (Santos, 2006) and role-play experiences (Lajoie & Maheux, 2013) were presented and discussed.

It is necessary to state, in agreement with Zehetmeier et al. (2015), that the issues of (self-)reflection are connected with other issues in teacher education, for example, teachers' practices; content and methods of PD courses; competency of didactical analysis; pedagogical content knowledge; didactic transposition; Anthropological Theory of Didactics; content representation; teaching experiments; common games; and childhood education.

5 Emerging issues and future studies on mathematics teacher education and professional development

Research on MTE and PD has been a stable focus over recent CERMEs. Research activities have been broad and varied from country to country due to different models of education and teacher education particularly. As mentioned by Zehetmeier,

> research has increasingly focused not only on the participating teachers, but also on the role of teacher educators and academic researchers. So far, the research community has attempted to develop theoretical and methodological frameworks to both describe and explain the complex topic of mathematics teacher education.
>
> *(Zehetmeier et al., 2015, p. 2730)*

Despite the huge variety of theories and methods, the results are usually convergent emphasising the specific character of mathematics; the crucial role of collaboration of different communities (consisting of future teachers, teachers and teacher educators/ researchers); and specific methods (for example, writing diaries, creation of portfolios, elaboration of analysis of video episodes) supporting deeper understanding of practice.

From reading the CERME papers, how theories affect the practical knowledge of teachers and how teachers perceive that their practice fits with theoretical issues emerged as important issues. The role of researchers, as theory-developers and mediators between theory and practice, university and schools, and research and teachers, needs further attention. Collaboration emerged as a convenient context for teacher development. Consideration of the papers in this chapter calls for the need to enhance the collaboration among researchers and teachers. In these contexts, as well as in any teacher development context, reflection plays a crucial role, but reflection needs to be guided and oriented toward outcomes that can be externally or internally characterised.

We have tried to give an account of the role of mathematics in the issues and papers selected in this chapter. It has not been an easy task, given that speaking of PD entails the consideration of dimensions that are more general.

The research community has formulated several issues to be dealt with in the future, especially:

How do we deal with different perspectives of groups of participants in teachers' education (mentors, teachers, prospective teachers, teacher educators)?

How can we assess effects (immediate and long-term) of supportive programmes for (prospective) teachers or teacher educators? How to cope with low sustainability of such programmes?

How can we actually observe identity development in terms of practice?
(Zehetmeier et al., 2015, pp. 2731–32)

We would add the role of new technologies in MTE and PD. As an example, Van Bommel and Liljekvist (2015) presented *Facebook* as a way to promote teachers' communication. They considered sharing ideas to be a good means for teachers to develop professionally. It is to be expected that the possibilities and availability of communication technologies will continue to increase and might represent a radical change in the approach to the teacher professionalisation process (for example: immediate sharing; availability of any material from anywhere).

MTE and PD are studied in other communities of researchers. PME devoted a chapter in an overview of research from 1976 to 2006 (Llinares & Krainer, 2006) to this issue, named 'Mathematics (student) teachers and teacher educators as learners'. The authors took a cognitive perspective and showed how research mapped the path from student to expert teachers. They regarded mathematics teachers' learning as a complex process and, to a large extent, influenced by personal, social, organisational, cultural and political factors. Ten years later, Lin and Rowland (2016) wrote, in another PME handbook, a chapter entitled 'Pre-service and in-service mathematics teachers' knowledge and professional development'. This overview of the research on MTE and PD discussed different aspects of teachers' knowledge gained in pre-service and in-service education. The authors mainly focused on studies that documented the relationship between a teacher's PD and the refinement of his/her practice. The authors organised the research results from the point of view of teachers' learning in different contexts (teaching; researching; and participating in a learning community).

It is obvious that teacher education and PD, which has become an increasingly frequent topic of research in mathematics didactics around the world since the end of the last century, is likely to be expanded. The role of a teacher changes as knowledge and communication technologies are promoted and become more influential. We suppose that our participation in the joint debates that characterise the CERME spirit strengthens our chances of deepening our understanding of what is happening in the mathematics classes in our countries, creating opportunities for positive change.

Note

1 Chapter 12 is devoted to teacher knowledge and beliefs, so we do not develop these issues further in this chapter.

References

Aizikovitsh-Udi, E., Clarke, D., & Star, J. (2013). Good questions or good questioning: an essential issue for effective teaching. *CERME8* (pp. 2908–2916).

Berg, C. V. (2007). Expressing generality: Focus on teachers' use of algebraic notation. *CERME5* (pp. 1837–1846).

Berg, C. V. (2011). Adopting an inquiry approach to teaching practice: The case of a primary school teacher. *CERME7* (pp. 2580–2589).

Bräuning, K., & Nührenbörger, M. (2010). Teachers' reflections of their own mathematics teaching processes. *CERME6* (pp. 944–953).

Brown, L. (2006). From practices to theories to practices . . . in learning to teach mathematics and learning mathematics. *CERME4* (pp. 1451–1461).

Carrillo, J., Climent, N., & Muñoz, C. (2006). The transition from initial training to the immersion in practice: The case of a mathematics primary teacher. *CERME4* (pp. 1526–1536).

Carrillo, J., Coriat, M., & Oliveira, H. (1999). Teacher education and investigations into teacher's knowledge. *CERME1* (pp. 99–146).

Climent, N., & Carrillo, J. (2002). Developing and researching professional knowledge with primary teachers. *CERME2* (pp. 269–280).

Cusi, A., & Malara, N. A. (2011). Analysis of the teacher's role in an approach to algebra as a tool for thinking: Problems pointed out during laboratorial activities with perspective teachers. *CERME7* (pp. 2619–2629).

da Ponte, J. P., Serrazina, L., & Fonseca, H. (2004). Professionals investigate their own practice. *CERME3*: www.mathematik.uni-dortmund.de/~erme/CERME3/Groups/TG11/TG11_Ponte_cerme3.pdf.

Dewey, J. (1910). *How we think*. Boston, MA: D. C. Heath & Co. Publishers.

Dias, P., & Santos, L. (2015). An assessment practice that teacher José uses to promote self-assessment of mathematics learning. *CERME9* (pp. 3002–3008).

García, M., Sánchez, V., Escudero, P., & Llinares, S. (2004). The dialectic relationship between theory and practice in MTE. *CERME3*: www.mathematik.uni-dortmund.de/~erme/CERME3/Groups/TG11/TG11_Garcia_cerme3.pdf.

Georget, J. (2007). Facilitate research activities at the primary level: Intentional communities of practice, teaching practices, exchanges about these practices. *CERME5* (pp. 1866–1975).

Goffree, F., & Oonk, W. (1999). A digital representation of "full practice" in teacher education: The MILE project. *CERME1* (pp. 187–199).

Guerreiro, A., & Serrazina, L. (2010). Communication as social interaction primary school teacher practices. *CERME6* (pp. 1744–1750).

Gunnarsdóttir, G., & Pálsdóttir, G. (2011). Lesson study in teacher education: A tool to establish a learning community. *CERME7* (pp. 2660–2669).

Gunnarsdóttir, G., & Pálsdóttir, G. (2013). New teachers' ideas on professional development. *CERME8* (pp. 3085–3094).

Helmerich, M. (2013). Competence in reflecting: An answer to uncertainty in areas of tension in teaching and learning processes and teachers profession. *CERME8* (pp. 3095–9104).

Hodgen, J. (2004). Reflection, identity and belief change in primary mathematics. *CERME3*: www.mathematik.uni-dortmund.de/~erme/CERME3/Groups/TG12/TG12_Hodgen_cerme3.pdf.

Hošpesová, A., & Tichá, M. (2006). Developing mathematics teacher's competence. *CERME4* (pp. 1483–1493).

Hošpesová, A., Tichá, M., & Macháčková, J. (2007). Differences and similarities in (qualified) pedagogical reflection. *CERME5* (pp. 1906–1915).

Jaworski, B. (1999). Teacher education through teachers' investigation into their own practice. *CERME1* (pp. 201–221).

Jaworski, B. (2003). Research practice into/influencing mathematics teaching and learning development: Towards a theoretical framework based on co-learning partnerships. *Educational Studies in Mathematics, 54*, 249–282.

Jaworski, B. (2004). Grappling with complexity: Co-learning in inquiry communities in mathematics teaching development. In *Proceedings of the 28th PME Conference* (Volume I, pp. 17–32). Bergen: Bergen University College.

Jaworski, B. (2007). Theoretical perspectives as a basis for research in LCM and ICTML. In B. Jaworski, A. B. Fuglestad, R. Bjuland, T. Breiteig, S. Goodchild, & B. Grevholm (Eds.), *Læringsfellesskapimatematikk – learning communities in mathematics*. Bergen: Caspar Forlag.

Jaworski, B. (2008). Building and sustaining inquiry communities in mathematics teaching development. In K. Krainer & T. Wood (Eds.), *Participants in MTE* (pp. 309–330). Rotterdam: Sense Publishers.

Jaworski, B. (2014). Communities of inquiry in mathematics teacher education. In S. Lerman (Ed.), *Encyclopedia of Mathematics Education* (pp. 76–78). New York: Springer.

Joubert, M., Back, J., De Geest, E., Hirst, C., & Sutherland, R. (2010). PD for teachers of mathematics: Opportunities and change. *CERME6* (pp. 1761–1770).

Karsenty, R., Arcavi, A., & Nurickp, Y. (2015). Video-based peer discussions as sources for knowledge growth of secondary teachers. *CERME9* (pp. 2825–2832).

Krainer, K. (1996). Some considerations on problems and perspectives of in service MTE. In C. Alsina et al. (Eds.), *8th International Congress on Mathematics Education: Selected Lectures* (pp. 303–321). Sevilla: SAEM Thales.

Krainer, K. (2002). Investigation into practice as a powerful means of promoting (student) teachers' professional growth. *CERME2* (pp. 281–291).

Lajoie, C., & Maheux, J. F. (2013). Richness and complexity of teaching division: Prospective elementary teachers' roleplaying on a division with remainder. *CERME8* (pp. 3155–3164).

Lin, F.-L., & Rowland, T. (2016). Pre-service and in-service mathematics teachers' knowledge and professional development. In Á. Gutiérrez, G. C. Leder & P. Boero (Eds.), *The second handbook of research on the psychology of mathematics education* (pp. 483–520). Rotterdam: Sense Publishers.

Llinares, S., & Krainer, K. (2006). Mathematics (student) teachers and teacher educators as learners. In A. Guriérrez & P. Boero (Eds.), *Handbook of research on the psychology of mathematics education* (pp. 429–459). Rotterdam: Sense Publishers.

Malara, N. A., & Navarra, G. (2007). A task aimed at leading teachers to promoting a constructive early algebra approach. *CERME5* (pp. 1925–1934).

Martignone, F. (2015). A development over time of the researchers' meta-didactical praxeologies. *CERME9* (pp. 2867–2873).

Martinho, H., & da Ponte, J. P. (2010). A collaborative project as a learning opportunity for mathematics teachers. *CERME6* (pp. 1961–1970).

Muñoz-Catalán, M. C., Carrillo, J., & Climent, N. (2007). The PD of a novice teacher in a collaborative context: An analysis of classroom practice. *CERME5* (pp. 1935–1944).

Muñoz-Catalán, M. C., Carrillo, J., & Climent, N. (2010). Analysis of interactions in a collaborative context of professional development. *CERME6* (pp. 2010–2019).

Nilsson, P., Sollervall, H., & Milrad, M. (2010). Collaborative design of mathematical activities for learning in an outdoor setting. *CERME6* (pp. 1101–1110).

Nunes, C., & da Ponte, J. P. (2010). Curriculum management in the context of a mathematics subject group. *CERME6* (pp. 1714–1723).

Nunes, C., & da Ponte, J. P. (2011). Teachers managing the curriculum in the context of the mathematics subject group. *CERME7* (pp. 2787–2797).

Oliveira, H., & Henriques, A. (2015). Characterizing one teacher's participation in a developmental research project. *CERME9* (pp. 2881–2887).

Pesci, A. (2007). From studies of cooperative learning practices towards a model of intervention on mathematics teachers. *CERME5* (pp.1945–1954).

Pesci, A. (2010). Developing mathematics teachers' education through personal reflection and collaborative inquiry: Which kinds of tasks? *CERME6* (pp. 1981–1990).

Reinup, R. (2010). Developing of mathematics teachers' community: Five groups, five different ways. *CERME6* (pp. 1831–1840).

Ribeiro, C. M., & Carrillo, J. (2011). Knowing mathematics as a teacher. *CERME7* (pp. 2817–2826).

Ribeiro, C. M., Aslan-Tutak, F., Charalambous, C., & Meinke, J. (2015). Introduction to the papers of TWG20: Mathematics teacher knowledge, beliefs, and identity: Some reflections on the current state of the art. *CERME9* (pp. 3177–3183).

Ribeiro, C. M., Monteiro, R., & Carrillo, J. (2010). Professional knowledge in an improvisation episode: The importance of a cognitive model. *CERME6* (pp. 2030–2039).

Sánchez, M. (2011). Concepts from mathematics education research as a trigger for mathematics teachers' reflections. *CERME7* (pp. 2878–2887).

Santos, L. (2006). The portfolio in teacher education. *CERME4* (pp. 1579–1588).

Santos, L., & Bento, A. (2007). The project work and the collaboration on the initial teacher training. *CERME5* (pp. 1974–1983).

Santos, L., Carrillo, J., Hošpesová, A., & Blanchard, M. (2010). From a study of teaching practices to issues in teacher education. *CERME6* (pp. 1688–1691).

Scherer, P., & Steinbring, H. (2004). The professionalisation of mathematics teachers' knowledge: Teachers commonly reflect feedbacks to their own instruction activity. *CERME3*: www.mathematik.uni-dortmund.de/~erme/CERME3/Groups/TG11/TG11_Scherer_cerme3.pdf.

Schön, D. A. (1983). *The reflective practitioner*. New York: Basic Books.

Seeger, F., & Steinbring, H. (Eds.) (1992). The dialogue between theory and practice in mathematics education: Overcoming the broadcast metaphor. *Proceedings of the Fourth Conference on Systematic Cooperation between Theory and Practice in Mathematics Education (SCTP). Brakel.* (IDM Materialien und Studien 38). Bielefeld: IDM Universität Bielefeld.

Skott, J. (2006). The role of the practice of theorising practice. *CERME4* (pp. 1598–1608).

Slavík, J. (2004). Profesionální reflexe a interpretace výuky jako prostředníkmeziteorií a praxí. (In Czech: Professional reflection and interpretation of education as a mediator between theory and practice.) In *Konference Oborové didaktiky v pregarduálním učitelském studiu*. Brno: PdF MUNI.

Somayajulu, R. V. (2013). Capturing pre-service teachers' mathematical knowledge for teaching. *CERME8* (pp. 3267–3276).

Tichá, M., & Hošpesová, A. (2011). Teacher competences prerequisite to natural differentiation. *CERME7* (pp. 2888–2897).

Van Bommel, J., & Liljekvist, Y. (2015). Facebook and mathematics teachers' professional development: Informing our community. *CERME9* (pp. 2930–2936).

Verhoef, N. C., & Terlouw, C. (2007). Training mathematics teachers in a community of learners (COL). *CERME5* (pp. 2014–2023).

Wenger, E. (1998). *Communities of practice: Learning, meaning and identity*. Cambridge, UK: Cambridge University Press.

Witterholt, M., & Goedhart, M. (2010). The learning of mathematics teachers working in a peer group. *CERME6* (pp. 1991–1999).

Wittmann, E. Ch. (2001). Developing mathematics education in a systemic process. *Educational Studies in Mathematics, 48*, 1–20.

Zehetmeier, S. (2010). The sustainability of professional development. *CERME6* (pp. 1951–1960).

Zehetmeier, S., Bosse, M., Brown, L., Hošpesová, A., Malara, N., & Rösken-Winter, B. (2015). Introduction to the papers of TWG18: MTE and professional development. *CERME9* (pp. 2730–2732).

14

MATHEMATICS EDUCATION AND LANGUAGE

Lessons and directions from two decades of research

Núria Planas, Candia Morgan and Marcus Schütte

1 Introduction

Several decades after the conceptualisation of language as a system of signs provided by the linguistic paradigm in language research, a range of questions about language and ways of tackling them have evolved inside and outside the field of mathematics education. In the midst of a diversity of premises of language, we know that language is a system of linguistic rules and texts, but also and importantly, an array of contexts of use for many kinds of rules and texts. In this chapter, we will argue that the progress of mathematics education and language research is taking place through a complex expansion rather than an overthrow of the linguistic paradigm, with an increase in the scope of the domain and in the spread of cultural and social claims. The questions addressed will be:

- What is the scope of the research on mathematics education and language?
- How can we map and link the newer approaches in the domain of mathematics education and language to classical approaches?
- What has been achieved in the last two decades of research?

Studying the progress of our knowledge of language in mathematics education research across the two decades of the European Society for Research in Mathematics Education (ERME) has led us to uncover classical themes regarding *the language of the learner, the language of the teacher/classroom* and *the language of mathematics*. In their contemporary forms, these are complementary themes, intertwined, either individually or in combination, with conceptualisations of *language as system, language as culture* and *language as discourse*. At the beginning of ERME in the working group entitled 'Social Interaction in Mathematical Learning Situations' and in the present Thematic Working Group (TWG), such themes and conceptualisations have been addressed primarily through *classroom-based research*. The study of language, inside and outside ERME, has mostly involved the study of mathematics classrooms as dynamic environments of interaction between students and teachers and between students and peers. The dominance of classroom studies and of the three themes regarding whose language is in focus suggests some continuity. Nonetheless, continuity is accompanied by a phenomenon of increasing complexity in the ways of understanding language. The postulation of the inseparability of language from cultural and social contexts (Morgan, 2013) has gained ground, along with interpretations of mathematics classrooms as communities of practice and configurations of discursive activity. The study of the language domain in ERME, therefore, points to a relationship between continuity and complexity. The sophistication in the ways of conceptualising language across major themes in classroom-based research inspires our overall characterisation of the ERME domain as *a continuum of complexity*.

After this introduction, in Section 2 we discuss what is involved in international research on mathematics education and language. In Section 3, we survey research reported at CERME since 1998 as a benchmark for assessing the phenomenon of increasing complexity as well as the relationship between continuity and complexity in the ERME domain. In Section 4, we map some gaps and directions for future research.

2 What does it mean (to) research on mathematics and language?

The review by Austin and Howson (1979) cited research into mathematics and language dating back to the 1940s, with a body of research beginning to establish itself in the early 1970s. Nearly 40 years later, the research developed within ERME and beyond still addresses the broad themes identified by these authors:

- *The language of the learner* (i.e. the language or languages and linguistic skills brought to the mathematics classroom by learners);
- *The language of the teacher and the classroom* (i.e. the language or languages and linguistic skills brought to the mathematics classroom by teachers);
- *The language of mathematics* (i.e. the language or languages and linguistic features of the texts that arise within the practice of mathematics).

We can see substantial development in the sophistication of these themes. There is more widespread and systematic engagement with theories of language and communication from psychology, sociology, linguistics, ethnology, semiotics and anthropology, as well as with specialised frames addressing the role of language in mathematics education. There has also been a growth in the diversity and complexity of the domain as researchers draw on a wider range of theoretical resources combined in new ways. One source of diversity is the fact that research on mathematics and language encompasses three main possible foci. The first takes language itself as the object of study, the second uses language as a vehicle for studying other phenomena and the third views participating in mathematical communication as learning mathematics itself. All three foci formulate descriptions of language-in-use in a mathematical context but analyse the descriptions in different ways. The description of language in some cases has been formed from 'common sense' knowledge about mathematics without a systematic theory of language or has drawn on tools from linguistics that do not fully serve the purpose of distinguishing characteristics of language use of interest to mathematics education. A major contribution toward more adequate description was the publication of Pimm's (1987) book, but there remains a need to develop greater rigour in the ways in which we define and distinguish between mathematical and 'everyday' language.

The significance of understanding what is specific in mathematical language appears stronger in the light of the development of theoretical understanding of mathematics itself as discursive activity. Recognising the distinctive nature of mathematical communication is a necessary element of any study of mathematical activity, whether one adopts the 'strong' discursive position that mathematical objects have no existence independent of the discursive means of communicating about them, or a less absolutist position that there is no direct material access to mathematical objects but the experience of them through some form of 'representation' or 'realisation'. These two terms reflect distinct ontological positions: speaking of representation of a mathematical object suggests that there exists an independent object, whereas speaking of realisation proposes that the communication about an object is what gives the object existence. In either of these positions, mathematical activity implies engagement in a form of discourse about real or discursive objects. Understanding such activity involves studying that discourse and its features.

Where language is the object of study, description of the language might be an end in itself, addressing the nature of *the language of mathematics*. Understanding the features of mathematical language enables us to describe and evaluate the mathematical discourse of teachers and students in classrooms, while principled description of mathematical language opens up many questions: What are the features of the mathematical discourse in which students are expected to participate? How do classroom activities induct students into (what kinds of) specialised mathematical discourse? To what extent are students engaging in specialised mathematical discourse? From a 'strong' discursive position, any study of mathematical knowledge and learning entails asking questions about *the language of the learner* and *of the teacher/classroom*, and how they change. However, the significance of language in mathematics education includes the use or function of language as well as its form. Paying attention to how language functions suggests questions about reasoning, argumentation, proof, mathematical objects and relationships. Communication in the classroom and in other contexts, including curriculum, assessment and policy, also has an interpersonal function, constructing positions for students and teachers and framing relationships between them and to the mathematics. Studying the interpersonal functioning of language, drawing on theoretical resources developed in fields such as pragmatics, social semiotics and conversation analysis, can contribute to understanding social aspects of mathematics education such as how teachers manage classroom interaction and how students from various groups experience mathematics education.

Drawing on theories that conceptualise language as constitutive, constructive or functional enables researchers to analyse what is achieved in a given context through language use, addressing *the language of the teacher and the classroom*. We see the use of linguistic data as a means of gaining insight into understanding and learning of mathematics. In the first meetings of the TWG, research drawing on social constructivist and social interactionist perspectives is strongly represented, starting from interactional approaches of interpretive classroom research. This research focuses on studying classroom interactions using interactional analysis (Krummheuer, 1999) in order to observe learners' collective negotiation of mathematical meaning. Likewise located in interpretive classroom research but with an additional focus on the special nature of mathematical knowledge, the work of Steinbring (2005) focuses on the interactive construction of mathematical knowledge through classroom interaction and signification. This line has been present at each CERME since the first, building, applying and adapting Steinbring's (2005) epistemological perspective on class interaction.

Although we have tended to refer to the focus of the TWG as 'language', it is relevant to recognise that mathematical communication uses a variety of modes, of which the linguistic is only one. There are specialised modes, especially suited to mathematical activity, including algebraic notation, Cartesian graphs, geometric diagrams and other symbolic and diagrammatic forms used in specific areas of mathematics. In addition to these, studies of face-to-face communication indicate the roles played by gesture and non-verbal language in doing mathematics. The study of multimodal communication has developed in recent years, stimulated in part by the transformations effected by the growth of new forms of communication technology. This development is reflected in the TWG, incorporating multimodal analyses of classroom communication and an as yet small number of studies looking at communication mediated by technologies.

3 What have we learned from mathematics and language research?

While complexity is relatively low near the origins of ERME with language viewed as a system of symbolic structures and a focus on classroom interactions, the complexity rises when researchers take account of the cultural and historical conditions of the researched environment, and it becomes higher when they attend to the social foundations of language and mathematical activity. Along the continuum, the social becomes less subordinated to the study of culture and cultural patterns. The ERME domain has thus experienced progress in parallel with the expansion of the social turn in the field and the understanding of mathematics classrooms as cultural and historical configurations. Drawing on the distinction between the

200 Planas, Morgan & Schütte

TABLE 14.1 Elements in the expansion of the ERME domain

ERME domain	Objects of study/themes	Conceptualisations of language
Mathematics and language research	The language of the learner The language of the teacher/class The language of mathematics	System → culture → discourse

language of the learner, the language of the teacher/classroom and the language of mathematics, we put each theme in relation to major conceptualisations of language as system, as culture and as discourse (see Table 14.1). This organisation allows us to articulate the complexity of the ERME domain in terms of the relationship between complexity and continuity over time. Each theme involves some continuum of complexity relative to the linguistic, cultural and social components progressively addressed.

Language as system refers to the focus on the semantic and therefore grammatical potential of pre-given linguistic systems brought into play in the interaction (e.g. Rowland, 2002). Language as culture challenges the attention to formal aspects and considers the relations between language and forms of action produced in a context (e.g. Edwards, 2007). Language as discourse further challenges the idea of locality to consider the relations between what we do with language in a context, our interpretation of that context and our reading of the social activity of the people in it (e.g. Morgan & Alshwaikh, 2010).

3.1 The language of the learner

We identify two lines of interest that have emerged through developing theoretical understanding of language as socially founded and of learning as discourse change. Complexity arises alongside discussion of the social and cultural conditions of mathematics learning in the classroom, and of how understanding these can contribute to understanding mathematics learning. Some of the papers pay attention to the diversity of languages involved in the learning process and negotiation of meaning, while others pay attention to the language-in-context of the learner. All, however, share an emphasis on the contextual conditions needed for mathematics learning to take place (Krummheuer, 1999). The learner is someone who needs to learn 'the language of mathematics', which requires access to and use of other languages and discourses of the classroom.

Discourse of the learner

By discourse of the learner, we mean the multiple uses of language that coincide in the learning process and through which the learner communicates realisations of this process. Within this frame, ERME studies differ not only in the notion of discourse they adopt, but also in the level of explicitness about their theoretical

Mathematics education and language **201**

tools and how these are used to produce methods for analysing discourse and discourse change. Some studies relate the idea of language-in-use to the interaction of the learner with the material world. Fetzer and Tiedemann (2015) examine how the discourse of the learner is made of discursive interactions with people and with objects. This implies redefinition of the social nature of the discourse of the learner to include objects as actors affecting the use of language for mathematics learning. Thus, discourse is more than what occurs between people in the form of verbal, written and other forms of symbolic communication. Mathematics learning emerges in the possibility of interacting with objects and abstracting from empirical realities. Although much research into language use still relies on analyses of written transcripts of recorded talk, these authors provide multimodal ways of transcribing video data for analyses of the interaction with objects.

Adopting an interactionist viewpoint, the critical correspondence between explicitness and implicitness in processes of developing conceptual understanding in the mathematics classroom has been investigated. Erath and Prediger (2015) address the question of how students learn to participate adequately in classroom mathematical practices through interaction regulated by explicit and implicit norms. Analysis of verbal interaction in the culture of the mathematics classroom reveals students who are involved in the performance of implicit norms about mathematical explanations. The discourse of the learner develops by participation in discursive practices, including ways of explaining, proving or defining mathematical concepts. It is interesting to note how most of the discursive practices in which the learner is expected to participate take place without detectable occurrence in spoken discourse. Implicitness thus appears as a condition of learning. Nevertheless, the learning opportunities vary depending on how and how much these discursive practices are communicated in visible ways in the discourse of the learner.

The non-verbal dimensions of language and the confluence of space and language in signed communication have been the focus in Krause (2017) regarding the discourse of the deaf learner in the mathematics classroom. The embodiment framework illustrates the interest in the analysis of non-verbal discourse and movement between verbal and non-verbal communication. Considering the discourse of the deaf learner, with signs and gestures produced in social interaction, opens up questions about the multimodal nature of the discourse of all learners. More generally, by understanding the learning processes of deaf learners, we may be in a better position to understand mathematics learning.

Although theories of orality and spoken languages in classroom-based research have dominated ERME research, the study of signs, gestures and particularly signed languages in the discourse of the learner has begun to come into focus. However, we find fewer papers centred on the written discourse of the learner and theoretical aspects of 'writtenness' in the mathematics classroom. One example is Schreiber's (2006) research into an internet-chat-based dialogue, which attends to differences between written data in the chat and spoken data collected during small group work. This experimental work suggests a way of interpreting the relationship between orality and writtenness as a social relationship with impact on mathematics learning. One finding is precisely that the concepts, theories, habits and competences of the participants are decisive for the emergent problem-solving and learning process.

Multilingualism in mathematics teaching and learning

In the early years of ERME, only a few papers addressed the issue of language diversity in mathematics teaching and learning. Where language diversity was an issue, most papers focused on linguistic aspects of mathematics that bilingual learners have to address. In the last decade, several papers have dealt with the experience of language diversity by the learner in more nuanced ways. Although there is not a unified theoretical approach to language diversity, recent work in sociolinguistics is present. Diversity refers to the languages of the learners as they interact with mathematics but also to the languages for communication: official languages of instruction, languages of teaching, and languages of thinking and learning. Learners of mathematics might switch from one language to another for different moments of communication in a lesson and combine aspects of these languages for different purposes. It is thus problematic to perpetuate discourses of monolingualism in the understanding of mathematics teaching and learning. Some studies are located in the transition between deficit perspectives on multilingual learners and views of language diversity as an asset for mathematics teaching and learning. The deficit perspective on multilingual learners is still present, though strongly contested nowadays, with language increasingly seen as an asset rather than a handicap. Far from focusing on obstacles for vocabulary, oral fluency and understanding in the language of instruction, we find studies centred on the resources that the languages of the learners bring to mathematics learning. Chronaki, Mountzouri, Zaharaki and Planas (2015) interrogate implications of the construction of the deficient multilingual mathematics learner. The case of a child whose dominant language differs from the language of the teacher reveals this child's participation in negotiation of numerical meanings. The support for flexible language use facilitates all children's engagement with diverse meanings for numbers. This study challenges taken-for-granted 'truths' about who is the competent learner of mathematics in the multilingual classroom, whose mathematics is valuable, and which discourses sustain language policies, curricular decisions and didactic actions.

Barwell (2015) also addresses the social dimension of language in studying multilingual learners in a way that challenges many common assumptions. This author draws on contemporary sociolinguistics of multilingualism to analyse the bilingual mathematics classroom, particularly on the notions of heteroglossia and orders of indexicality. The diversity of languages and the social diversity of speech types within any language, translated as heteroglossia, are stressed. Barwell suggests that the construction of mathematical learning in multilingual settings is often guided by views about languages and their speakers, rather than views of mathematical competence, performance and achievement. Other authors have developed from focusing on language forms and devices in the multilingual mathematics classroom to considering the social dimension of multilingual mathematics learning. This is the case of Poisard and colleagues, who have expanded their initial psycholinguistic frames of language. In Poisard, Ní Ríordáin and Le Pipec (2015), we find a move toward recognition of the relevance of other influences, such as the culture of the mathematics classroom and the discourses at large in society. There is reflection on some of the compensatory responses in interpreting the needs of students whose home languages are different from the language of instruction. They note that some research has shown the positive pedagogic effect of using the languages of the learners in the multilingual mathematics classroom. This is in line with views of language as pedagogic resource and language use as cultural and social practice.

3.2 The language of the teacher/classroom

The previous section examined mathematics education and language in relation to learners; this section provides a change in perspective. Mathematics learning takes place in different social settings. Often in the interaction, one or more participants have more advanced skills, for example teachers or parents. In this context, the focus falls upon the language of such individuals (here briefly called teachers, even if including kindergarten teachers and others) and upon the language in the classroom or kindergarten. Studying the interpersonal functioning of language can contribute to understanding social aspects of mathematics education (Steinbring, 2005), including how teachers manage classroom interactions and how students from various social and cultural groups gain opportunities for mathematics learning in the 'learning spaces' structured by teachers and peers.

The standpoint that mathematical activity is socially originated and developed is central to most of the research concerned with classroom language. The earlier expansion of the linguistic paradigm was brought into clearer focus within the discussion of studies using interactionistic approaches of interpretive classroom research. These studies were distanced from the previously dominant view that learning was merely an internal psychological phenomenon. Thereafter, the inclusion of interactionistic aspects of learning and teaching meant a shift of focus from the structure of objects to the structures of learning processes, and from the individual learner to the social interactions between them. The transformed understanding of learning led to the development of theories that regard meaning, thinking and reasoning as cultural products of social activity. Based on the assumption that meaning is negotiated in interactions between individuals and that social interaction is thus to be understood as fundamental for learning processes, language can no longer be understood only as the medium in which meaning is constructed. Rather, speaking about mathematics in collective argumentations is to be seen as the doing of mathematics and the development of meaning. Thus, language acquires central significance in the building of mathematical knowledge and mathematical thought.

204 Planas, Morgan & Schütte

One can find numerous studies from the early days of ERME that focus on children's participation in classroom interaction. This focus is connected to the aim of these works to primarily understand, rather than change, children's learning processes. Krummheuer (1999) examines the relationship between students' participation in argumentative processes and their individual content-related development. Using transcripts from two research projects to reconstruct aspects of narrativity in interactional processes in the classroom, he emphasises a 'folk psychology' of learning, where learning is conceptualised as a social process of cultural co-creation. In Price (1999), there is equal emphasis on understanding and change, indicating opportunities for the teacher to support mathematical learning. Price addresses the social nature of learning by analysing a transcript of a simple addition exercise in a group of children aged 4–5. She shows the importance of teachers using examples from everyday experience to promote children's learning, pointing out that although mathematical concepts such as addition are essentially abstract, they should not be taught only in an abstract way. Rowland (2002) also adopts the interactionistic focus on language in mathematics teaching. He examines utterances of two 10-year-old pupils discussing a problem with a teacher and notes that language has an interactional function, expressing both social relationships and inner attitudes. He argues that linguistic means can be used to analyse social and affective factors in mathematics teaching. Edwards (2007) focuses on participation in classroom mathematics learning and places the emphasis on learning in small groups. Reporting on collaborative classroom group work, her findings suggest that groups self-selected by pupils on the basis of friendship and trust produce dialogical reasoning and exploratory talk. This supports the idea of social interaction as a means toward cognitive change.

Jung and Schütte (2015) investigate to what extent the linguistic discourse in kindergarten and primary school gives children the opportunity to achieve mathematics-specific discursive competences that allow them to participate in the discourse of the mathematics classroom. This contribution illustrates an increasing trend to focus on the potential of improving conditions for learning mathematics. The teacher and the teacher's language become increasingly central in studies toward more optimal conditions for mathematics learning. Schütte (2006) analyses the linguistic accomplishment of instruction in a class. His results support a hypothesis of limited learning opportunities for a multilingual pupil body in classes because the linguistic accomplishment of the teacher orients itself toward perceptions of unity of a monolingual 'normal' child and the diversity is barely considered.

With the change of focus from the learner to the language design of classroom interaction of the teacher and the interactive interdependence between all participants, starting in the early 2000s, special emphasis has been given not only to the description of learning processes but also to demonstration of potential change or even initiation of these changes. Tatsis (2011) shifts the focus from the identities of learners supported by teachers to those of the teachers. He looks at the importance of language in the narratives that define teachers' identities, arguing that these identities are useful in understanding teachers' relationship to their actual practice

Mathematics education and language   **205**

and to the practice that they would expect to perform in the future. Through observation and analysis of teachers' participation in a training course, he finds that their identities and stories emerge from first-, second- and third-person narratives in verbal and written contexts.

Because of the increasing diversity of student populations, all places of learning inside, outside or before school – whether with an individual with advanced skills in the interaction, as in the classroom conversation, or in small groups without such an individual – will increasingly be characterised by a plurality of interpretations in negotiations of meaning. It is of particular importance to note that mathematical language is itself diverse.

3.3 The language of mathematics

In this section, we review work that has addressed the relationship between language and mathematics, describing both the forms and functions of language-in-use. A source of complexity in this area is the developing breadth and sophistication of the conceptualisation of language itself. It has long been recognised that any consideration of mathematical language needs to take account of the specialised forms of communication distinctive of written mathematics, in particular algebraic notation (Pimm, 1987). So-called 'natural' language has been an object of study throughout the period, both in oral interactions and in written texts. In the early years of ERME, the orientation toward classroom interaction meant that the majority of research focused on spoken language. While transcriptions of classroom episodes sometimes included mention of gestures, artefacts or writing, these tended to be treated as contextual information and their roles in mathematical communication were not analysed. Reflecting the development of fuller theorisation of multimodal communication in the fields of linguistics and semiotics as well as in mathematics education, the scope of the group has come to incorporate a wider range of communicative modes, including gestures, diagrams and the multiple modes offered by new technologies. While forming rigorous descriptions of non-linguistic modes has been an essential part of expanding the conceptualisation of the language of mathematics, the main focus of research has been on how (multimodal) language functions in the construction of mathematical knowledge, and how use of various modes of communication contribute to support mathematical reasoning. Bjuland, Cestari and Borgersen (2007) studied how students and teacher combine their use of gestures and verbal language while interpreting a Cartesian diagram. They distinguished pointing and sliding gestures and identified how students integrated these with verbal language as they reasoned about the mathematical situation, using discursive strategies such as comparison or coordination.

The adoption of discourse perspectives on language has introduced further complexity. Within such perspectives, language (including multiple modes) is not conceptualised merely as a means of communication or as a tool for doing mathematics but as constitutive of the mathematics itself. Analysis of language use in a classroom interaction or a written text can thus illuminate the nature of the mathematics that is made available for students to experience. One distinction between types of school mathematics discourse focuses on how students might construe their position with regard to mathematical activity: whether they are invited to engage in creative intellectual activity and to see mathematics as involving making decisions and choices, or whether they are subject to an external authority that presents mathematical knowledge to be received as unquestionable. This distinction is made in Stamou and Chronaki's (2007) analysis of a mathematics magazine for lower secondary students. The authors note the interdiscursivity of the texts they study – that is, the way that linguistic characteristics typical of one discourse, in this case the 'traditional' authoritative discourse, are incorporated into texts that appear to be within another, 'progressive', discourse. They identify this as a possible source of confusion rather than providing students with access to mathematics. Interdiscursivity, the mixing of resources from different discursive practices, is the focus of another strand of interest, albeit not always addressed from an explicitly discourse theoretic standpoint: the movement between 'everyday' and mathematical forms of language or, to use sociolinguistic terms, between colloquial and literate registers.

Functioning of language in the construction of mathematical knowledge and reasoning

The study of how language functions mathematically has varied from analysis of single signs and their use to studies of the qualities and purposes of whole genres, such as Misfeldt's (2007) study of the roles of different genres of writing in the practices of mathematicians. The work of Steinbring and researchers influenced by his epistemological perspective on classroom interaction provides insight into the roles specific words, symbols or diagrams play in children's construction of new mathematical concepts. This perspective emphasises that relations between representations and concepts are mediated by the 'reference context', including knowledge and experiences of the children, and hence may vary between individuals with different prior knowledge and may change as the reference context develops to include new knowledge. Nührenbörger and Steinbring (2007) explain how the interpretations by a teacher and two children of the decomposition of 8 into 4 + 4 varied because of differences in their reference contexts. Although the teacher used mathematical principles to explain why 4 + 4 should appear only once in a list of decompositions, the younger child persisted in interpreting two occurrences of 4 + 4 as distinct, referring to differences in the notation used rather than to the mathematical objects. Steinbring's framework has served various analyses of classroom episodes involving children working on tasks. While this epistemological perspective provides insight into the role that signs play in forming children's mathematical concepts, other studies have revealed the power of communicational modes to transform mathematical reasoning. Consogno (2006) introduces the notion of the semantic-transformational function of written language to argue that the dynamic process of production and reinterpretation of a text contributes to mathematical reasoning. She shows how, while writing their solutions to a problem, a process of linguistic expansion leads students to associate new words and meanings with the key words of the problem situation, thus shaping the direction of their reasoning. Using a discursive perspective, Morgan and Alshwaikh's (2010) multimodal analysis of an episode of problem solving in a technologically rich environment demonstrates how students' use of language and other modes of communication affect their approach to the problem. The variety of perspectives in the study of how language functions in mathematical activity provides a range of explanatory frameworks but also achieves strong evidence and a powerful consensus that language communicates mathematical activity and influences its trajectory.

A related strand of research, as yet under-represented in ERME, addresses differences between the structures of various 'national' languages and how these might influence the mathematical activity of speakers of these languages. International interest in this topic has tended to focus on non-European languages. The work of Ní Ríordáin (2013) with bilingual students in Ireland begins to address this issue by relating variation in mathematical performance to characteristics of the English and Irish languages. While differences between European languages are generally less than those between the languages of Europe and of Asia, Africa and the indigenous languages of Australasia and the Americas, there is nevertheless scope for further research in this area. This might be of particular importance in light of the increased significance to educational policy of international comparisons based on tests translated into multiple languages.

Distinguishing mathematical language from everyday language

As discussed earlier, mathematics can be considered a discursive activity, using and manipulating specialised discursive resources (language, notations, diagrams, etc.) in distinctive ways. Mathematics education, however, is a hybrid activity, involving pedagogic and mathematical communication. The objective of mathematics education can be seen as induction of students into mathematical activity (and mathematical ways of communicating) rather than as simply doing mathematics. The language used in mathematics education thus inevitably includes non-mathematical and mathematical characteristics. This phenomenon is not unique to mathematics; learning in any specialised practice involves learning to use the specialised language of the practice. Distinguishing mathematical from non-mathematical forms and studying how these function in mathematics classrooms has been a strand of ERME research.

Pedagogic strategies frequently involve making connections between mathematics and familiar 'everyday' artefacts or problem situations. Whether these connections are intended as concrete support for developing mathematical concepts and procedures, as motivation for engaging in mathematics or as a form of application of mathematics through modelling and problem solving, the combination and coordination of the everyday and the mathematical also involves using a mixture of everyday and mathematical language. This juxtaposition might appear as a source of confusion and difficulty or as a means by which mathematical knowledge comes to be constructed. During a lesson in which primary school children were measuring and mixing ingredients to make waffles, Rønning's (2010) semiotic analysis of talk about fractions, decimals and measurements of volume suggests that the numbers and measurements given in the written recipe and marked on artefacts such as milk cartons and measuring jugs were interpreted differently by the teacher and by the children. For the teacher, marks such as '¼ litre' and '15 dl' formed a connected chain of signs, linking the practical activity to the mathematical activity. The children did not make connections between these signs but instead found practical solutions to the problem of mixing a batter of the right consistency, solutions that did not necessitate use of numerical measurements or calculations.

Connections and disconnections between everyday and specialised mathematical language can also occur when specific words or other communicative elements have potential to be used for making either everyday or mathematical meanings. Some of the authors have discussed differences between teacher and student use of apparently similar words and gestures in the context of the description and construction of mathematical objects in the classroom. Albano, Coppola and Pacelli (2015) use the general distinction between colloquial and literate registers (originating in functional linguistics) as a lens to analyse and discuss errors made by university students on a task involving graphs and analytic properties of functions. The components of the written answers could be said to be elements of specialised mathematical language, but were frequently used in ways characteristic of a colloquial rather than a literate register. The students, for example, evoked the local context of situation rather than general conventions of mathematical notation and treated graphs as iconic rather than symbolic representations. While use of the literate mathematical register is necessary to support mathematical thinking, the colloquial register also plays an essential role in supporting conceptual development. They conclude that the skill of moving between colloquial and literate registers needs to be developed and fostered from an early age by planned teaching activities – a conclusion echoed by other researchers in the field. Studies such as these provide insights into sources of apparent difficulties, misunderstandings and errors, locating these in the structures of mathematical and non-mathematical activities and the properties of language associated with those activities, rather than seeing them as arising from deficiencies in the students. The delineation of lexical, grammatical and structural characteristics of mathematical language developed by researchers involved in the TWG contributes to the knowledge required to underpin teaching that will help students develop skills in distinguishing between everyday and mathematical forms of language and moving between them.

We have exemplified papers situated within the different conceptualisations of language. For example, Krause (2017) and Ní Ríordáin (2013) mostly conceptualise language as system through the respective foci on the potential of a sign language structure and on two oral grammars. In Jung and Schütte (2015) and in Rønning (2010), we see the conceptualisation of language as culture in the respective foci on the relations between language and forms of talking mathematics in kindergarten and early primary school, and between language and forms of talking fractions in the resolution of a problem with everyday artefacts. Finally, in Chronaki et al. (2015) and in Tatsis (2011), we see the conceptualisation of language as discourse in the respective foci on how either learners or teachers view their contexts of language use and the people engaged in the activity there.

4 What more could we learn in the next decades?

We have discussed the progress and vitality of the ERME domain of research in mathematics education and language. Nonetheless, little is still known about many other aspects, e.g. how language is influenced by new technologies that enable new discourse practices (oral, chat, computer-mediated graphics, gestures . . .) and give rise to new questions: Do the newer tools change the ways people speak and write? Do they reflect established patterns of verbal interaction? How do we conceptualise the relationship between conventional forms of verbal interaction and communication and those mediated by new technologies? Little is also known about the ways in which methods and findings from the domain can be applied to mathematics teacher education and professional development. Past research has established the connection between teachers' pedagogical knowledge and experiences of professional development with little attention to issues of language responsiveness in teaching. Working with practitioners who have successfully integrated multilingual and multimodal practices in their classrooms would help.

There is also energy needed to address some practices within our research community. In a domain where language is at the core of the agenda, the ethics and practices of power involved in the use of language by researchers remain surprisingly under-examined. Knowing what we know now, in a period in which global information flows in English, we cannot expect that the question of English does not affect the domain. Researchers from a small number of countries conduct a majority of international work and English is the language with official status for this. The quality of the research experience is framed by how different languages and codes of communication are accepted, represented and acknowledged, particularly those of the participants, which might not even be known by the researchers. Analysis of how this situation influences empirical work is fundamental.

References

Albano, G., Coppola, C., & Pacelli, T. (2015). Reading data from graphs: A study on the role of language. *CERME9* (pp. 1326–1332).

Austin, J. L., & Howson, A. G. (1979). Language and mathematical education. *Educational Studies in Mathematics*, *10*(2), 161–197.

Barwell, R. (2015). Linguistic stratification in a multilingual mathematics classroom. *CERME9* (pp. 1333–1339).

Bjuland, R., Cestari, M. L., & Borgersen, H. E. (2007). Pupils' mathematical reasoning expressed through gesture and discourse. *CERME5* (pp. 1129–1139).

210 Planas, Morgan & Schütte

Chronaki, A., Mountzouri, G., Zaharaki, M., & Planas, N. (2015). Number words in 'other' languages: The case of little Mariah. *CERME9* (pp. 1347–1353).

Consogno, V. (2006). The semantic-transformational function of written verbal language in mathematics. *CERME4* (pp. 810–820).

Edwards, J.-A. (2007). The language of friendship: Developing sociomathematical norms in the secondary school classroom. *CERME5* (pp. 1190–1199).

Erath, K., & Prediger, S. (2015). Diverse epistemic participation profiles in socially established explaining practices. *CERME9* (pp. 1374–1381).

Fetzer, M., & Tiedemann, K. (2015). The interplay of language and objects in the mathematics classroom. *CERME9* (pp. 1387–1392).

Jung, J., & Schütte, M. (2015). Discourses in kindergarten and how they prepare for future decontextualised learning of mathematics. *CERME9* (pp. 1414–1420).

Krause, C. (2017). DeafMath: Exploring the influence of sign language on mathematical conceptualisation. *CERME10* (pp. 1316–1323).

Krummheuer, G. (1999). The narrative character of argumentative mathematics classroom interaction in primary education. *CERME1* (pp. 331–341).

Misfeldt, M. (2007). Idea generation during mathematical writing: Hard work or a process of discovery? *CERME5* (pp. 1240–1249).

Morgan, C. (2013). Language and mathematics: A field without boundaries. *CERME8* (pp. 50–67).

Morgan, C., & Alshwaikh, J. (2010). Mathematical activity in a multi-semiotic environment. *CERME6* (pp. 993–1002).

Ní Ríordáin, M. (2013). A comparison of Irish and English language features and the potential impact on mathematical processing. *CERME8* (pp. 1576–1585).

Nührenbörger, M., & Steinbring, H. (2007). Students' mathematical interactions and teachers' reflections on their own interventions. *CERME5* (pp. 1250–1269).

Pimm, D. (1987). *Speaking mathematically: Communication in mathematics classrooms*. London: Routledge.

Poisard, C., Ní Ríordáin, M., & Le Pipec, E. (2015). Mathematics education in bilingual contexts. *CERME9* (pp. 1468–1474).

Price, A. J. (1999). It is not just about mathematics, it is about life. *CERME1* (pp. 364–374).

Rønning, F. (2010). Tensions between an everyday solution and a school solution to a measuring problem. *CERME6* (pp. 1013–1022).

Rowland, T. (2002). Pragmatic perspectives on mathematics discourse. *CERME2* (pp. 408–419).

Schreiber, C. (2006). Semiotic processes in a mathematical internet-chat. *CERME4* (pp. 903–912).

Schütte, M. (2006). Interaction structures in primary school mathematics with multilingual student body. *CERME4* (pp. 913–923).

Stamou, A. G., & Chronaki, A. (2007). Writing mathematics through dominant discourses: The case of a Greek school mathematics magazine. *CERME5* (pp. 1311–1320).

Steinbring, H. (2005). *The construction of new mathematical knowledge in classroom interaction: An epistemological perspective*. New York: Springer.

Tatsis, K. (2011). Language as a shaping identity tool: The case of in-service Greek teachers. *CERME7* (pp. 1376–1385).

15

DIVERSITY IN MATHEMATICS EDUCATION

Guida de Abreu, Núria Gorgorió and Lisa Björklund Boistrup

1 The emergence and development of a Thematic Working Group on diversity in mathematics education

This chapter reviews and reflects the development of the CERME Thematic Working Group (TWG) on 'Diversity in Mathematics Education'. The name of this group has been transformed and extended over the years, as a reflection of the change and expansion of the interests of its members. Thus 'Teaching and Learning Mathematics in Multicultural Classrooms' at CERME 3 (proceedings published in 2004) has been progressively transformed into its present name 'Diversity in Mathematics Education: Social, Cultural and Political Challenges'. To illustrate this development, this section summarises how the interests of the group have expanded throughout the years.

The centrality of culture in the doing, thinking, learning, and teaching of mathematics has been discussed by many scholars in CERME meetings since they started. Already in the proceedings of CERME 1, before the creation of the group, we find many references that consider several aspects related to culture, from mathematics as a cultural product (Arzarello, Dorier, Hefendehl-Hebeker, & Turnau, 1999), to mathematical learning as being co-constructed by culture, and to the culture of mathematical classrooms (e.g. Krummheuer, 1999). Similarly, in his keynote address in CERME 1, Jeremy Kilpatrick (1999) pointed out that the increased multicultural and multilingual composition of many classrooms in many countries called for new research. At the next congress, CERME 2, the challenges associated with multicultural, multiethnic, multilingual aspects of mathematics education were addressed in several papers, for example Krummheuer (2002) who discussed the challenges in relation to both theory and methods.

Bishop, Clarkson, FitzSimons, and Seah (2002) also contributed to the discussion, stressing the importance of values at personal, institutional, social, and

212 Abreu, Gorgorió & Björklund Boistrup

cultural levels stating that "at the cultural level, the very sources of knowledge, beliefs, and language, influence our values in mathematics education. Further, different cultures develop different values" (p. 370). Around the same time, the sudden increase in levels of migration in many European countries contributed to the visibility of the cultural, ethnic, and linguistic diversities in mathematics classrooms, and several research projects focusing on these issues emerged (e.g. Favilli, Oliveras, & César's (2004) research in Italy, Spain, and Portugal; Gorgorió, Planas, & Vilella's (2002) research in Catalonia; and Alrø, Skovsmose, and Valero's (2004) in Denmark). This provided the impetus for the foundation of the TWG 'Teaching and Learning Mathematics in Multicultural Classrooms' to become a forum for European researchers involved with the topic area to share, discuss, and reflect on the challenges and types of research being carried out.

One of the key aspects discussed in CERME 3, and which persisted throughout the different meetings, is the realisation that in many European countries, teachers could expect to work with students from ethnic, linguistic, and cultural groups distinct from their own. Cultural, linguistic, political, and social issues, which are often seen as specific to using, teaching, and learning mathematics, for addressing situations where students are from cultures other than those regarded as mainstream have become central to many European classrooms. This has been reflected in papers presented from several European countries, such as Denmark (Alrø, Skovsmose, & Valero, 2004), Holland (Elbers & de Haan, 2004), Italy, Spain, and Portugal (Favilli, Oliveras, & César, 2004), and Germany (Kaiser, 2004). Another key aspect noted early on, was that 'multicultural classrooms' were not the only space for research: instead, other settings, and the transitions between those settings, had to be included (e.g., between the school and educational policies, the home mathematical practices, workplaces, etc.). This resulted in a change of the title to 'multicultural settings' (CERMEs 4 & 5; see Abreu, César, Gorgorió, & Valero, 2006), and of titles that have explicitly included diversity in mathematics education from social, cultural, and political perspectives since CERME 7.

The inclusion of 'political perspectives' in the title reflects the interests of researchers presenting their work in this group. Political aspects have been part of the TWG's work from the beginning, and have become a prominent theme in the last four CERMEs. This position rejects the naive idea of research as politically neutral, providing objective data that is used to guide policy making on a supposedly rational basis (Pais, Crafter, Straehler-Pohl, & Mesquita, 2013; see also Valero, 2013). In CERME 5, an example of this position was introduced by Stentoft (2007) who addressed methodologies of research in multicultural mathematics classrooms from the perspective of power relations between actors in the research. A recent example is Fyhn, Meaney, Nystad, and Nutti (2017) at CERME 10, who addressed culturally responsible teaching of mathematics in relation to indigenous (Sámi) teachers' self-determination. Political perspectives focus on how the broader political context of mathematics education (taken in a broad systemic sense, including more than mathematics classrooms) affects the teaching and learning of mathematics. Two examples derive from Sweden, where Bagger (2015) addressed the effects of national testing in grade three for students in 'special needs' classes, and where Boistrup and Keogh (2017) addressed 'workplace mathematics' and institutional norms in a nationwide in-service programme.

Political perspectives may also focus on how diversity among learners has consequences in terms of unequal access to the learning of mathematics. This research might include critical investigations of socioeconomic backgrounds, among other factors, of students as grounds for unequal mathematics education (e.g., Doğan & Haser, 2013), where one consequence is the sorting of students due to their socioeconomic backgrounds (e.g., Turvill, 2015). Another focus of research is on examining the tensions between 'official discourses' (positing inclusion and equity as fundamental goals) and the actual practices of mathematics teaching (which might actually perpetuate inequities) (Straehler-Pohl & Pais, 2013).

2 Meanings attached to diversity in mathematics education

In the social sciences, it is acknowledged that the challenges that social and cultural diversity poses to education have many facets, and these have been studied from different approaches (Abreu, 2014; Abreu & Crafter, 2016; De Haan & Elbers, 2008). Conceptions of the role of the social and the cultural in processes of learning inspire these different approaches, and consequently the different meanings of diversity explored in research. This is also the case for mathematics learning and education.

Thus, despite diversity being of interest to the members of the group, the meanings attached to it have been multiple from the beginning of the group, and remain so. Key meanings of diversity that reflect the patterns of the research presented in CERME include:

- cultural, ethnic, social, and linguistic backgrounds of school students and their school experiences taking into account: (a) increased numbers of students with immigrant backgrounds in schools, and classrooms which have changed from mono- to multicultural, multiethnic, and multilinguistic composition of the classroom population; and increasing gaps between schools in terms of socioeconomic factors; (b) differences in school levels of achievement of students from non-mainstream ethnic, social, economic, and cultural backgrounds (e.g. some minority groups achieving significantly lower grades, and sometimes higher grades, than the majority group, as reflected in the education statistics in several countries); (c) other forms of student diversity such as gender, level of achievement in school mathematics, and their like or dislike of school mathematics;
- perspectives and experiences of diversity, taking into consideration teachers' students', and parents' perspectives as well as school/institutional perspectives and policy and political perspectives;
- the focus of discourses used to discuss diversity: (a) diversity as a problem; (b) diversity as a resource.

A working definition of diversity that includes these aspects was introduced by Valero, Crafter, Gellert, and Gorgorió (CERME 7, 2011), and further elaborated by Boistrup, Meaney, Mesquita, and Straehler-Pohl (CERME 9, 2015). In this definition of diversity, they included:

- *diversity of people*: of students, teachers, parents, and many other participants in mathematics education – with the diversity even more refracted through aspects such as gender, ethnicity, culture, language, social and socioeconomic status, disability, qualification, life opportunities, aspirations, career possibilities, etc., that shape their acting, interacting, valuing and identities, – affected and framed in and by various contexts;
- *diversity of contexts which both frame and affect all actors*: this includes the formation of policies informing mathematical education, the sites where mathematics education takes place, and the differences in the organisation and structure of practice in such contexts – in schools, homes, workplaces, etc.;
- *diversity and possibilities of practice*: due to the concrete situations of mathematics education in which multiple diversities may intersect, posing challenges to actual learning and teaching practices, as well as a basis for rethinking what is possible.

This definition of diversity thus includes considerations for research around theoretical approaches to diversity (What is diversity in the context of mathematics education and mathematics education research?), and around the engagement with diversity in educational practice and research (What are the challenges and possibilities emerging from increasing levels of diversity?).

Another key observation is that, despite the multiple sources and perspectives in the study of diversity, this TWG is united in rejecting views and practices of 'diversity as a problem' or 'diversity as a deficit', and in working toward developing ways of addressing diversity as a resource. The focus on addressing diversity as a resource takes many forms in the group's research. For example, some studies examine the discourses of diversity embedded in the practices and policies of schooling; some studies examine subjective views and experiences of diversity by learners, teachers, and parents; and other studies examine the possibilities of new school practices that draw on diversity as a resource.

3 How different diversities have been theorised and empirically addressed

The multiplicity of understandings of diversity also impacts on the way it has been theorised and analysed by TWG participants. It is not only that diversity is polyhedral in itself, but that it can also be seen from different perspectives that make visible particular faces such as the cultural, the social, the political, and the linguistic.

Diversity in mathematics education **215**

3.1 Theorising when researching diversity

When the group began, the theorising was dominated by approaches drawing on the cultural nature of mathematics (Bishop, 1988), cultural psychology (Cole, 1996), critical mathematical education (Skovsmose, 2014), and ethnomathematics (D'Ambrosio, 1985). As any one of these approaches was already too broad in itself, while at the same time the number of European researchers working in each area was relatively small, it was difficult at times to foster productive dialogue within the group.

However, over time, despite the different ways of theorising diversity, it was apparent that the group shared an interest in understanding the key processes in learning and teaching in the context of diversity. These include, for example, an interest in understanding identities (e.g., Abreu, 2006; Black, Solomon, & Radovic, 2015), agency (e.g., Andersson, & Norén, 2011), social representations of mathematical knowledge (e.g., Abreu & Gorgorió, 2007; Gorgorió & Prat, 2011), cultural representations of mathematical knowledge (e.g., Crafter, 2010; Mukhopadhyay & Greer, 2015), discourses of diversity (Alrø, Skovsmose, & Valero, 2006; César & Favilli, 2006), home and school mathematical practices (Abreu et al., 2006), and transitions between mathematical practices (e.g., Abreu, Crafter, Gorgorió, & Prat, 2013).

This focus on the processes brings some unity to a group that is truly multidisciplinary, and benefits from drawing on and developing a sophisticated theoretical understanding in the field of mathematics education, and social sciences more generally. Focusing especially on the most recent CERMEs, for example CERMEs 7, 8, and 9, we notice that the theorising has drawn on a variety of theoretical approaches:

- *Sociocultural psychology*: sociocultural theories of learning and development evolving from Vygotsky and European social representations theory is one family of theoretical approaches informing many studies (e.g., Abreu & Gorgorió, 2007; Crafter & Abreu, 2011; Newton & Abreu, 2011). These also include cultural historical activity theory which is one strand of sociocultural theory that has evolved from the work of Vygotsky (e.g., Gebremichael, Goodchild, & Nygaard, 2011), the dialogical self (Abreu et al., 2013; Newton & Abreu, 2013), and dialogue – drawing on Bakhtin's ideas of dialogism (Rangnes, 2011).
- *Discursive and sociological approaches*: the notion of discourse in a sociological sense, along with other concepts, has been adopted to explore the social construction of what counts as mathematical knowledge, identity positioning, and issues of equity. For example, Lange and Meaney (2011) examined how public discourse may construct disadvantage; Gellert and Straehler-Pohl (2011) draw on Bernstein's differences between horizontal and vertical discourse, where the concepts of discourse and knowledge are closely interrelated. Johansson

216 Abreu, Gorgorió & Björklund Boistrup

and Boistrup (2013) use Bourdieu's concepts habitus and field to investigate signs of mathematical aspects of a person's workplace competence; and Turvill (2015) uses Bourdieu's notions of social and cultural capital to explore the way mathematics education systematically disadvantages particular groups of children.

- *Culture and mathematics education*: notions from ethnomathematics (e.g., Stathopoulou, François, & Moreira, 2011; Mukhopadhyay & Greer, 2015), and critical mathematics education (e.g., Alrø et al., 2004, 2006; Hauge et al., 2015) were concerned with the sociopolitical dimension of mathematics education that also informs many studies. Some of the participants based their research within one of these two approaches (e.g., Domite & Pais, 2010), whereas others sought an articulation between the two (e.g., François & Pinxten, 2013).

3.2 Researching diversity

Similarly, and for the same reasons, the ways diversities have been addressed empirically are multiple. The settings of these studies were varied, including classrooms, schools, communities, institutions, and countries. Within these settings, the focus was on learners, teachers, parents, and professionals from the perspective of the role that various diversities play in the construction of mathematical learning, teaching, practices, and uses.

The main way these questions have been pursued was essentially through qualitative approaches (see Seah, Davis, & Carr, 2017 for an exception), drawing on interpretive frameworks. In many studies, the approach is described as qualitative, with the focus on interpretation. However, other studies clearly situate their approaches within social sciences traditions, including:

- *Ethnographic approaches*: Ethnography is a popular approach that reflects the shared interest of the group in research that fundamentally aims at uncovering the meanings and experiences of diversity located in sociocultural contexts. A methodological approach developed by anthropologists as means to understand and describe 'other cultures', it was adopted by sociologists to investigate cultures perceived as other within western societies, and by social and cultural psychologists to investigate the role of culture and social contexts on psychological functioning. In this way, it is an approach that has been combined with different theoretical approaches (see, e.g., CERME 7 papers by Andersson (2011), Crafter & Abreu (2011), Díez-Palomar & Ortin (2011), and Stathopoulou et al. (2011); also CERME 9 papers by Bagger (2015), Parra-Sánchez (2015), and Radovic, Black, Salas, & Williams (2015)).
- *Discourse approaches*: Interest in discourse approaches has been increasing in recent years and reflects an interest in a methodological stance that uncovers the social and political constructions of what counts as appropriate mathematical practices, issues of inclusion and exclusion, and processes of identity

Diversity in mathematics education   **217**

development and positioning (see, e.g., papers presented in CERME 7 by Andersson & Norén (2011); Gellert & Straehler-Pohl (2011); Lange & Meaney (2011) and the papers presented at CERME 9 by Bagger (2015), Radovic et al. (2015), and Montecino & Valero (2015)).

- *Dialogical and narrative approaches*: These approaches are emerging from an interest in understanding the way the person (student, teacher, parent, actors in workplaces, etc.) develops their participation in mathematical practices, and by examining the dialogues between identity positions (associated with different times and past, present, and future identities; different settings, such as home and school; or different roles, such as teacher and parent) (see, e.g., the papers presented at CERME 8 by Abreu, Crafter, Gorgorió, & Prat (2013); and Newton & Abreu (2011)).

Within these various research approaches, strategies to collect data have included a variety of methods, such as observations, interviews, questionnaires, and narratives. The review of methods clearly shows a lack of longitudinal studies (an exception is Bagger, 2015) as well as surveys. This confirms the fact that the methods and strategies used not only reflect the users' interpretations of diversity, but also their constructions of its meaning. Moreover, it gives a clear image of how funding (or the lack of it) affects certain research domains.

4 Conclusions and looking toward the future

Overall, this review shows that, on a superficial level, there have been times in this TWG when the research presented gave the impression of a collection of papers more like a patchwork than a single united piece. In fact, the difficulty in establishing coherence between the papers presented in the group was noted in several of the coordinators' reports. This retrospective review, however, reveals that, on a deeper level, the research presented over these years does reflect a shared common interest in the directions of the theorisation of mathematics learning and the related empirical research; namely, as a human activity and sociocultural practice located in historical and political contexts.

We suggest that three specific turns unite the research on diversity in mathematics education presented at CERME. The *first turn* focused on establishing the cultural nature of mathematics knowledge and learning, and it was informed both by ethnomathematical and sociocultural approaches to mathematical cognition. This resulted in an emphasis on understanding *differences* in mathematical practices, such as differences between in-school and out-of-school mathematics, and the situated nature of mathematical cognition. One key contribution of this turn was to acknowledge *diversity* as part of both, with an emphasis on psychological functioning and also an account of the uses and learning of mathematics as socioculturally and politically located in the context of specific practices. Evidence from research has shown a *discontinuity* between the ways a person has learned or expressed mathematical competence in school and in their out-of-school practices. Research has shown that being competent in one practice, such as school mathematics, does not always predict how the individual will perform in another practice (such as in out-of-school practices). This evidence has been crucial in informing initial studies when the research work focused on multicultural classrooms, and attempted to explore reasons for the barriers in learning experienced by students of immigrant backgrounds without attributing these to any form of individual deficit.

Having achieved a more sophisticated understanding of the cultural diversity of mathematical knowledge, the researchers turned their attention to the social and political aspects of learning. This marks the *second turn in the research*, and is revealed in the focus on the role of the social and political contexts, such as social interactions, social representations, social institutions, power relationships, public discourses, etc. This added focus moved the understanding of constructions and experiences of diversity forward. Theoretically, this turn resulted in an interest in sociological and socio-psychological perspectives. Thus, as shown in many of the papers presented in the group, social valorisations, and social representations of what counts as school mathematics, embedded in dominant institutional discourses and in practices, might play a key role in the way diversity is experienced. In particular, this turn resulted in exploring the process of mathematical learning in terms of participation in mathematical practices, which involved both the psychological re-construction of forms of mathematical knowledge and skills (cultural tools), and in terms of constructions of identities. These approaches also highlighted tensions between 'official discourses', such as mandated policies and educational practices. A key insight from considering the social and the political is the realisation of the possibilities of forms of participation that construct diversities as a resource, and enable experiences of continuity, bridging, dialogue, and negotiation between practices and identities.

The conceptual clarity achieved with the examination of the roles played by the cultural and the social opened the path to a *third turn in researching diversity* in mathematics education. This third turn focuses on the person as a participant in multiple mathematical practices (Abreu & Crafter, 2016). This research is interested in exploring the trajectories of participation, and the mediating roles of identities, as they interact in social, cultural, and political contact zones. This includes participation across practices that co-exist in time (e.g., home–school) and over time (school–university, moving countries).

Finally, looking toward the future, we expect that research along the lines summarised above will continue. However, we also expect that new emerging themes will take priority. These could include the ethics of doing research on diversity, the problematisation of diversity and a critical reflection of diversity-focused research practices.

The ethics of doing research in relation to diversity of various forms has begun to be addressed (e.g., Eikset et al., 2017). Here, diversity might concern students, children, parents, teachers, classrooms, etc., and might be about culture, achievement, ethnicity, gender, social class, values, histories, or the like. This TWG is united in its striving for social justice, inclusivity, and variety. One consequence of engagement

Diversity in mathematics education **219**

in ethical considerations is reflexivity in research, where the researcher's acts are also critically observed and analysed.

The problematisation of diversity as a concept that is socially constructed is also a key theme that must continue informing research. During CERME 10, diversity as a concept, and the connotations thereof, were problematised (Boistrup, Bohlmann, Díez-Palomar, Kollosche, & Meaney, 2017; Roos, 2017). One aspect here is that the concept of diversity itself might arise from the assumption that there is something normal from which, for example, diverse students deviate, whereas for this TWG the concept of diversity is likely to be viewed as the norm itself. We expect that there will be more problematisations of diversity as a concept in future CERMEs. A related matter here are words that are similar to diversity, but perhaps carry other connotations such as difference, heterogeneity, multiplicity, or variety; also connected words such as democracy, segregation/integration, inclusion/exclusion, or empowerment.

A final key theme that deserves to be addressed is the critical reflection of research practices, including forms of collaboration and methodologies, and the promotion of new and innovative ways of data collection emerging from new technologies and means of communication. The group could benefit from collaborative research of a wider dimension which could provide new insights. In relation to this wider dimension, we are thinking, for example, in terms of comparative qualitative case studies across countries, and in terms of longitudinal studies. These combined could contribute to an understanding of the impact of the diversity of educational contexts, and trajectories of development over time, and could incorporate the complementary perspectives of the co-construction of the social, cultural, and political with the psychological.

Acknowledgement

We are grateful to the feedback from Gail FitzSimons in a previous version of this chapter.

References

Abreu, G. de, (2006). Cultural identities in the multi-ethnic mathematical classroom. *CERME4* (pp. 1131–1140).

Abreu, G. de (2014). Cultural diversity in mathematics education. In S. Lerman (Ed.), *Encyclopedia of mathematics education* (pp. 125–129). Dordrecht: Springer.

Abreu, G. de, César, M., Gorgorió, N., & Valero, P. (2006). Issues and challenges in researching mathematics education in multicultural settings. *CERME4* (pp. 1125–1130).

Abreu, G. de, & Crafter, S. (2016). Mathematics learning in and out of school: Towards continuity or discontinuity? In L. English & D. Kirshner (Eds.), *Handbook of international research in mathematics education* (3rd ed.) (pp. 395–415). London, UK: Routledge.

Abreu, G. de, Crafter, S., Gorgorió, N., & Prat, M. (2013). Understanding immigrant students' transitions as mathematical learners from a dialogical self-perspective. *CERME8* (pp. 1648–1655).

220 Abreu, Gorgorió & Björklund Boistrup

Abreu, G. de, & Gorgorió, N. (2007). Social representations and multicultural mathematics teaching and learning. *CERME5* (pp. 1159–1566).

Alrø, H., Skovsmose, O., & Valero, P. (2004). Communication, conflict and mathematics education in the multicultural classroom. *CERME3*: www.mathematik.uni-dortmund. de/~erme/CERME3/Groups/TG10/TG10_Alro_cerme3.pdf.

Alrø, H., Skovsmose, O., & Valero, P. (2006). Culture, diversity and conflict in landscapes of mathematics learning. *CERME4* (pp. 1141–1152).

Andersson, A. (2011). Interplays between context and students' achievement of agency. *CERME7* (pp. 1399–1408).

Andersson, A., & Norén, E. (2011). Agency in mathematics education. *CERME7* (pp. 1389–1398).

Arzarello, F., Dorier, J.-L., Hefendehl-Hebeker, L., & Turnau, S. (1999). Mathematics as a cultural product. *CERME1* (pp. 70–77).

Bagger, A. (2015). Pressures and positions of need during the Swedish third-grade national test in mathematics. *CERME9* (pp. 1558–1563).

Bishop, A. (1988). *Mathematical enculturation: A cultural perspective on mathematics education.* Dordrecht: Kluwer.

Bishop, A. J., Clarkson, P., FitzSimons, G., & Seah, W. T. (2002). Studying values in mathematics education: Aspects of the VAMP Project. *CERME2* (pp. 368–376).

Black, L., Solomon, Y., & Radovic, D. (2015). Mathematics as caring: The role of 'others' in a mathematical identity. *CERME9* (pp. 1564–1570).

Boistrup, L. B., Bohlmann, N., Díez-Palomar, J., Kollosche, D., & Meaney, T. (2017). Introduction to the papers of TWG10: Diversity and mathematics education – social, cultural and political challenges. *CERME10* (pp. 1301–1304).

Boistrup, L., & Keogh, J. (2017). The context of workplaces as part of mathematics education in vocational studies: Institutional norms and (lack of) authenticity. *CERME10* (pp. 1337–1344).

Boistrup, L. B., Meaney, T., Mesquita, M., & Straehler-Pohl, H. (2015). Introduction to the papers of TWG10: Diversity and mathematics education – social, cultural and political challenges. *CERME9* (pp. 1534–1537).

César, M., & Favilli, F. (2006). Diversity seen through teachers' eyes: Discourses about multicultural classes. *CERME4* (pp. 1153–1164).

Cole, M. (1996). *Cultural psychology.* Cambridge, MA: The Belknap Press of Harvard University Press.

Crafter, S. (2010). Parental resources for understanding mathematical achievement in multiethnic settings. *CERME6* (pp. 1453–1461).

Crafter, S., & Abreu, G. de. (2011). Teachers' discussions about parental use of implicit and explicit mathematics in the home. *CERME7* (pp. 1419–1429).

D'Ambrosio, U. (1985). *Socio-cultural bases for mathematics education.* Campinas, Brasil: Unicamp.

De Haan, M., & Elbers, E. (2008). Diversity in the construction of modes of collaboration in multiethnic classrooms. In B. van Oers, W. Wardekker, E. Elbers, & R. van der Veer (Eds.), *The transformation of learning: Advances in cultural-historical activity* (pp. 219–241). Cambridge, UK: Cambridge University Press.

Díez-Palomar, J., & Ortin, S. T. (2011). Socio-cultural roots of the attribution process in family mathematics education. *CERME7* (pp. 1491–1500).

Doğan, O., & Haser, Ç. (2013). The gap between mathematics education and low income students' real life: A case from Turkey. *CERME8* (pp. 1697–1704).

Domite, M. do C., & Pais, A. S. (2010). Understanding ethnomathematics from its criticisms and contradictions. *CERME6* (pp. 1473–1483).

Eikset, A., Fosse, T., Lange, T., Lie, J., Lossius, M. H., Meaney, T., & Severina, E. (2017). (Wanting to do) Ethical research in a shifting context. *CERME10* (pp. 1353–1360).

Elbers, E., & de Haan, M. (2004). The construction of word meaning in a multicultural classroom: Talk and collaboration during mathematics lessons. *CERME3*: www.mathematik.uni-dortmund.de/~erme/CERME3/Groups/TG10/TG10_Elbers_cerme3.pdf.

Favilli, F., Oliveras, M. L., & César, M. (2004). Maths teachers in multicultural classes: Findings from a Southern European project. *CERME3*: www.mathematik.uni-dortmund.de/~erme/CERME3/Groups/TG10/TG10_Favilli_cerme3.pdf

François, F., & Pinxten, R. (2013). Multimathemacy. *CERME8* (pp. 1735–1743).

Fyhn, A., Meaney, T., Nystad, K., & Nutti, Y. (2017). How Sámi teachers' development of a teaching unit influences their self-determination. *CERME10* (pp. 1361–1368).

Gebremichael, A. T., Goodchild, S., & Nygaard, O. (2011). Students' perceptions about the relevance of mathematics in an Ethiopian preparatory school. *CERME7* (pp. 1430–1439).

Gellert, U., & Straehler-Pohl, H. (2011). Differential access to vertical discourse: Managing diversity in a secondary mathematics classroom. *CERME7* (pp. 1440–1449).

Gorgorió, N., Planas, N., & Vilella, X. (2002). Immigrant children learning mathematics in mainstream schools. In G. de Abreu, A. Bishop, & N. Presmeg (Eds.), *Transitions between contexts of mathematical practice* (pp. 23–52). Dordrecht: Kluwer.

Gorgorió, N., & Prat, M. (2011). Mathematics teachers' social representations and identities made available to immigrant students. *CERME7* (pp. 1450–1459).

Hauge, K. H., Sørngård, M. A., Vethe, T. I., Bringeland, T. A., Hagen, A. A., & Sumstad, M. S. (2015). Critical reflections on temperature change. *CERME9* (pp. 1577–1583).

Johansson, M. C., & Boistrup, L. B. (2013). It is a matter of blueness or redness: Adults' mathematics containing competences in work. *CERME8* (pp. 1744–1753).

Kaiser, G. (2004). Learning mathematics within the context of linguistic and cultural diversity: An empirical study. *CERME3*: www.mathematik.uni-dortmund.de/~erme/CERME3/Groups/TG10/TG10_Kaiser_cerme3.pdf.

Kilpatrick, J. (1999). Ich bin Europäisch. *CERME1* (pp. 49–68).

Krummheuer, G. (1999). Introduction (Social interactions in mathematical learning situations). *CERME1* (pp. 305–307).

Krummheuer, G. (2002). The comparative analysis in interpretative classroom research in mathematics education. *CERME2* (pp. 339–346).

Lange, T., & Meaney, T. (2011). Becoming disadvantaged: Public discourse around national testing. *CERME7* (pp. 1470–1480).

Montecino, A., & Valero, P. (2015). Statements and discourses about the mathematics teacher: The research subjectivation. *CERME9* (pp. 1617–1623).

Mukhopadhyay, S., & Greer, B. (2015). Cultural responsiveness and its role in humanizing mathematics education. *CERME9* (pp. 1624–1629).

Newton, R., & Abreu, G. de (2011). Parent-child interactions on primary school-related mathematics. *CERME7* (pp. 1481–1490).

Newton, R., & Abreu, G. de (2013). The dialogical mathematical 'self'. *CERME8* (pp. 1784–1791).

Pais, A., Crafter, S., Straehler-Pohl, H., & Mesquita, M. (2013). Introduction to the papers and posters of WG10: Cultural diversity and mathematics education. *CERME8* (pp. 1820–1824).

Parra-Sánchez, A. (2015). Dialogues in ethnomathematics. *CERME9* (pp. 1644–1650).

Radovic, D., Black, L., Salas, C., & Williams, J. (2015). The intersection of girls' mathematics and peer group positionings in a mathematics classroom. *CERME9* (pp. 1651–1657).

Rangnes, T. E. (2011). Moving between school and company. *CERME7* (pp. 1501–1510).

222 Abreu, Gorgorió & Björklund Boistrup

Roos, H. (2017). Diversity in an inclusive mathematics classroom: A student perspective. *CERME10* (pp. 1433–1440).

Seah, W. T., Davis, E. K., & Carr, M. (2017). School mathematics education through the eyes of students in Ghana: Extrinsic and intrinsic valuing. *CERME10* (pp. 1441–1448).

Skovsmose, O. (2014). *Critical mathematics education*. Dordrecht: Springer.

Stathopoulou, C., François, K., & Moreira, D. (2011). Ethnomathematics in European context. *CERME7* (pp. 1511–1520).

Stentoft, D. (2007). Multiple identities in the mathematics classroom: A theoretical perspective. *CERME5* (pp. 1597–1608).

Straehler-Pohl, H., & Pais, A. (2013). To participate or not participate? That is not the question! *CERME8* (pp. 1794–1803).

Turvill, R. (2015). Number sense as a sorting mechanism in primary mathematics education. *CERME9* (pp. 1658–1663).

Valero, P. (2013). Mathematics for all and the promise of a bright future. *CERME8* (pp. 1804–1813).

Valero, P., Crafter, S., Gellert, U., & Gorgorió, N. (2011). Introduction to the papers of WG 10: Discussing diversity in mathematics education from social, cultural and political perspectives. *CERME7* (pp. 1386–1388).

16

COMPARATIVE STUDIES IN MATHEMATICS EDUCATION

Eva Jablonka, Paul Andrews, David Clarke and Constantinos Xenofontos

1 Introduction

1.1 History of the group

The group 'Comparative Studies in Mathematics Education' was established in 2007 at CERME 5, reflecting a growing interest of researchers and regional school authorities in comparative and international dimensions of education, not least in response to cross-national achievement studies carried out by supranational organisations. The group has been truly multilinguistic since its inception; and more recently it has attracted participants from locations far beyond geographical Europe or ERME affiliated regions, such as for example, Ghana, South Africa, China, Japan, Australia, Vietnam, Mexico, Chile and the USA.

Immediate access to colleagues from contexts that participants had chosen for their cross-cultural comparisons proved invaluable to better mutual understanding of cultural, social and political contexts of mathematics education. At each CERME, participants included international postgraduate research students from outside Europe who worked with supervisors at European universities and individual researchers who had moved between contexts. Hence the meetings provided an excellent forum for productive interaction of insiders' and outsiders' perspectives and exchange of experiences in support of the development of culturally sensitive research designs and analyses.

The number of presenting participants remained relatively small throughout the group's history, with many opportunities for sustained and focused discussion in an inclusive atmosphere, greatly aided by the reservoir of languages available for communication and the help provided by many participants who were fluent in a couple of these.

1.2 Developing aims and frameworks for comparative studies

One of the group's initial aims was to overcome simplistic identifications of culture with country and associated misinterpretations of comparative studies. In acknowledgement of cultural complexities within many countries or administrative units, the scope of the work has not been restricted to comparative studies across national educational systems. Consequently, the group adopted an eclectic perspective in its interpretation of comparison as referring to any study that documents, analyses, contrasts or juxtaposes cross-cultural or cross-contextual similarities and differences across all aspects and levels of mathematics education. The notion of culture adopted in the discussion refers to participation in practices in which members share a discourse or identities, so that their membership in a particular group or sub-group is not an outcome of a categorisation only employed as a unit of comparison by the researcher.

Besides sharing outcomes of empirical studies, the group aimed at developing and defining research methodologies specific to comparative studies and developing better understanding of how various theoretical approaches and conceptual frameworks shape the goals and the design of comparative research in mathematics education.

One important issue for the shaping of comparative studies was seen in the development of general strategies for undertaking 'culturally sensitive research' (Tillman, 2006). The group highlighted negotiation between cultural insiders and outsiders throughout the research process as an important strategy. At CERME 7, the potential for cultural biases as a consequence of the theoretical frameworks chosen, was a major topic for discussion. This issue was raised in relation to the appropriateness of using particular cognitive psychological frameworks (such as for students' motivation or engagement) developed in one context, for comparison with students or teachers from different cultural contexts. However, the group also discussed examples in which the use of analytical frameworks stemming from an outsider's cultural heritage could be considered a strength (Jablonka & Andrews, 2012).

While understanding the role of culture in the construction of mathematics teaching and learning remained a shared interest of the groups' participants, studies accepted for presentation in the working group often included much smaller units of 'culture'. Consequently, rather than assuming or seeking coherence in the field of comparative studies in mathematics education, the working group invited contributions based on a broad range of methodologies and analytical frameworks that reflected the diversity of aims and concerns of researchers working in different contexts.

Yet, comparisons of aspects of curriculum and pedagogy across (national) education systems remained the focus of many studies presented and discussed in the group. While agreement was reached that comparison itself cannot constitute the goal of such studies, comparison was seen as being always of interest in providing a new 'lens' and making the familiar look unfamiliar. The often very vivid discussions revealed how comparison across systems can afford insights beyond the reach of research confined to one context. Further, cross-system comparison of 'distant' cultures was discussed as a research strategy because contrasting cases might have substantial theoretical and methodological bearing.

2 Foci of comparison

In this section, we discuss the foci of comparison of the CERME studies. These foci are either implicitly or explicitly presented in the papers. Briefly, we identify three themes underlying the comparative nature of the studies. First, we examine the aspects of mathematics education that are addressed and compared. Second, we look into the choice of the systemic contexts for comparison. Finally, we consider studies that make comparisons beyond 'culture'. Interestingly, no studies adopting a historically comparative perspective have been identified.

2.1 Aspects of mathematics education addressed and compared in empirical studies

A majority of cross-system studies addressed aspects of curriculum and pedagogy, with some also including assessment practices and policy. Curriculum studies included analyses of curriculum goals and 'standards' in official documents (i.e. An, Mintos & Yigit, 2013; Bjarnadóttir, 2007; Gosztonyi, 2015), representation in textbooks and teacher guides (i.e. Bofah & Hannula, 2011; Cabassut & Ferrando, 2013; da Ponte & Marques, 2007; Xenofontos & Papadopoulos, 2015), as well as curriculum enactment in classroom practice (Clarke & Xu, 2010). Many of these studies focused on the teaching of selected mathematical topics, some on broader competencies, such as number sense (Andrews, Sayers & Marschall, 2015), problem solving and modelling (Saeki, Matszaki, Kawakami & Lamb, 2015), or on specific aspects of pedagogy, such as language use and modes of questioning (Hommel & Clarke, 2015) and assessment methods (Brown, 2007).

Serving teachers were the focus of some cross-system studies examining participants' professional motives (Andrews, 2010), reactions to classroom interventions and professional development activities (Cabassut & Villette, 2011), teachers' knowledge (Tchoshanov, Quinones, K. Shakirova, Ibragimova & L. Shakirova, 2017), and views on 'good' examination tasks (Peng, Sollervall, Stadler, Shang & Ma, 2015). Other studies were concerned with prospective teachers, looking at issues such as mentor–prospective teacher relationships (Knutsson, Hemmi, Bergwall & Ryve, 2013), the use of metaphors for describing mathematics (Kiliç, 2011), their mathematical knowledge for teaching (Andrews & Xenofontos, 2017; Kingji-Kastrati, Sajka & Vula, 2017; Xenofontos & Andrews, 2017), and problem solving in pre-service teacher training (Kuzniak, Parzysz, Santos-Trigo & Vivier, 2011).

Students also featured commonly in cross-system studies. Examples comprise investigations of students' mathematical understanding and meaning construction (Nguyen & Grégoire, 2013; Vollstedt, 2007), examinations of factors that influence learners' mathematical literacy (Törnroos, 2007), explorations of emotional aspects and affective co-productions of mathematics learning (Jablonka, 2013;

226 Jablonka et al.

Nosrati & Andrews, 2017; Tuohilampi, Hannula, Giaconi, Laine & Näveri, 2013), and identities of 'elite' students (Saari, 2010).

The group recognised the difficulties with generalisations at the level of 'culture', particularly in contexts where cultural homogeneity of school students or teachers cannot be assumed, and agreed that a more nuanced articulation of 'culture' is necessary. Some participants foregrounded social relations in their conceptualisation of 'culture', while others focused more on shared traditions and systems of values.

2.2 Systemic contexts chosen for comparison

A variety of reasons were provided by the authors of the cross-system papers, explicitly or in some cases implicitly, in order to justify their preferences for the comparisons of specific systemic contexts. It should be noted, however, that the reasons presented below are not mutually exclusive, and that for every study more than one reason might apply. The majority of studies located their choices within an argument of 'cultural proximity versus cultural distance'. In regard to the former, we find examples like those of da Ponte and Marques (2007), comparing data from Portugal, Brazil and Spain; of Hannula, Lepik, Pipere and Tuohilampi (2013), comparing data from Estonia, Latvia, and Finland; of Nostrati and Andrews (2017), with their study in Norway and Sweden; and of Vula, Kingji-Kastrati and Podvorica (2015), comparing data from Albania and Kosovo. Other studies emphasise the different educational traditions of the systems under scrutiny, such as that of Saari (2010) who compares data from Finland and the USA; of Kuzniak et al. (2011) and their study in France and Mexico; as well as of Nguyen & Grégoire (2013), highlighting the cultural and linguistic differences between the French Belgium and Vietnam. A few studies comment on the geographical proximity of the compared contexts, such as Bjarnadóttir (2007), examining issues regarding Iceland, Denmark, Norway and England.

Co-authors' locations appear to be another feature that, implicitly or explicitly, enables cross-system comparisons. For instance, Xenofontos and Papadopoulos (2015) talk explicitly about their respective locations in Cyprus and Greece, while in the cases of Cabassut and Ferrando (2013), comparing issues in France and Spain, of Saeki et al. (2015) and their study concerning Australia and Japan, and of Modeste and Rafalska (2017) with a study regarding France and Ukraine, this is implied by the authors' affiliating institutions.

In a few studies, typically single-authored, the choice of a comparative approach is justified by claims of the author's familiarity with the systems under examination (see, for instance, Cabassut, 2007; Gosztonyi, 2015; Kiliç, 2011). A relatively large number of papers draw on data from larger cross-system projects, and compare findings from typically two to four of the participating countries. Such examples constitute the studies of An et al. (2013), Andrews (2011), Clarke and Xu (2010) and Jablonka (2013). Finally, a couple of studies explain how the choice of a comparative cross-national approach was adopted for testing the cross-cultural affordances of frameworks or theoretical models under development (see Andrews et al., 2015; Andrews & Xenofontos, 2017; Bofah, 2015).

Comparative studies in maths education **227**

2.3 Comparisons beyond 'culture'

In addition to studies that used the location in a particular school system for defining their unit of comparison, a couple of presentations reported investigations with other comparative foci. These included comparisons across school sectors and between students at different levels of attainment or in different streams, or belonging to an ethnic minority. In regards to comparisons between students at different attainment levels, we find studies like those of Maréchal (2011), who talks about 'praxeologies' in different classes for 'ordinary' or other students in Geneva, Schäfer and Winkler (2011), investigating metacognitive actions used by high-achieving and low-achieving German students chosen from different streams, and Sajka and Rosiek (2015), who report an experimental study of eye tracking of high- and low-achieving students and their problem-solving competencies in Poland.

Another focus of comparison concerned school sectors within a single educational system. For example, Nilsen (2010) compares Norwegian lower- and upper-secondary teachers' beliefs, while Larson and Bergsten (2013) explore 'praxeologies' in lower- and upper-secondary classrooms in Sweden. Finally, we see studies comparing settings with demographic differences in school intake, as, for example, Eisenmann and Even (2007) who follow the same teacher, applying different pedagogies in different schools, and Koljonen (2017) who examines the use of Finnish teaching materials by a Swedish teacher; studies comparing differently qualified teachers, such as Vantourout (2007); and longitudinal comparisons within the same context, such as that of Branchetti and her colleagues (2015), who follow the same cohort of Italian students and compare their test performance in grade 6 and grade 8.

These studies were comparative in so far as differences at the individual level were taken as contributing to the representation of one group in comparison with another group. A methodological question discussed in this context was whether these groups exist independently of the comparative research; if not, then the notion of a coherent culture or micro-culture that is being compared is somewhat problematic. These questions initiated a continuous methodological discussion.

3 Developing methodology and theory

3.1 Methodological approaches

The methodological stances adopted by the working group's members have shown considerable variation, although there have been many more qualitatively focused than quantitative. In broad terms, they have fallen into five forms, summarised in Table 16.1.

The purpose of the following review of methodologies is primarily descriptive. It will help to identify trends and assist researchers to find papers employing a particular strategy and evaluate its usefulness.

228 Jablonka et al.

TABLE 16.1 Summary data of the forms of study presented to the working group

	2007	2009	2011	2013	2015	2017	Total
Case study	3	5	4	3	5	3	23
Small survey		1	4	3	1	6	15
Documentary analysis	3	1	2	3	3	1	13
Large survey	1				2		3
Position paper	1	1		2	2		6

Case studies

The most frequently occurring methodological tradition was case study in various forms, although the majority of studies identified below were not reported as case studies by their authors. Typically, the case element of the study was left implicit with authors attending to details pertaining to other methodological aspects such as theoretical perspectives or the articulation of an analytical procedure. In some respects, acknowledging the brevity of a CERME paper, this is probably not surprising. In categorising case studies, choices have to be made between, say, whether to categorise by approaches to data collection and analysis or by the nature of the cases under scrutiny. In this instance we have chosen to categorise first by the nature of the case, as it better represents the diversity of the working group's papers.

In broad terms, case studies as identified in this manner fell into four categories in which the cases were countries, classrooms, teachers or students. The most common of these, accounting for 14 of these 23 studies, involved the country as the case. These included analyses of lessons focused on particular topics and students of various ages (Andrews, 2011; Andrews et al., 2015; Back, Sayers & Andrews, 2013; Navarra, Malara & Ambrus, 2010; Saeki et al., 2015), comparisons of particular teacher education interventions (Asami-Johansson, Attorps & Laine, 2017; Knutsson et al., 2013; Kuzniak et al., 2011), analyses of classroom interactions and their meaning for participants (Clarke & Xu, 2010; Hommel & Clarke, 2015; Vollstedt, 2007), teachers' perceptions of the nature of good problems (Peng et al., 2015), the different mathematics pedagogical vocabularies used by teachers (Clarke, Mesiti, Cao & Novotna, 2017) and a study of elite students' experiences of mathematics learning (Saari, 2010).

Eight of the remaining nine studies were undertaken in single cultural contexts in which different groups within a particular setting were compared. Case studies focused on the student included differently conceptualised comparisons of the mathematical behaviour of high- and low-achieving students (Sajka & Rosiek, 2015; Schäfer & Winkler, 2011), examinations of minority students' mathematical resilience (Mulat & Arcavi, 2010) and a comparison of upper- and lower-secondary students' perceptions of school mathematics (Larson & Bergsten, 2013). Two studies exploited students' socioeconomic affiliation as the case, as in Eisenmann and Even's (2007) comparison of grade 7 algebra lessons taught by

Comparative studies in maths education **229**

the same teacher in different schools and Maréchal's (2011) investigation of the teaching of addition to grade 1 students in three structurally different classrooms. Two cases involved the teacher: one of these compared pre-service teachers' approaches to the assessment of fictitious student solutions (Vantourout, 2007), while the other compared lower- and upper-secondary teachers' beliefs about the teaching of mathematics (Nilsen, 2010). Finally, one case study compared the use of curriculum materials in two countries, one in which the materials were developed and another (Koljonen, 2017).

Typically, case studies drew on qualitative data, whether derived from interview, observation transcripts or both, although the means of analysis varied considerably. A number exploited the constant comparison process of the grounded theorists. In some papers, this was explicit (Andrews, 2011; Vollstedt, 2007; Peng et al., 2015) and in others it was implicit (Knutsson et al., 2013; Mulat & Arcavi, 2010; Saari, 2010). In other studies, established theoretical frameworks were applied to the researchers' qualitative data (Asami-Johansson et al., 2017; Eisenmann & Even, 2007; Koljonen, 2017; Larson and Bergsten, 2013; Maréchal, 2011; Schäfer & Winkler, 2011), while others applied frameworks derived from a larger project of which the reported paper was part (Andrews et al., 2015; Back et al., 2013; Saeki et al., 2015). Finally, in this section, Clarke et al. (2017) adopted an open coding approach to their investigation of mathematics pedagogical vocabulary, while (Nilsen, 2010) offered an unspecified analysis of qualitative data derived from lesson observations and interviews.

A smaller number of studies exploited quantitative analyses of their different data. Two, due to the ways in which data were focused on particular tasks, undertook analyses of time on various aspects of the tasks: Sajka and Rosiek's (2015) eye-tracking study, and Vantourout's (2007) examination of teacher education students' approaches to the assessment of fictitious students' problem solutions. Two studies offered more conventional analyses of teacher education student or school student responses to mathematical tasks, respectively (Kuzniak et al., 2011; Navarra et al., 2010). Finally, two studies, both drawing on data from The Learner's Perspective Study (LPS), offered quantitative summaries of differently focused data derived from codes applied to lesson transcripts (Clarke & Xu, 2010; Hommel & Clarke, 2015; for the LPS see, for example, Clarke, Emanuelsson, Jablonka & Mok, 2006).

Of course, alternative interpretations might not have presented many of the studies above as case studies, but our view is that the privileging of, typically, two groups for comparison created particular cases.

Small surveys

Fifteen studies were construed as small surveys. In this respect it is important to explain how we distinguished, on the one hand, between small surveys and large surveys and, on the other hand, small surveys and case studies. For the former, the distinction was based less on sample size, although this was not ignored, than on

230 Jablonka et al.

the source of the data analysed. Studies classed as large surveys drew their data from large-scale assessments of achievement, whether national or international, while small surveys exploited instruments developed by the project team or similar. In the latter case, small surveys were construed as requiring sufficient data units to achieve thematic saturation (O'Reilly & Parker, 2013), which case studies typically do not seek to achieve. Three forms of small survey were identified, one reflecting the former case above and two the latter.

The 15 studies in this broad categorisation of small survey could be categorised in various ways, some conventional and others not so. Three studies addressed different aspects of mathematics-related beliefs, including a survey of teachers' beliefs about mathematical modelling undertaken in France, Germany, Hungary and Spain (Cabassut & Villette, 2011), a study of teachers' beliefs in Estonia, Latvia and Finland (Hannula et al., 2013) and an examination of Chilean and Finnish grade 3 students' mathematics-related affect structures (Tuohilampi et al., 2013). Five studies exploited test-like surveys to evaluate various aspects of pre-service and in-service mathematical knowledge for teaching. Studies focused on pre-service teachers included Andrews and Xenofontos' (2017) and Xenofontos and Andrews' (2017) task-based studies of Greek and Cypriot students' didactical understanding of linear equations and Kingji-Kastrati and colleagues' (2017) comparisons of Polish and Kosovan preservice teachers' understanding of fractions. Studies focused on serving teachers included Jakobsen, Fauskanger, Mosvold and Bjuland's (2011) adaptation of an American instrument to examine Norwegian teachers' MKT and Tchoshanov et al.'s (2017) TIMSS-related comparison of American and Russian teachers. A sixth study exploited a battery of project-developed test items to examine the impact of the number-name system on third-grade French-speaking Belgian and Vietnamese students' understanding of the role of zero in their understanding of number in various forms (Nguyen & Grégoire, 2013).

Three studies were construed as interview surveys. Andrews (2010) exploited semi-structured interviews, which were subjected to a constant comparison analysis, to examine English and Hungarian teachers' professional motivations, Nosrati and Andrews (2017) undertook semi-structured interviews with Norwegian and Swedish upper-secondary students to elicit their views on the purpose of school mathematics, while Jablonka (2013) analysed a representative sample of LPS interviews to examine student perspectives on boredom in the mathematics lessons of Germany, Hong Kong and the United States. The remaining four studies were less straightforwardly categorised. These included an atypical social media distributed survey to examine interested parties' perspectives on mathematics education in South Africa and England (Joubert, 2015), and an eye-tracking comparison of the problem-solving behaviours of different mathematically elite groups (Sajka, 2017). Finally, in this section, one study was construed as a written small survey in that Turkish and Belgian teacher education students were invited to describe any metaphors they had for describing mathematics (Kiliç, 2011). In all cases, analyses were conducted according to both project goals and the nature of the instruments used.

Documentary analyses

Thirteen papers, falling into three categories, were construed as invoking some form of documentary analysis. The largest of these three categories comprised six studies comparing different classroom-related texts. Of these, five focused on comparisons of how school textbooks from different countries facilitate students' learning of different topics. These included an analysis of the opportunities to learn calculus in Finland and Ghana (Bofah & Hannula, 2011), proportion in middle-school mathematics in Portugal, Brazil, Spain and the USA (da Ponte & Marques, 2007), fractions in Albania and Kosovo (Vula et al., 2015), the history of mathematics in Cyprus and Greece (Xenofontos & Papadopoulos, 2015) and a summary of various studies involving England, France and Germany (Pepin, 2010).

The second group of studies comprised comparisons of curricular documents in various ways. These included the introduction of quadratic equations in the Caribbean, China, Turkey and the USA (An et al., 2013), mathematical modelling in France and Spain (Cabassut & Ferrando, 2013), curricular approaches to algorithmics in France and Ukraine (Modeste & Rafalska, 2017) and the graphical calculator in the examinations of Denmark, Australia and the International Baccalaureate (Brown, 2007). The third group comprised two papers summarising the history of the 'New Math' movement in Iceland, Denmark, Norway and Sweden (Bjarnadóttir, 2007) and France and Hungary (Gosztonyi, 2015). Finally, there was one paper focused on document-based comparison of two mathematics teaching and learning traditions – the metaphor and the *Grundvorstellungen* (Soto-Andrade & Reyes-Santander, 2011).

From an analytical perspective studies invoked a variety of approaches. Two exploited a project-derived framework for supporting quantitative analyses (An et al., 2013, Vula et al., 2015). Four exploited external frameworks in different ways. For example, both Cabassut and Ferrando (2013) and Modeste and Rafalska (2017) presented an ATD-based qualitative analysis, while da Ponte and Marques (2007) offered a categorical analysis based on PISA's assessment framework. Two studies offered, essentially, categorical analyses based on frameworks drawn from elsewhere in the literature (Hemmi, Koljonen, Hoelgaard, Ahl & Ryve, 2013; Bofah & Hannula, 2011). Three studies offered qualitative analyses based on frameworks derived from the vocabulary of the analysed documents themselves (Bjarnadóttir, 2007; Brown, 2007; Gosztonyi, 2015), while just one invoked the traditions of constant comparison (Xenofontos & Papadopoulos, 2015).

Large surveys

Finally, three studies presented novel analyses of large extant data sets. One of these, Branchetti et al. (2015), compared the same cohort of students on two iterations of the Italian national test with the aim of characterising their understanding of fractions. The other two studies, Törnroos (2007) and Bofah (2015) drew on

232 Jablonka et al.

data from PISA 2003 and TIMSS 2011, respectively, to analyse the impact of non-mathematical factors on students' mathematics performance. Törnroos' study involved students from Finland and Sweden, while Bofah's exploited data from five diverse African countries: Botswana, Ghana, South Africa, Morocco and Tunisia. In all three studies, data were analysed statistically, the latter two, in particular, exploiting confirmatory factor analyses to identify the structural properties of the entities under scrutiny.

Position papers

The six studies categorised as position papers, unsurprisingly, followed similar formats in their drawing on literature to support the authors' proposed arguments. However, their foci fell into two distinct categories. First, there were papers with a methodological emphasis for comparative mathematics education research, such as Cabassut (2007), Clarke (2013, 2015), Jablonka (2015), and Knipping and Müller-Hill (2013). All such papers offered significant insights into the conduct of comparative mathematics education research. Second, a single study (Xenofontos, 2010) drew on literature to propose an agenda for future research on mathematical problem solving.

3.2 Theories and analytical framings

As comparative studies are dependent on various theoretical interests and changing methodological standards established in research in general, delineation will always be in flux. Yet, the view was shared that in studies that involve comparison across educational systems there are two branches, one deriving from documentation, description and analysis of classroom and school practices, and another stemming from large-scale international achievement surveys. As is apparent from the overview of methodologies, researchers affiliated with the first tradition constituted a clear majority in the working group. However, it was noted that within this group studies with a broader conception of the enacted mathematics curriculum, which aim at illuminating issues of power, identity and subjectivity, were still rare (Jablonka, 2015).

Members of the group employed and developed a wide range of theories and analytical framings. These included literature-based frameworks for the teaching/learning of particular mathematical topics (An et al., 2013; Andrews, 2011; Andrews et al., 2015), mathematical knowledge for teaching (Jakobsen et al., 2011), conceptual metaphors and *Grundvorstellungen* (Kiliç, 2011; Soto-Andrade & Reyes-Santander, 2011), cognitive psychology and theories of mathematical abstraction (Schäfer & Winkler, 2011), social cognitive theory (Mulat & Arcavi, 2010), particular theories from educational psychology (Bofah, 2015; Andrews, 2010; Tuohilampi et al., 2013; Vollstedt, 2007; Saari, 2010), Anthropological Theory of Didactics (Cabassut & Ferrando, 2013; Gosztonyi, 2015; Larson & Bergsten, 2013; Maréchal, 2011), Theory of Didactic Situations (Gosztonyi, 2015; Kuzniak et al., 2011), neo-Vygotskian theories (Hemmi et al., 2013; Pepin, 2010), theory of speech acts (Schäfer & Winkler, 2011), Bernstein's theory of pedagogic discourse (Larson & Bergsten, 2013) and semiotics (Branchetti et al., 2015). Some of the studies aimed at theory-networking or complementary analysis (Jablonka, 2013; Larson & Bergsten, 2013; Xenofontos, 2010).

The diversity of approaches and theoretical lenses aroused interest in the role of the cultural origins of didactical principles and related issues pertaining to conceptual equivalence of categories used for comparative analyses. For instance, a couple of studies (Kiliç, 2011; Soto-Andrade & Reyes-Santander, 2011) triggered a discussion of the role of metaphors in a culture. Group members recommended investigating whether there are structural differences in the metaphors used in different contexts. Given the unavoidable cultural authorship of many theoretical frameworks, awareness of blind spots as well as of the danger of imposing categories and taxonomies that arise in one curriculum and teaching tradition onto another was recognised as important if the (mathematics) education community is to overcome unquestioned assumptions of a shared view on the nature of the subject being compared. A more general issue examined in this context was whether we can successfully use a 'home-grown' theory from one culture to ask questions about another. The use of the French Anthropological Theory of Didactics in settings with different institutional layers and curriculum traditions, or the appropriateness of tools developed in the West for use in the Global South served as examples.

Most group members agreed that comparisons are undertaken between constructed representations, whose structure and attributes are reflective as much of the value system of the researcher as of the objects being represented for the purposes of comparison. Any cross-cultural comparative analysis faces the challenge of honouring the separate cultural contexts, while employing an analytical frame that affords reasonable comparison. In particular, in an international comparative study, any evaluative aspect is reflective of the cultural authorship of the study. More specifically, seven 'dilemmas' were identified by Clarke (2013) to reveal some of the contingencies under which international comparative research might be undertaken. Each dilemma can also serve as an interrogatory instrument: a tool directing the researcher's attention to salient characteristics that, while presenting impediments to comparison, simultaneously provide insight into nuances of meaning and practice.

4 Trends and directions

At each CERME, the group identified a number of practical challenges, methodological issues and complexities that comparative studies need to address, and shared experiences how this can be done. An increase in reflective awareness of methodological and theoretical issues is evidenced in the number of position papers with methodological and theoretical contributions submitted to the group. It was concluded that comparison can assist in basic, fundamental research that opens up new issues and that the role and potential of comparative approaches in theory building needs to be explored further.

234 Jablonka et al.

It proved useful to consider comparative research from the perspective of boundary crossing. Recognition of the significance of acts of comparison in both boundary crossing and boundary construction foregrounds comparison as a key tool in the essential act of boundary deconstruction.

Notions of culture were examined at all meetings with the initial aim of developing a more differentiated view of culture than is typically evidenced in large-scale surveys. Group members contended that there are different levels and dimensions of culture, and limitations of 'zooming in' or 'zooming out' were highlighted. In the case of employing psychological constructs, a potential 'ecological fallacy' by taking aggregated scores of individuals to represent a culture's score was acknowledged. Further, it was concluded that cultural homogeneity cannot be assumed, even if there are shared political institutions. Hence it was suggested that researchers should employ more specific notions of identity for the purpose of analytical comparison rather than cultural affiliation, in recognition of individuals with multiple identities.

A major concern at all CERMEs was the development of culturally sensitive research that provides valid comparison. Rather than developing a set of fixed criteria, conceptualising methodological decisions in terms of balancing tensions was seen as useful. Some unavoidable 'dilemmas' in comparative studies were discussed in more depth, such as the advantages or disadvantages of exploiting too inclusive or too distinctive categories in the pursuit of valid comparisons. A study might risk sacrificing validity in the interest of comparability (equivalence of constructs assumed) or limit comparability by the use of culturally specific categories for cross-cultural description or evaluation. It was also noted that when imported instruments fail to be reliable in new contexts, this is an important research outcome. In order to overcome a preoccupation with adapting existing frameworks, openness toward the empirical ('letting the data speak') was seen as a condition for allowing the voice in the new context to be heard.

Further, it was acknowledged that all data can be seen as constructed rather than given, and that in cross-cultural research criteria for what constitutes data must hold legitimately across settings. Although matching a research question to an appropriate unit of analysis is a major decision in all research studies, the group concluded that the choices are more visible in comparative studies. Consequently, methodological questions relating to the unit of analysis in comparative studies deserve more discussion, in particular with respect to appropriate theoretical framings for different levels or units of comparison. Another direction to be pursued arises from the need to better understand power relations between participants during data collection, and to acknowledge culturally and micro-culturally constructed interpersonal relationships, as for example in characterising and comparing classroom instruction from the vantage points of an ostensibly 'neutral' observer, the teacher and the learner.

A question resulting from the diversity of systemic contexts chosen for comparisons pertained to the tension between the 'cultural proximity' and 'cultural distance' of the researchers involved. Generally, the group acknowledged the value of both the insider's and outsider's perspectives, and noted that insiders share blind spots due to cultural blindness. Also, a distinction between form and function proved productive; while insiders can easily recognise the function of events in a culture, outsiders first of all attend to the form and might be more open to alternative interpretations. Similarly, the group also considered examples in which the use of theoretical frameworks linked to the researcher's cultural heritage could be viewed as a strength, and contrasting examples where it might be a limitation to be overcome by cooperation and exchange between researchers, such as through the group meetings at CERME.

References

An, T., Mintos, A., & Yigit, M. (2013). A cross-national standards analysis: Quadratic equations and functions. *CERME8* (pp. 1825–1834).

Andrews, P. (2010). Comparing Hungarian and English mathematics teachers' professional motivations. *CERME6* (pp. 2452–2462).

Andrews, P. (2011). The teaching of linear equations: Comparing effective teachers from three high achieving European countries. *CERME7* (pp. 1555–1564).

Andrews, P., Sayers, J., & Marschall, G. (2015). Developing foundational number sense: Number line examples from Poland and Russia. *CERME9* (pp. 1681–1687).

Andrews, P., & Xenofontos, C. (2017). Beginning teachers' perspectives on linear equations: A pilot quantitative comparison of Greek and Cypriot students. *CERME10* (pp. 1594–1601).

Asami-Johansson, Y., Attorps, I., & Laine, A. (2017). Comparing the practices of primary school mathematics teacher education: Case studies from Japan, Finland and Sweden. *CERME10* (pp. 1602–1609).

Back, J., Sayers, J., & Andrews, P. (2013). The development of foundational number sense in England and Hungary: A case study comparison. *CERME8* (pp. 1735–1844).

Bjarnadóttir, K. (2007). Development of the mathematics education system in Iceland in the 1960s in comparison to three neighbouring countries. *CERME5* (pp. 2403–2412).

Bofah, E. A. (2015). Reciprocal determinism between students' mathematics self-concept and achievement in an African context. *CERME9* (pp. 1688–1694).

Bofah, E. A., & Hannula, M. (2011). The case of calculus: Comparative look at task representation in textbooks. *CERME7* (pp. 1545–1554).

Branchetti, L., Ferretti, F., Lemmo, A., Maffia, A., Martignone, F., Matteucci, M., & Mignani, S. (2015). A longitudinal analysis of the Italian national standardized mathematics tests. *CERME9* (pp. 1695–1701).

Brown, R. G. (2007). Policy change, graphing calculators and 'High stakes examinations': A view across three examination systems. *CERME5* (pp. 2413–2422).

Cabassut, R. (2007). Examples of comparative methods in the teaching of mathematics in France and in Germany. *CERME5* (pp. 2423–2432).

Cabassut, R., & Ferrando, I. (2013). Modelling in French and Spanish syllabus of secondary education. *CERME8* (pp. 1845–1854).

Cabassut, R., & Villette, J.-P. (2011). Exploratory data analysis of a European teacher training course on modelling. *CERME7* (pp. 1565–1574).

Clarke, D. J. (2013). The validity-comparability compromise in crosscultural studies in mathematics education. *CERME8* (pp. 1855–1864).

236 Jablonka et al.

Clarke, D. J. (2015). The role of comparison in the construction and deconstruction of boundaries. *CERME9* (pp. 1702–1708).

Clarke, D. J., & Xu, L. H. (2010). Spoken mathematics as a distinguishing characteristic of mathematics classrooms in different countries. *CERME6* (pp. 2463–2472).

Clarke, D. J., Emanuelsson, J., Jablonka, E., & Mok, I. A. (Eds.) (2006). *Making connections: Comparing mathematics classrooms around the world*. Rotterdam: Sense Publishers.

Clarke, D. J., Mesiti, C., Cao, Y., & Novotna, J. (2017). The lexicon project: Examining the consequences for international comparative research of pedagogical naming systems from different cultures. *CERME10* (pp. 1610–1617).

da Ponte, J. P., & Marques, S. (2007). Proportion in school mathematics textbooks: A comparative study. *CERME5* (pp. 2443–2452).

Eisenmann, T., & Even, R. (2007). Types of algebraic activities in two classes taught by the same teacher. *CERME5* (pp. 2433–2442).

Gosztonyi, K. (2015). The 'New Math' reform and pedagogical flows in Hungarian and French mathematics education. *CERME9* (pp. 1709–1716).

Hannula, M., Lepik, M., Pipere, A., & Tuohilampi, L. (2013). Mathematics teachers' beliefs in Estonia, Latvia and Finland. *CERME8* (pp. 1865–1874).

Hemmi, K., Koljonen, T., Hoelgaard, L., Ahl, L., & Ryve, A. (2013). Analyzing mathematics curriculum materials in Sweden and Finland: Developing an analytical tool. *CERME8* (pp. 1875–1884).

Hommel, M., & Clarke, D. J. (2015). Reflection and questioning in classrooms in different cultural settings. *CERME9* (pp. 1717–1723).

Jablonka, E. (2013). Boredom in mathematics classrooms from Germany, Hong Kong and the United States. *CERME8* (pp. 1885–1894).

Jablonka, E. (2015). Why look into mathematics classrooms? Rationales for comparative classroom studies in mathematics education. *CERME9* (pp. 1724–1730).

Jablonka, E., & Andrews, P. (2012). CERME7: Comparative studies in mathematics education. *Research in Mathematics Education, 14*(2), 203–204. doi: 10.1080/14794802.2012.694291.

Jakobsen, A., Fauskanger, J., Mosvold, R., & Bjuland, R. (2011). Comparison of item performance in a Norwegian study using U.S. developed mathematical knowledge for teaching measures. *CERME7* (pp. 1575–1584).

Joubert, M. (2015). The perceived causes of the (assumed) mathematics problems in England and South Africa: A social media experiment. *CERME9* (pp. 1731–1737).

Kiliç, Ç. (2011). Belgian and Turkish pre-service primary school mathematics teachers' metaphorical thinking about mathematics. *CERME7* (pp. 1585–1593).

Kingji-Kastrati, J., Sajka, M., & Vula, E. (2017). Comparison of Kosovan and Polish pre-service teachers' knowledge of fractions. *CERME10* (pp. 1618–1625).

Knipping, C., & Müller-Hill, E. (2013). The problem of detecting genuine phenomena amid a sea of noisy data. *CERME8* (pp. 1895–1904).

Knutsson, M., Hemmi, K., Bergwall, A., & Ryve, A. (2013). School-based mathematics teacher education in Sweden and Finland: Characterising mentor–prospective teacher discourse. *CERME8* (pp. 1905–1914).

Koljonen, T. (2017). Finnish teaching materials in the hands of a Swedish teacher: The telling case of Cecilia. *CERME10* (pp. 1626–1633).

Kuzniak, A., Parzysz, B., Santos-Trigo, M., & Vivierr, L. (2011). Problem solving and open problems in teachers' training in the French and Mexican modes. *CERME7* (pp. 1594–1604).

Larson, N., & Bergsten, C. (2013). Comparing mathematical work at lower and upper secondary school from the students' perspective. *CERME8* (pp. 1915–1924).

Maréchal, C. (2011). What kind of teaching in different types of classes? *CERME7* (pp. 1605–1613).

Modeste, S., & Rafalska, M. (2017). Algorithmics in secondary school: A comparative study between Ukraine and France. *CERME10* (pp. 1634–1641).

Mulat, T., & Arcavi, A. (2010). Mathematical behaviors of successful students from a challenged ethnic minority. *CERME6* (pp. 2473–2482).

Navarra, G., Malara, N. A., & Ambrus, A. (2010). A problem posed by J. Mason as a starting point for a Hungarian-Italian teaching experiment within a European project. *CERME6* (pp. 2483–2494).

Nguyen, H. T. T., & Grégoire, J. (2013). Re-examining the language supports for children's mathematical understanding: A comparative study between French and Vietnamese languages. *CERME8* (pp. 1925–1934).

Nilsen, H. K. (2010). A comparison of teachers' beliefs and practices in mathematics teaching at lower secondary and upper secondary school. *CERME6* (pp. 2494–2503).

Nosrati, M., & Andrews, P. (2017). Ten years of mathematics education: Preparing for the supermarket? *CERME10* (pp. 1642–1649).

O'Reilly, M., & Parker, N. (2013). 'Unsatisfactory saturation': A critical exploration of the notion of saturated sample sizes in qualitative research. *Qualitative Research, 13*(2), 190–197. doi: 10.1177/1468794112446106.

Peng, A., Sollervall, H., Stadler, E., Shang, Y., & Ma, L. (2015). Swedish and Chinese teachers' views on what constitutes a good mathematical test task: A pilot study. *CERME9* (pp. 1738–1744).

Pepin, B. (2010). Mathematical tasks and learner dispositions: A comparative perspective. *CERME6* (pp. 2504–2512).

Saari, J. v. R. (2010). Elite mathematics students in Finland and Washington: Access, collaboration, and hierarchy. *CERME6* (pp. 2513–2522).

Saeki, A., Matszaki, A., Kawakami, T., & Lamb, J. (2015). Examining the heart of the dual modelling cycle: Japanese and Australian students advance this approach. *CERME9* (pp. 1745–1751).

Sajka, M. (2017). Visual attention while reading a multiple choice task by academics and students: A comparative eye-tracking approach. *CERME10* (pp. 1650–1657).

Sajka, M., & Rosiek, R. (2015). Solving a problem by students with different mathematical abilities: A comparative study using eye-tracking. *CERME9* (pp. 1752–1758).

Schäfer, I., & Winkler, A. (2011). Comparing the construction of mathematical knowledge between low-achieving and high-achieving students: A case study. *CERME7* (pp. 1614–1624).

Soto-Andrade, J., & Reyes-Santander, P. (2011). Conceptual metaphors and Grundvorstellungen: A case of convergence? *CERME7* (pp. 1625–1635).

Tchoshanov, M., Quinones, M. C., Shakirova, K., Ibragimova, E., & Shakirova, L. (2017). Cross-national study of lower secondary mathematics teachers' content knowledge in USA and Russia. *CERME10* (pp. 1658–1665).

Tillman, L. C. (2006). Researching and writing from an African-American perspective: Reflective notes on three research studies. *International Journal of Qualitative Studies in Education, 19*(3), 265–287.

Törnroos, J. (2007). Factors related to students' mathematical literacy in Finland and Sweden. *CERME5* (pp. 2453–2462).

Tuohilampi, L., Hannula, M., Giaconi, V., Laine, A., & Näveri, L. (2013). Comparing the structures of 3rd graders' mathematics-related affect in Finland and Chile. *CERME8* (pp. 1935–1944).

Vantourout, M. (2007). A comparative study of assessment activity involving 8 pre-service teachers: What referent for the assessor? *CERME5* (pp. 2463–2472).

Vollstedt, M. (2007). The construction of personal meaning: A comparative case study in Hong Kong and Germany. *CERME5* (pp. 2473–2482).

Vula, E., Kingji-Kastrati, J., & Podvorica, F. (2015). A comparative analysis of mathematics textbooks from Kosovo and Albania based on the topic of fractions. *CERME9* (pp. 1759–1765).

Xenofontos, C. (2010). International comparative research on mathematical problem solving: Suggestions for new research directions. *CERME6* (pp. 2523–2532).

Xenofontos, C., & Andrews, P. (2017). Explanations as tools for evaluating content knowledge for teaching: A cross-national pilot study in Cyprus and Greece. *CERME10* (pp. 1666–1673).

Xenofontos, C., & Papadopoulos, C. (2015). The history of mathematics in the lower secondary textbook of Cyprus and Greece: Developing a common analytical framework. *CERME9* (pp. 1766–1773).

17

HISTORY AND MATHEMATICS EDUCATION

Uffe Thomas Jankvist and Jan van Maanen

1 Introduction

When considering mathematics education from a historical perspective one may follow two distinct strands: history (as a subject or a didactical tool) *in* mathematics education; and history *of* mathematics education. Within ERME both strands have been represented since CERME 6 (Lyon, 2009). In this chapter we sketch the goals, methods and structure of both research strands as well as their main results, while providing illustrative examples of both strands from the work within the CERME Thematic Working Group (TWG) on 'History in Mathematics Education'. At the same time we take into account the cooperation between researchers with different backgrounds, collaboration between experienced researchers and starting researchers as well as by relating to the general communication between the TWG's researchers.

We first review the period before CERME 6, as far as history in mathematics education is concerned, and then describe the creation of the TWG. The TWG's work is represented in two ways. We outline the contents of six representative papers, which show the various theoretical frameworks and methodologies applied in the TWG. Also the collaboration between starting and established researchers can be recognised in this subset of papers. Then the distribution of the papers along the research questions set by the organisers is surveyed. Finally, we address the work carried out in the TWG and relate this to the three Cs of ERME.

One specific feature of the TWG has been the thematic discussions, which, starting from the contributions of the participants, addressed various general issues related to conducting research in the two strands. While the history *in* mathematics education strand addresses actual uses of history of mathematics in teaching and learning, the other strand addresses the very history *of* mathematics teaching and learning. We will argue that this distinction and the thematic discussions of both strands ended up

240 Jankvist & van Maanen

being a strength of the work carried out in the TWG, not least in terms of the three ERME Cs of Communication, Cooperation and Collaboration.

2 The pre-ERME period

Early interest in the connection between the history of mathematics and mathematics education goes back to the nineteenth century. Fauvel (1991) and Tzanakis and Arcavi (2000, p. 202) present some early cases. Fauvel quotes George Heppel, who in 1893 presented three conditions for 'The use of history in teaching mathematics': (1) history should be auxiliary, (2) history should assist the student in learning mathematics and (3) history should not be tested on exams. Heppel feared that history would overload the 'ordinary schoolboy or schoolgirl', a fear that was also expressed by many later authors. A broad perspective was sketched by Otto Toeplitz, who introduced in mathematics education the so-called genetic principle, already used by biologists. In 1927 he pleaded for taking the historical development of mathematical ideas as a guide for teaching (cf. Fried & Jahnke, 2015).

History of mathematics, either as a topic to be taught or as an inspiration and background for mathematics teaching, became *en vogue* in several countries in the 1960s and 1970s. Widely distributed and influential was the 31st yearbook of the National Council for Teachers of Mathematics (NCTM) on 'historical topics for the mathematics classroom' (NCTM, 1969; updated 1989). The 1989 Preface describes it as "a pioneering effort to assist in the teaching of mathematics from a historical perspective" (p. vii). Also in the early 1970s, the executive committee of the International Commission on Mathematical Instruction (ICMI) decided to start with a so-called Study Group, affiliated with ICMI, which received the name International Study Group on Relations between History and Pedagogy of Mathematics; now better known as the HPM Study Group (see Fasanelli & Fauvel, 2006). The next phase can be characterised as a period of combined efforts. Joint work of teachers and academics within the French Instituts de Recherche sur l'Enseignement des Mathématiques (IREMs) was brought to the UK thanks to Evelyne Barbin and John Fauvel. IREMs were set up in the 1970s for research into mathematics teaching in France, which was carried out by practising teachers (more in Fauvel 1990, p. 139). The introduction in the UK started with a conference of the British Society for the History of Mathematics (BSHM) in 1998 in Leicester. Fauvel organised in 1990 the first *History in Mathematics Education* (HiMEd) conference, for which he made the French IREM materials available in English (Fauvel, 1990). Until 1996 there were HiMEd conferences of the BSHM every other year (Nottingham 1992, Winchester 1994, Lancaster 1996). After 2000 the question of the effect of the actual teaching with historical material became more and more important, also due to the influence of ERME.

2.1 Research in the pre-ERME period

Again, we distinguish between history *in* mathematics education and history *of* mathematics education. The second one speaks for itself: like art and politics, mathematics education has a history. There was a time when Calculus had not yet been found out, and a year (1696) when its first textbook appeared. There are teachers who integrate history of mathematics in their lessons. In teaching the logarithm, for example, they show how the logarithm in the 17th century speeded up big calculations by transforming time-consuming multiplications into additions. Doing so, they hope that their students see that $\log(a \cdot b) = \log(a) + \log(b)$ is not just a rule, but a rule to remember, because it changed the world. Such integration of historical elements is here referred to as history *in* mathematics education.

Research studies on history in mathematics education gradually appeared. Journals brought thematic issues, such as *For the Learning of Mathematics* (1991), *Educational Studies in Mathematics* (2007), and *Science & Education* (2014). Collections of papers published in the *MAA Notes* illustrate well the scope of the field (e.g. Katz, 2000; Katz & Tzanakis, 2011). An overarching view of the field is given in the ICMI-Study (Fauvel & van Maanen, 2000). A comprehensive account of empirical studies on history in mathematics education is available in Jankvist (2012), which also provides an overview of the academic fora of the field.

Interest in the history of mathematics education goes back to the 19th century. In the *Handbook on the History of Mathematics Education* (Karp & Schubring, 2014) Schubring reviews the historiography of teaching and learning mathematics. He locates early interest especially in Germany, with books about one particular Prussian Gymnasium (1843) and about methods for teaching arithmetic (1888). By the end of the century historical studies also appeared in several other countries. An important stimulus came from international cooperation within the Internationale Mathematische Unterrichts-Kommission (IMUK, 1908), which since 1952 continued as the ICMI. IMUK and ICMI ordered several broad international studies that generally concerned parts and aspects of the mathematics education of that period. Schubring also presents a number of studies about mathematics education in specific countries (Germany, England, USA and Canada, Finland, Russia).

Research about the history *of* mathematics education was explicitly put on ICMI's agenda at ICME10 (Copenhagen 2004). The goals then were to "gather the researchers working in this field [. . .] and develop research programmes, which enhance international perspectives and the study of the 'general' within national specific histories". The initiative connected to existing national activities, and had a considerable follow-up. A specialised *International Journal for the History of Mathematics Education* appeared in 2006, but publication terminated in 2016. Biennial conferences started in 2009 (Reykjavik, base of the important initiator Kristín Bjarnadóttir), followed by Lisbon (2011), Uppsala (2013), Turin (2015) and Utrecht (2017). Also, participation in the history TWG since CERME 6 much stimulates researchers in this field.

3 The creation of the ERME TWG on history

The creation of the CERME TWG should be seen as an attempt to create a forum and a platform for fostering empirical studies in the field of history in/of mathematics education and to also better link research in this field with research in mathematics education in general. This is reflected in the bullets of the 'call for papers/posters' that the TWG has operated with from CERME 6 through CERME 9:

1. Theoretical, conceptual and/or methodological frameworks for including history in mathematics education;
2. Relationships between (frameworks for and empirical studies on) history in mathematics education and theories and frameworks in other parts of mathematics education [this point featured only from CERME 7 onwards];
3. The role of history of mathematics at primary, secondary, and tertiary level, both from the cognitive and affective points of view;
4. The role of history of mathematics in pre- and in-service teacher education, from cognitive, pedagogical, and/or affective points of view;
5. Possible parallelism between the historical development and the cognitive development of mathematical ideas;
6. Ways of integrating original sources in classrooms, and their educational effects, preferably with conclusions based on classroom experiments;
7. Surveys on the existing uses of history in curricula, textbooks, and/or classrooms in primary, secondary, and tertiary levels;
8. Design and/or assessment of teaching/learning materials on the history of mathematics;
9. The possible role of history of mathematics/mathematical practices in relation to more general problems and issues in mathematics education and mathematics education research.

As can be seen from bullet 9 above, due to CERME's focus on present day mathematics education, a constraint was made that studies on the history of mathematics education should relate their historical results to current teaching and learning practice. On average, the TWG counts around 20 participants, but not always the same 20 people. Also, it is a rather mixed crowd. There is an overlap with the HPM community, but the TWG has also managed to attract newcomers. Some participants come from the history of mathematics while others are more rooted in mathematics education or mathematics proper. In hindsight, the TWG has been rather successful in combining young and established researchers. One of the main things expressed when the TWG has been evaluated is its friendly, inclusive and productive atmosphere. In 2011 one young participant said:

> A week ago . . . I didn't know how everyone in the working group would react to my work and my opinions (if I had enough courage to express them). Today I have in my memory the best conference I ever attended: a fantastic working group that made me desire for more opportunities to work with everyone.

Here we see the importance of the TWG for researchers: those in their starting phase; those from countries in which there is not a strong HPM community; and

History and mathematics education **243**

those who even have difficulties finding relevant sources and other documentation. For the following two sections we selected six papers from the output of the TWG from CERMEs 6 to 9. In making the selection we tried to represent both starting and experienced researchers, who come from different geographical backgrounds.

4 Three selected papers on history *in* mathematics education

4.1 Paper 1: Development of learning strategies and historical awareness

Kjeldsen (2011), whose background is in the history of mathematics, suggests a theoretical framework for discussing "how history benefits students' learning of mathematics, and develops students' historical awareness" (p. 1700). She draws on a framework by the Danish historian B. E. Jensen, who in 2010 outlined a broad approach to history. Jensen distinguishes between several approaches to history: a pragmatic vs. a scholarly approach to history; observer history vs. actor history; identity neutral vs. identity concrete history writing; and finally, a so-called 'living history' approach. Kjeldsen addresses mainly the first two pairs. In a pragmatic approach the use of history is guided by the idea that through history we are to gain knowledge of the world today. History is studied from a utility perspective. This is contrasted to historians' critical distance to past events in a more scholarly approach. From another perspective, if history is used to orient oneself and act in a present context, then it is referred to as actor history and history helps to intervene. If the past, however, is used retrospectively with a purpose to enlighten rather than to act or intervene, then Jensen refers to it as a use of the past from an observer perspective.

As an illustration, Kjeldsen argues that if the focus is on developing students' mathematical competencies (e.g. Niss & Højgaard, 2011), a pragmatic approach from an actor perspective may be considered. However, mathematical competencies may also be developed alongside the development of historical overview and awareness, in which case the weight should be on a scholarly approach, e.g. with an observer perspective. The paper ends with an empirical example of an in-service teacher who "used different approaches to history and used past episodes from various perspectives for different purposes" so that history "was used in ways in which students gained genuine historical insights, developed learning strategies, and enhanced their mathematical problem solving skills even though they worked on mathematics that might not be part of the core curriculum" (p. 1708).

4.2 Paper 2: Teaching the concept of tangent line using original sources

Mota, Ralda and Estrada (2013) report about a teaching experiment in two consecutive years of upper secondary school (grades 11 and 12), in which the concept of

244 Jankvist & van Maanen

tangent line is part of the mathematics curriculum. The authors wanted to determine the benefits and disadvantages of the introduction of the concept via the discussion of historical texts, chosen from Greek authors via 17th-century French mathematicians and Leibniz to a 1790 Portuguese textbook. In Grade 11, 21 students worked in groups on tasks about five texts from antiquity to Leibniz. Each text focused on aspects of the tangent. Typical tasks for students were: to answer the question 'Is Euclid's definition suitable for Archimedes' spiral?' and 'Using Fermat's method, determine the extremes of a 2nd degree polynomial function'. In Grade 12, half of the group of 22 students had also participated in Grade 11. The students appeared to be more receptive for the digestion of the different historical definitions via notes that each student wrote as a preparation for a classroom debate.

The paper is a fair example of the transitional phase that studies about history (as a tool) in mathematics education have been undergoing since 2000. Teacher and researcher, who often coincide in one person, have a strong positive attitude toward using history in mathematics education. In Mota et al.'s paper this is reflected for example in the paper's final paragraph, which starts with the words "it is our conviction". Also, the teaching material receives a clear and complete display. But no real evidence is presented to support the conviction that contact with the historical evolution of the tangent was helpful for students. Instead of presenting evidence, the authors give a list of the obstacles that they encountered during the experiment (for example, students arguing that mathematics is not history). The next step in the development of a paper like this would be to pay more attention to the learning of the students, and to step from 'conviction' to the 'evaluation of evidence'. This paper is part of a larger Portuguese PhD project and it might well be that a later phase of the project concentrated more on evidence about the learner.

4.3 Paper 3: History of mathematics and mathematical thinking competency

Being an American researcher in the field of HPM, Clark was, among other sources, introduced to the Danish mathematics competencies framework (Niss & Højgaard, 2011) through the TWG. As one of eight distinct, yet interrelated, competencies, this framework describes the mathematical thinking competency as

> being able to *recognise, understand and deal with the scope of given mathematical concepts* (as well as their limitations) and their roots in different domains; *extend the scope of a concept by abstracting* some of its properties; *understand* the implications of *generalising* results; and be able to *generalise* such results to larger classes of objects.
>
> *(emphasis in original; Niss & Højgaard, 2011, p. 53)*

Clark's thesis is that the history and philosophy of mathematics holds promising potential for having students think about essential mathematical concepts. Clark (2015) states that it is her hope that "The results of the present research may enable instructors of history and philosophy of mathematics (HPhM) courses to consider the power of the course on improving mathematical understanding and could find it beneficial to revise courses to capitalise on this implication" (p. 1805).

More precisely, the primary goal of the study is to link college students' experiences from a course on HPhM to potential changes in their mathematical thinking competency with regard to three selected fundamental mathematical concepts/constructs: infinity; the complex number system; and the axiomatic structure in mathematics. Through a method of 'think-aloud task interviews' Clark is able to address selected students' progression of the thinking competency in relation to the structure of the number fields (up to complex numbers). In general, students' possession and development of the mathematical thinking competency has not been researched to the same extent as some of the other competencies (e.g. modelling, problem tackling, reasoning). Hence, also from a non-HPM perspective Clark's study brings something new to the table.

5 Three selected papers on history *of* mathematics education

5.1 Paper 4: Arithmetic in Brazilian primary school in late nineteenth century

Da Costa's (2010) paper shows a valuable feature of the CERME conferences: CERME assembles researchers from a variety of cultural and national backgrounds. The author was a doctoral student from Brazil, who worked for a year in Paris in a collaborative team of doctoral students. The Brazilian-French connection is clearly recognisable in the structure of the paper, which researches Brazilian arithmetic teaching around 1900 from a French perspective, especially Chervel's 1998 framework of the history of school disciplines. Central in Chervel's framework is the *vulgata*, the standard set of educational principles and materials shared by most of the teachers. Part of the *vulgata* are representative teaching materials. For arithmetic education in Brazil, da Costa describes the *Cartas de Parker* (Parker's Cards, named after Francis Wayland Parker, 1837–1902) and their introduction in Brazil. Parker connected with his cards, on which numbers and relations between them were represented geometrically, to the ideas of Grube and followers in Germany. The introduction of the cards in Brazil was much stimulated by a series of positive articles in a pedagogical journal. The teaching method was heuristic, discovery by the pupils was stimulated and the strict order of teaching (addition → subtraction → multiplication → division) was given up, as was the importance of memorising.

The paper is a fair example of the power of crossing borders: national borders in the 19th century and also in the research community of the 21st century.

5.2 Paper 5: Evaluation and design of historical mathematics curricula

As mentioned, participants at CERME come from different backgrounds. Quite common is the educational professional who on a part-time basis researches the area in which she or he works. Several teachers of mathematics research their teaching practice supported by an academic supervisor. Such research has produced interesting contributions to CERME and sometimes also leads to a doctorate. The paper by Krüger and van Maanen (2013) is one written against the background of professional experience, in this case in curriculum development. It presents the research questions and structure of a PhD project, which could also be positioned within the area of curriculum studies. The main aim is to understand the process of curriculum development from a historical perspective. Three historical mathematical curricula are studied, from the 17th, 18th and 19th centuries, respectively, and in each case the factors and actors are identified that influence to a high degree the content of the new curriculum. In each case there appeared to be socioeconomical influences next to important personal influences.

The analysis of the historical cases produces valuable conclusions for current-day curriculum development, a comment that was also made in the reviews of two fellow participants. Comparison of the cases shows that through time there is a considerable shift in the heart of the curriculum. In the curricula of the 17th and 18th centuries there was only a very global indication of the mathematical subjects, whereas much attention was paid to the circumstances in which the students had to learn and live, and to work after they had left the school. In the 19th century the mathematical subjects were described in more detail, and nowadays the curriculum, at least in the Netherlands, is primarily a description of subjects to the minutest detail. One of the reviewers concludes: "This is a study . . . which gives us things to think about, in particular when re-defining the procedures for defining the curriculum."

5.3 Paper 6: The first periodical on mathematical sciences in Ottoman Turkey

The paper by Alpaslan, Schubring and Günergun (2015) is an example of collaboration between a young PhD student (Alpaslan) and two experienced researchers. They discuss the periodical, *Mebahis-i İlmiye*, which was published between 1867 and 1869 in Ottoman Turkey. The periodical presented a wide variety of pure and applied mathematics, often related to physics and engineering, and including also vocational mathematics. The specific questions that the authors seek to answer are: (1) For what reasons did *Mebahis-i İlmiye* provide 19th-century Ottoman Turkey society with mathematics education? How were the reasons addressed in its content? (2) On which mathematical traditions did the periodical rely? These questions are addressed through the three categories of fundamental reasons for mathematics education elicited by Niss (1996): (i) the technological and socioeconomic development of society; (ii) the political, ideological and cultural maintenance and development of society, and finally (iii) providing individuals with prerequisites that might help them to cope with life in general. Another theoretical construct used is transmission of mathematical knowledge as a dissemination process of mathematical ideas from the scientifically established 'metropolis' countries to the not yet scientifically productive countries in the 'periphery' (Schubring, 2000):

Findings indicate that *Mebahis-i İlmiye* addressed all the three kinds of reasons for mathematics education (Niss, 1996) to a certain degree. The authors utilised transformation of the recent knowledge of both pure and applied mathematics from Europe, mainly from France, as the 'metropolis' of the time (Schubring, 2000). Reception occurred in the difficult social setting of conflicts between modernisers and traditionalists, and within the already existing culture of Islamic mathematics. An important aspect of this transmission was the development of a terminology for the modern mathematics in Ottoman Turkish language, since the traditional mathematics did not provide terms for the new developments in the field.

(Alpaslan, Schubring & Günergun, 2015, p. 1788)

The study also illustrates the importance of local (national) driving forces in society, in this case three such individuals: Yusuf Ziya Bey (1826–1882), who conceived the idea of the periodical; Ahmet Muhtar Bey (1839–1919), a science teacher at the military academy; and Hüseyin Tevfik Bey (1832–1901), a remarkable mathematician trained both in the context of traditional Islamic mathematics and European mathematics, thus being the "appropriate person for transmitting European mathematics into the Ottoman context of mathematics" (ibid., p. 1783).

6 The distribution of research topics in papers and posters

In an attempt to indicate what topics received the main interest from CERMEs 6–10, we have categorised all papers and posters in the TWG according to the subdivision in the call for papers, discussed above, keeping in mind that a paper might belong to more than one category. This survey reveals to the following conclusions:

- One-third of the papers and also one-third of the posters are on the history *of* mathematics education, against two-thirds on history *in* mathematics education.
- Teaching and learning materials, such as textbooks, receive more attention than historical teaching practice. The main interest in the history *of* mathematics education was on its social, cultural and political aspects.
- In the field of history *in* mathematics education the main interest was on the way history is applied in primary, secondary, tertiary education and in teacher education. Frequent also were papers about teaching materials (showing how history can be used in mathematics teaching) and theoretical frameworks for including history in mathematics education.
- The activities of one research group is well recognisable in CERME 7, when four out of five posters came from a Portuguese group of PhD students, who worked on related research topics. Over all five conferences there were 13 posters.

248 Jankvist & van Maanen

This survey does not reveal important aspects of research, such as the chosen methodology, or a possible research paradigm, or to what extent a paper focuses on the learner. For this, we refer to the six selected papers and the thematic discussions.

7 Development of the thematic discussions in the TWG

Through the years the work of the TWG underwent a notable development. The focus shifted from content of the individual papers to the discussion of aspects that the papers had in common. After a more traditional organisation of the TWG sessions in CERME 6 (Furinghetti, Dorier, Jankvist, van Maanen & Tzanakis, 2010), thematic discussions and shortened presentations of individual papers were introduced for CERME 7. Themes were found among the general issues that had remained un(der-)discussed in the previous CERMEs. These themes were also fitted to the specific composition of the TWG, which in CERME 7 consisted of a relatively high number of younger (beginning) researchers. In particular, further reflection upon aspects of methodology and upon connections with mathematics education research more generally were brought into play. This is seen also from the four general themes discussed at CERME 7: (1) research questions and relevance of the research; (2) use of HPM theoretical constructs vs. mathematics education theoretical constructs; (3) methods, data and analysis; and (4) validity, reliability and generality of the reported research results. For the younger researchers the discussions of theme 1 had the outcome that several of them would "reconsider their research aim(s), formulate questions, refine formulations of existing questions, or expand their research perspectives" (Jankvist, Lawrence, Tzanakis & van Maanen, 2011, p. 1638). Among the experienced researchers the following key issues were identified in relation to theme 2: "[to] *provide some order in the wide spectrum of research and implementations* [. . .], *to somehow check the efficiency of introducing a historical dimension*, [. . .] *to convince the target population* [. . .]; *and to develop appropriate conditions for designing, realising, and evaluating our research*" (ibid., p. 1638, italics in original). At CERME 7 it was decided that in the future an effort should be made to make it easier for the poster-presenters, who oftentimes are the young researchers, to join in the thematic discussions by providing them with short timeslots for a brief and early presentation of their posters. When practised at CERME 8, this indeed had the intended effect.

For CERME 8 five themes for thematic discussions were identified based on the submitted papers: (1) interdisciplinarity; (2) theoretical frameworks for history *of* mathematics education; (3) history in high school/upper secondary school mathematics education; (4) history in pre-high school mathematics education; (5) history of mathematics in teacher education. New in CERME 8 was that discussion questions related to the themes were ready and sent out prior to the congress, which too was based on previous experience. As an example, for theme 2 questions were: What is in this respect the difference between 'story' and 'history'? A supplemental suggestion was made that the difference might very well be related to the use of various (theoretical) frameworks. Another question asked which theoretical frameworks are actually available for research in this strand? And yet another question raised the issue of to what extent the history of mathematics education requires study of primary or original sources, documents, etc.? Among the participants in the TWG, there seemed to be some agreement

> about *story* being something narrative, whereas *history*, although it may contain narratives (or stories), is structured by theoretical frameworks, the purpose of which includes being able to see benefits or limitations, to communicate results, and to enable the researchers to organise and present findings, assertions, etc. [. . .] The participants, for example, point to constructs from history research, e.g., those of more externalist historiography of studying factors crucial to the development of institutions, etc.
> *(Jankvist, Clark, Lawrence & van Maanen, 2013, pp. 1947–1948)*

The issue of narrative and story vs. history came up also for theme 4, which among others was centered around questions of special challenges when using history in primary school, kindergarten, etc. and how to stay true to history, i.e. non-Whig, when applying it at pre-high school levels? In the report it is noted that

> in practice when using history at younger age levels there is a need for compromise, also in order to make the mathematics itself more accessible to the children [. . .] there may be the need for narratives in the form of telling stories of mathematics, rather than confronting them with the actual history of mathematics. But as one of the subgroups state in its report: 'You have to tell stories, but the knowledge of history enables you to tell *true* stories.'
> *(ibid., p. 1948)*

In CERME 9 a set of discussion themes were again identified (Jankvist, Chorlay, Clark & Lawrence, 2015). This time an attempt was made to draw attention to the circumstance that most often HPM research is not concerned with large-scale quantitative studies or even large-scale historical surveys. In regard to this local/ global tension the following questions were phrased: Can large-scale surveys (e.g. history of algebra, notion of proof from Euclid to Hilbert, evolution of the concept of function, etc.) go beyond the 'bird's eye view'? Can we elicit necessary conditions for such large-scale surveys to make any sense? Another theme addressed was that of 'meta-level or methodological reflections' exemplified through questions such as: What (if any) is or could be the role assigned to epistemological/historical reflection in some major mathematics education theoretical frameworks, e.g. *Theory of Didactical Situations* (TDS), *Mathematical Knowledge for Teaching* (MKT), etc.? As for the outcome of discussions related to these themes, "several participants shared the view that using a general survey of history (i.e. 'global view') helps to create a cultural landscape, which includes and accommodates multiple tools, concepts, and ideas – and which establishes a meaningful lens to use from the outset" (ibid., pp. 1780–1781). Yet a theme, which is ever present within the

TWG, is the role of history of mathematics in teacher training. This theme was explicitly taken up during the thematic discussions of CERME 9 by means of the question: What minimal/satisfactory level of command of history of mathematics can we reasonably attempt to achieve in teacher training? During discussions, this question was addressed through two 'sub-issues': criteria for being an *able reader* (such as having the ability to assess a primary source with a critical mind) and the *epistemological tool-box* (which contains the descriptive/analytical concepts we wish to provide to teachers). Finally, it was discussed to what extent we shall expose (future)-teachers to elements of history of mathematics which have no direct connections with classroom contents (in particular to enrich their 'image' of the parts of higher mathematics, which they studied but will not teach). Reflections, also methodological ones, on actual teacher-training modules were made. The discussions led to further questions, e.g. how to go beyond the mere 'raise awareness' objective, how to objectify the impact on student-teachers, and not least how to stabilise any potential impact.

The history TWG at CERME is usually one that covers aspects of the entire educational system, i.e. from primary school to higher education and teacher training, which was the case during CERME 10 in Dublin as well (Chorlay, Clark, Gosztonyi & Lawrence, 2017). At this meeting some of the themes for thematic discussions included: innovative teaching experiments; original sources or original problems, techniques, or experiments; as well as history *of* mathematics education. In relation to the latter the following question was proposed: 'A provocative rephrasing of the theme: Is the history of mathematics education a part of general history or a part of mathematics education? Which side do you fall on and why?' Some of the topics that were picked up for further discussion during the TWG sessions were: connections with the didactics of physics (kinematics); connections with current trends in the philosophy of mathematics; investigation of teacher-practices when using historical sources; connections between history of education and history of theoretical frameworks in mathematics education research. At CERME 10, 11 of the participants of the TWG were newcomers to CERME, which hopefully indicates a still growing interest in the HPM topics, in particular among younger researchers.

What might we conclude from the papers and discussions in five CERME conferences since CERME 6? Most remarkable is the shift from individual contributions to shared insight. Already in CERME 7 common aspects of contributions were discussed, such as how to select productive research questions, what methodology to use and what data to collect and analyse. Exchange of ideas and experience fostered the construction of common standards for sound research in the subfields of history *in/of* mathematics education, much in line with the spirit of ERME.

8 The Cs of ERME

One characteristic of ERME is its three Cs. Former ERME president Ferdinando Azzarello used to recite *Communication, Cooperation* and *Collaboration* – stressing their importance for ERME's activities. Much of the *Communication* within the history TWG took place during the thematic discussions. These exchanges contributed considerably to forming the 'identity' of the TWG and over time some 'common research standards'. One example of this is paper 3 (Section 4.3) by Clark, in which a use of history in the US is addressed through the Danish framework of mathematical competencies (Niss & Højgaard, 2011). So, not only is this a result of more general mathematics education research informing HPM research, it is also an example of European frameworks finding their way across the ocean to inform research studies carried out in a somewhat different cultural setting. Another example is paper 6 (Section 5.3) by Alpaslan, Schubring and Günergun, where frameworks developed by a German researcher (Schubring, 2000) and a Danish researcher (Niss, 1996) structure a cultural and historical case from the Ottoman Empire. This is an example of theory from general mathematics education informing not history *in* but history *of* mathematics education. Paper 6 is also a fair illustration of *Cooperation/Collaboration* between Alpaslan as a young and promising researcher and two rather experienced researchers. Alpaslan participated in the history TWG since CERME 7, when still a Master's student. At CERMEs 7 and 8 he presented papers related to history *in* mathematics education. No doubt, due to his participation in the TWG, Alpaslan had become acquainted with the history *of* mathematics education, so he began researching the history of mathematics education in Turkey, which led to this joint paper 6. Furthermore, the topic of the paper was not the core research topic of Alpaslan; his PhD work was concerned with the use of primary historical sources in teacher education and how teachers could apply aspects of these in their own practice (Alpaslan & Haser, 2015). Sadly, due to the tragic passing away of both Alpaslan and his wife in a traffic collision on 31 July 2015, this work remains unfinished.

Related to *Communication*, in order to convince the mathematics education community of the benefits of history for mathematics education, the 'history people' need to 'speak' the established language of this community. In this sense, the TWG can be seen as a spearhead in getting the HPM 'hinterland', which has its own idiom and frameworks, to cooperate toward being understood by the educational researchers. Also, the TWG has provided a perfect habitat for testing which general frameworks, notions, etc. would apply to history and mathematics education. Paper 2 (Section 4.2), for example, reports about classroom experiments. The authors investigated if studying and discussing historical appearances of the concept of tangent would help students in their concept building. Another aspect of *Communication* is that CERME gave the first author of paper 2, who was preparing a PhD, the possibility to present and discuss her work with several experts in her field. This also holds for paper 3 (Section 4.3). CERME gave its author, who is from Brazil and who was working within the French research culture, the chance to communicate with researchers from other research cultures. Paper 5 (Section 5.2) is an example of cooperation between a professional curriculum developer who was preparing a PhD about historical mathematics curricula (Krüger) and the thesis supervisor (van Maanen). The wide range of findings led to the possibility of asking and answering a number of general questions.

The history TWG reflects the diversity of the HPM community. This diversity has fostered rich discussions and reflections. An example is paper 1 by Kjeldsen (Section 4.1). Inspired by the Danish historian B. E. Jensen, Kjeldsen proposed to distinguish between a pragmatic and a scholarly approach to history, and between observer history and actor history. At the time of CERME 7, this fostered intense discussion among the more experienced researchers, who asked from which other fields history in mathematics education gets its inspiration in terms of frameworks, theoretical constructs, etc. The conclusion was that it does so from a wide range of fields (from mathematics to the social sciences, etc.). Another reflection was that, in order to 'do well' in relation to research on history *in* or *of* mathematics education, one must possess a profound knowledge of the history of mathematics – and/or the history of mathematics education – which indeed requires a firm mathematical background, and at the same time one must be well read in mathematics education. Since this 'trinity' (history, mathematics, and education) is not always easy to come by, the field of history and mathematics education is a perfect place for *Cooperation/Collaboration* between researchers with different backgrounds.

CERME's history TWG is important as a platform where starting and established researchers are open to meet each other, and where research cultures meet. Next, the rules on how to behave on this platform, especially the requirement to submit a paper, to peer-review papers and to read the papers before the conference, give participants a fair chance to be successful. For us, this indicates once again the value of historical knowledge and research for mathematics education.

References

Alpaslan, M., & Haser, Ç. (2015). Selecting and preparing original sources for pre-service mathematics teacher education: The preliminary of a dissertation. In E. Barbin, U. T. Jankvist, & T. H. Kjeldsen (Eds.), *History and Epistemology in Mathematics Education. Proceedings of the Seventh European Summer University ESU 7, Copenhagen, Denmark 14–18 July 2014* (pp. 451–463). Copenhagen, Denmark: Danish School of Education, Aarhus University.

Alpaslan, M., Schubring, G., & Günergun, F. (2015). 'Mebahis-i İlmiye' as the first periodical on mathematical sciences in the Ottoman Turkey. *CERME9* (pp. 1783–1789).

Chorlay, R., Clark, K., Gosztonyi, K., & Lawrence, S. (2017). Introduction to the papers and posters of TWG12: History in mathematics education. *CERME10* (pp. 1681–1684).

Clark, K. (2015). The contribution of history of mathematics on students' mathematical thinking competency. *CERME9* (pp. 1804–1810).

da Costa, D. A. (2010). Arithmetic in primary school in Brazil: End of the nineteenth century. *CERME6* (pp. 2712–2721).

Fasanelli, F., & Fauvel, J. (2006). History of the international study group on the relations between the History and Pedagogy of Mathematics: The first twenty-five years, 1976–2000. In F. Furinghetti, S. Kaijser, & C. Tzanakis (Eds.), *Proceedings of HPM 2004 & ESU 4* (pp. x–xxviii). Iraklion, Greece: University of Crete.

Fauvel, J. (1990). *History in the mathematics classroom: The IREM papers.* Leicester, UK: The Mathematical Association.

Fauvel, J. (1991). Using history in mathematics education. *For the Learning of Mathematics 11*, 3–6.

Fauvel, J., & van Maanen, J. (Eds.) (2000). *History in mathematics education: The ICMI Study*. Dordrecht: Kluwer Academic Publishers.

Fried, M. N., & Jahnke, H. N. (2015). Otto Toeplitz's 1927 paper on the genetic method in the teaching of mathematics. *Science in Context, 28*, 285–295.

Furinghetti, F., Dorier, J. L., Jankvist, U. T., van Maanen, J., & Tzanakis, C. (2010). Introduction: Theory and research on the role of history in mathematics education. *CERME6* (pp. 2679–2681).

Jankvist, U. T. (2012). A first attempt to identify and classify empirical studies on 'History in mathematics education'. In B. Sriraman (Ed.), *Crossroads in the history of mathematics and mathematics education* (pp. 295–332). The Montana Mathematics Enthusiast Monographs 12. Charlotte, NC: Information Age Publishing.

Jankvist, U. T., Chorlay, R., Clark, K., & Lawrence, S. (2015). Introduction to the papers and posters of TWG12: History in mathematics education. *CERME9* (pp. 1779–1781).

Jankvist, U. T., Clark, K., Lawrence, S., & van Maanen, J. (2013). Introduction to the papers and posters of WG12: History in mathematics education. *CERME8* (pp. 1945–1950).

Jankvist, U. T., Lawrence, S., Tzanakis, C., & van Maanen, J. (2011). Introductory report from WG12: History in mathematics education. *CERME7* (pp. 1636–1639).

Karp, A., & Schubring, G. (Eds.) (2014). *Handbook on the history of mathematics education*. New York: Springer.

Katz, V. J. (Ed.) (2000). *Using history to teach mathematics: An international perspective*. Washington, DC: Mathematical Association of America.

Katz, V. J., & Tzanakis, C. (Eds.) (2011). *Recent developments on introducing a historical dimension in mathematics education*. Washington, DC: Mathematical Association of America.

Kjeldsen, T. H. (2011). Uses of history in mathematics education: Development of learning strategies and historical awareness. *CERME7* (pp. 1640–1649).

Krüger, J., & van Maanen, J. (2013). Evaluation and design of mathematics curricula: Lessons from three historical cases. *CERME8* (pp. 2030–2039).

Mota, C., Ralda, M. E., & Estrada, M. F. (2013). The teaching of the concept of tangent line using original sources. *CERME7* (pp. 2048–2057).

NCTM Yearbook (1969). *Historical topics for the mathematics classroom*. Reston, VA: NCTM. (Updated edition 1989).

Niss, M. (1996). Goals of mathematics teaching. In A. J. Bishop, M. A. Clements, C. Keitel, J. Kilpatrick, & C. Laborde (Eds.), *International handbook of mathematics education* (pp. 11–48). Dordrecht: Kluwer Academic.

Niss, M., & Højgaard, T. (2011). *Competencies and mathematical learning ideas and inspiration for the development of mathematics teaching and learning in Denmark*. Danish edition 2002, English edition, October 2011. IMFUFA tekst no. 485. Roskilde, Denmark: Roskilde University.

Schubring, G. (2000). The first international curricular reform in mathematics and the role of Germany: A case study in the transmission of concepts. In A. Gagatsis, C. P. Constantinou, & L. Kyriakides (Eds.), *Learning and assessment in mathematics and science* (pp. 265–287). Nicosia, Cyprus: Department of Education, University of Cyprus.

Tzanakis, C., & Arcavi, A. (2000). Integrating history of mathematics in the classroom: An analytic survey. In J. Fauvel & J. van Maanen (Eds.), *History in mathematics education: The ICMI Study* (pp. 201–240). Dordrecht: Kluwer Academic Publishers.

18

THEORETICAL PERSPECTIVES AND APPROACHES IN MATHEMATICS EDUCATION RESEARCH

Ivy Kidron, Marianna Bosch, John Monaghan and Hanna Palmér

1 Introduction

Dealing with different theoretical perspectives and approaches in mathematics education is in accord with the spirit of ERME, especially with regard to communication, cooperation and collaboration. At the time of writing this chapter, seven Thematic Working Groups (TWGs) on theoretical perspectives have been organised. The first was held at CERME 4 in 2005. This chapter offers a reflection on the development of the TWG since then. Besides the TWG, two plenary sessions have been devoted to the theme of theories, one at CERME 5 in 2007 and one at CERME 6 in 2009. We are aware that the perspectives about research and the place of theory in research, as well as the metalanguage that accompanies these perspectives come with different connotations and baggage. Nevertheless, we will use the terms 'theory', 'theoretical approach', 'perspective' and 'theoretical framework' interchangeably.

The chapter is structured as follows. Section 2 discusses the discourse on theories in mathematics education research (MER) until 2005. This short overview serves as background for the next sections. Sections 3 and 4 include sustained development of ideas by the CERME TWG on theoretical perspectives over a number of CERME conferences. They are the core of the chapter. They are presented as separate sections but the issues are closely interrelated. Section 3 deals with the diversity of theoretical perspectives and networking theories, understanding how theories can be successfully connected while respecting their underlying conceptual and methodological assumptions. In Section 4, questions concerning the role of theory, the relationships between theory and practice and the dynamic character of theory are discussed with regard to the diversity of theoretical approaches and with networking in the background. Section 5 deals essentially with ideas that arose within CERME and initiated or influenced, at least partially, developments

Theoretical perspectives and approaches **255**

in mathematics education beyond ERME. In the last section, we reflect on what has been achieved within the TWG and outline a vision for the future.

2 Theoretical perspectives circa 2005

The CERME TWG on theoretical perspectives was formed in 2005 but it was not formed in a vacuum. The purpose of this section is to outline the 'theoretical context' the TWG was born into. Due to word length constraints we do not explain the theoretical perspectives to which we refer but expositions of these frameworks can be found in the *Encyclopaedia of Mathematics Education* (Lerman, 2014).

Cognitive perspectives dominated MER in the 1980s and the influence of Piaget's developmental psychology was strong. In anglophone MER this influence can be seen in radical constructivism (von Glasersfeld, 1995); in francophone MER this influence can be seen in the Theory of Didactic Situations (Brousseau, 1997). But this influence was not uniform. For instance, as Radford points out, the dialectic between cognition and the social realm differ:

> For constructivist approaches . . . the social realm is considered as a mere facilitator of the individual's development of these structures. In the Theory of Didactic Situations, cognition is also conceived of in an adaptive manner, but the social realm is thematized as a 'milieu' and a game that the individual plays with it.
>
> *(Radford, 2008, p. 320)*

By the end of the millennium many researchers were influenced by 'the social turn', "the emergence into the mathematical education community of theories that see meaning, thinking, and reasoning as products of social activity" (Lerman, 2000, p. 23). The social turn represented a wide variety of theoretical perspectives (Activity Theory, the Anthropological Theory of the Didactic, Situated Cognition, Social Constructivism and more). The period around the end of the millennium also witnessed increased communication between mathematics education researchers across (and beyond) Europe, coupled with mutual attempts to understand each other's theoretical perspective (which falls under what we call now 'networking theoretical perspectives'). An example of this is the forum in 2002 of the *International Group for the Psychology of Mathematics Education* (PME) devoted to mathematical abstraction in which representatives of different theoretical perspectives were invited to engage in critical dialogue.

Turning our attention to CERME we note that theoretical perspectives were explicitly considered before 2005. One example is in the TWG on 'Research paradigms and methodologies and their relationship to questions in mathematics education' (Hejny, Shiu, Godino & Maier, 1999) at CERME 1. The focus of the group was "the nature, roles and function of theories in mathematics education" (p. 212) and questions discussed were: What is a theory and how do theories arise? What is the role and function of theory in the research process? Is it feasible and useful to distinguish

256 Kidron et al.

between different kinds of theories? How can theories in mathematics education be evaluated? This TWG also noted that the research paradigms discussed seemed to be bound to the cultural or national contexts in which they were developed.

Theoretical perspectives were also raised in CERME discussions on empirical issues in MER, for example, in the TWG 'Inter-relating theory and practice in mathematics teacher education' (Jaworski, Serrazina, Koop & Krainer, 2004) at CERME 3. This TWG focused on the relationship between theories and practice and questions discussed included: What is the difference between applying theory and practising theory? Does a theoretical framework allow us to make more sense of what we see from reflection on and analysis of practice? To what extent should theories be explicit or implicit for researchers and teachers? And is practice a theory generator and does theory impact on existing practice?

So, we see that, by the time of CERME 4, there was an emergent community with a special interest in the role and function of theories as well as on how to deal with diversity of theories.

3 Plurality of theoretical perspectives: networking theories

In this section, we analyse the progression of knowledge through the work of the TWG during the different CERMEs. The following subsections are presented as questions that motivated the work of the TWG. We describe attempts to address these questions as well as outcomes from these attempts.

3.1 Diversity of theories

The discussion in the TWG started with the questions:

Why diversity? Is diversity a necessity in the field or the sign that the field is in an immature state?

At CERME 4, Artigue, Bartolini-Bussi, Dreyfus, Gray and Prediger (2006) observed that the plurality of theories results from the influence of theoretical constructions and approaches initially developed outside the field of MER (psychology, sociology, anthropology, mathematics, linguistics and epistemology) and that progressively become genuine constructions of mathematics education. They also observed "a more intrinsic diversity linked to the diversity of educational cultures, and to the diversity of the institutional characteristics of the development of the field of mathematics education in different countries" (ibid., p. 1240). In her plenary lecture at CERME 6, Bikner-Ahsbahs (2010) argued that the diversity of theories is scientifically necessary. Already at CERME 4, the TWG participants agreed that diversity is both a source of richness and a source of fragility for research if we do not make scientific efforts to counterbalance the difficulties that stem from communication. This position was widely accepted in the TWG in the next CERMEs and led to new questions:

Theoretical perspectives and approaches **257**

How to deal with diversity? How to interact and to know each other?

Diversity is considered as a challenge for a community that intends to enable communication and progress between researchers of different theoretical frameworks. Therefore, the question how to deal with the diversity, complexity and richness of theoretical perspectives in mathematics education has been the core of the work of the TWG since CERME 4 and led to the idea of networking.

3.2 The idea of networking

Raising the question of dealing with the diversity of approaches was a first step. The expression 'networking' soon appeared in the TWG as a way to highlight the question without directly answering it. Networking was first seen as an opportunity to become aware of and compare theoretical standpoints. This was initially done by: separate analyses of the same data via methods appropriate to different theoretical frameworks and then considering and reflecting on each analysis; or by analysing the interactions of different theories as they are applied to the same empirical study. Important questions arose from various TWG discussions:

Why is networking important? What are the aims of connecting theories? What are the problems which arise in the efforts to network theories? To what extent does the networking depend on the type of theories that are considered?

There was a consensus in the TWG on the importance of networking as a first step on the "path towards establishing mathematics education as a scientific discipline" (Artigue et al., 2006, p. 1242). Motivation for networking resulted from the fact that learning, especially in classrooms, is a complex phenomenon. Different theoretical approaches have, to some extent, arisen from a dominant focus on one or more aspects of this complex phenomenon. But since learning, whatever it is, goes on (or not) regardless of how we might analyse it, the question of connections between theoretical approaches to understand this learning arises. Networking tries to provide tools to address this question. The benefits and problems that arise in the efforts to network theories are described in the following subsections.

3.3 Explicitness and awareness

To what extent do we share the same notion of theory? Do we share the same degree of explicitness and awareness of the theoretical assumptions?

Explicitness and awareness of the underlying assumptions of each theory serve as a basis toward networking:

258 Kidron et al.

> It would be useful to make explicit the level at which a theory operates . . . Only on the basis of such awareness (of the theoretical assumptions), can a discussion on the possible coherence of underlying assumptions begin to take place so that a common language supporting such networking can be developed.
>
> *(Artigue et al., 2006, p. 1243).*

Among the underlying assumptions of each theory, "the assumptions on the nature of mathematical knowledge appear to be one key point for the analysis of similarities and differences between approaches" (Arzarello, Bosch, Lenfant and Prediger, 2007, p. 1622). The nature of knowledge and learning as individually or socially constituted was discussed in the first TWG as a difference that might be an obstacle in connecting theories. But by the second TWG cases of networking were analysed that demonstrate attempts at new conceptualisations of the individual/social interplay. For example, Kidron, Lenfant, Bikner-Ahsbahs, Artigue and Dreyfus (2007) compare the different roles that the social interaction can play within three different theories. New questions arose such as what can be connected and what are the difficulties encountered in connecting theories

3.4 Aims and limitations of networking

What does networking mean? Is networking possible? What are the different aims of networking? What are the characteristics of the different cases of networking theories?

Even at the outset, in the CERME 5 TWG, comparing different initial strategies for networking made it clear that networking was important. Different aims in the efforts to network theories were differentiated. In some cases, the researchers were interested in the complementary insights that are offered when given data or an empirical phenomenon is analysed with different theories. In other cases, the interest in the rich diversity of theories was to explore the insights offered by one theory to the other. In Arzarello et al. (2007, pp. 1625–1626) examples of different profiles of networking are presented; it also includes an initial attempt to understand the use of different theoretical perspectives in research on technology enhanced learning in MER. The next step was to explore the limits of such networking efforts (Kidron et al., 2007).

3.5 Different kinds of dialogues/interactions between theories

What do researchers do when they use more than one theory? Do different approaches use the same words with the same meanings?

In analysing the source of problems which arise in efforts to network theories, new discussions took place in the TWG at CERME 6. The participants distinguished

Theoretical perspectives and approaches **259**

between different kinds of dialogues between theories. The word 'dialogue' was used because networking does not necessarily imply changes in the theories at stake. The word 'dialogue' was also used not only to describe that which enables mutual understanding in the way we communicate our theories but also to emphasise differences in the use of language. The TWG observed that even 'neighbouring approaches' (theoretical approaches that were born in the same educational and didactic culture) do not use the same words with the same meanings. This fact was observed in the different kinds of dialogues between neighbouring approaches that are offered in the papers by Ligozat and Schubauer-Leoni (2010), by Sensevy (2010) and by Artigue, Bosch, Gascón and Lenfant (2010). These dialogues were explored in the TWG and important differences between theories were observed.

3.6 Toward a theory of networking theories

What is the methodology for 'networking theories'?

The idea of a "theory of networking theories" emerged at CERME 5 (Arzarello et al., 2007, p. 1627). The challenging task of networking "approaches for networking theories" was an important part of the discussion in the TWG at CERME 7. A central term that emerged from the TWG at CERME 7 was 'transformation'. After networking, researchers are often transformed and are able to approach phenomena which they could not see before. Conditions which permit fruitful dialogues were discussed in the TWG. Artigue, Bosch and Gascón (2011) consider the potential offered by the Anthropological Theory of the Didactic (ATD), with its central notion of praxeology, for addressing issues of networking between theories. Other approaches for networking were proposed in the TWG and an effort of networking "approaches for networking theories" started (Kidron, Bikner-Ahsbahs, Monaghan, Radford & Sensevy, 2011, p. 2377). It was also noted that strategies for networking depend to an important extent on how close (or far) the networked theories are located in the semiosphere (Radford, 2008). At CERME 8, the work continued with a focus on theory and methodology. Strategies for networking theories as described in Radford's semiosphere (2008) and Artigue et al. (2011) were discussed. At this stage of work, the TWG had reached a certain maturity, allowing it to go more deeply into the epistemological question.

3.7 Epistemology and networking

What is the role of epistemology in networking?

The claim that "the study of any didactic phenomena needs to question common epistemological models of mathematics" (Bosch, Chevallard & Gascón, 2006, p. 1256) was already present at CERME 4. The maturity of thinking in the TWG brought back these questions at the heart of the reflection. Epistemological principles about how to interpret mathematical knowledge were discussed at

CERME 6. At CERME 7, awareness of underlying assumptions of different theories also concerned the nature of mathematical objects (for example, Font, Malaspina, Giménez & Wilhelmi, 2011). Epistemological aspects of theories were further developed in the TWG. The construct 'reference epistemological model' was presented at CERME 8 to highlight this dimension (Ruiz-Munzón, Bosch & Gascón, 2013) and used at CERME 9 in Florensa, Bosch and Gascón (2015). At CERME 9, we reflected on epistemology and networking of theories. Kidron (2015) analysed how the epistemological analysis is tightly linked to the cultural dimension and provided an example of networking that demonstrates how the social dimension might influence the epistemological analysis.

In the following, we observe the progression that has been achieved through the collective work in the TWG. The diversity in the nature of the results the TWG has produced is especially striking. We observe how networking was first interpreted as the ability to reflect on one's own work from the perspectives of others. As a consequence, it enables the awareness of one's own assumptions and insights offered by other theories. Then different aims of networking were introduced. Networking strategies were presented as well as difficulties in networking, especially the difficulties that accompany the use of language. A remarkable result consists of the methodological creativity the TWG has nurtured and structured during the different CERMEs. The place of methodology in theoretical frameworks was carefully analysed as well as the relation of this issue to the methodology for networking theories. Another important result of the work of the TWG resides in the analysis of the role of epistemology in networking. We end this section by highlighting the fact that the TWG contributed to building and disseminating a new culture, the networking culture. The TWG on theoretical perspectives at CERME 10 continued the work on networking but shifted attention to design research and tensions between home-grown and borrowed theories. Networking within a design research perspective raises more complex questions at the level of the conception of the design. "What is special about design research is that the justification of an educational goal requires normative theories, and the ways in which means are implemented to reach the goal – for example in design principles – require prescriptive theories" (Bikner-Ahsbahs, Bakker, Haspekian & Maracci, 2017, p. 2686). This, in turn, leads to a challenge for future TWGs, to develop "argumentative grammars for types of research that explicitly have a normative and/or prescriptive element, such as design research" (ibid., p. 2689).

4 Theory

In all the work related to networking theoretical approaches, deepening into the notion of theory appears as a crucial issue and it became important to explicitly state what is considered (or used) as a theory in mathematics education (from a local model to a well-established theory). The first call for papers at CERME 4 asked "to concentrate the discussion on research paradigms and/or theories within the context of their effect on empirical research" (Artigue et al., 2006, p. 1239). After

Theoretical perspectives and approaches **261**

the conference, a focus of discussion within the TWG team included the need to "make explicit the level at which a theory operates" and "to have an awareness of the underlying assumptions of each theory" (ibid., p. 1243). The discussions at CERME 5 and CERME 6 made evident that 'theory' is also a metonymy of research activities and that methodologies are intimately inseparable from theoretical perspectives. An important result shared by the group is the conviction that using theories is unavoidable in any research (the limit case being 'common-sense theories'), which raised the question of the use of 'metatheories' for networking. At all the conferences, some papers and many discussions dealt with these questions, making the reflection 'reflective' on our own research work. This led us to re-examine some of the questions the diversity of theoretical approaches brings about.

What is a theory? How do theoretical frameworks shape MER?

One of the first statements that came into play when comparing or just talking about theories is the heterogeneity of what is considered as a theoretical framework in MER and the consequent possible incommensurability of the investigations that are carried out in different theories. Diversity appeared as a main issue at CERME 5 (Arzarello et al., 2007) and, at CERME 6, Prediger, Bosch, Kidron, Monaghan & Sensevy (2010) called for concrete discussions. It appeared as more productive to avoid talking about the 'essence' of theories and, instead, consider the role theories play in research methodologies, especially in the delimitation of the 'reality' to study (the 'unit of analysis') and the kind of distinctions made about it. Differences soon appeared between approaches questioning the way students learn/acquire knowledge and those starting from the kind of knowledge that is taught and learnt, some of which were clearly stated during the work at CERME 7 (Kidron et al., 2011).

It is evident that, behind the foci and implicit assumptions of theories, are people – researchers – and collectives – research communities – that create, adapt, develop and share theories. The problem of how to incorporate this issue in the group work was raised in the plenary session on networking theories at CERME 6 (Dreyfus, 2010; Monaghan, 2010) and showed how the reflection on our own research cannot be separated from the models used to analyse teaching and learning processes. In consonance, at CERME 7 some researchers made proposals for general models to describe the dynamics and social dimension of MER, using for instance the notions of 'semiosphere', 'shifts of attention', 'theoretical genesis' or 'research praxeologies' (Kidron et al., 2011).

How to extend the scope of a given theory to embrace a wider and more complex phenomenon such as the relationships between teaching/learning practices and research?

This question was raised at CERME 5 and the papers by Goodchild (2007) and Jaworski (2007) provide interesting answers. Teaching and learning difficulties are the main motivation of MER and are thus usually located at the core of the

262 Kidron et al.

construction of theories. Reciprocally, the influence of MER in teaching and learning practices is one of the main concerns of our community. "Exploring issues in networking theories is not a purely academic exercise; through this work we aspire to better understand the construction of mathematical knowledge" (Kidron, Bosch, Monaghan and Radford, 2013, p. 2788). The interactions between MER and mathematic teachers' communities, in both empirical research and in teacher professional development, were analysed (Goodchild, 2007; Jaworski, 2007). Some ambiguities in the use of theoretical guidelines for teaching appeared (Bingolbali & Bingolbali, 2015), as well as strong divergences (Roos & Palmér, 2015). Related to this, the issue of the tools or technologies used in teaching practices has also been a recurrent theme, not least because the implementation of new practices using digital tools raises new theoretical issues in our understanding of mathematical activities (Lagrange and Monaghan, 2010). The role of different theoretical perspectives in approaching technology was also a topic of discussion (e.g. Cerulli, Pedemonte & Robotti, 2006) and a plenary lecture on digital technologies as a window on theoretical issues in mathematics education was given at CERME 5 (Artigue, 2007).

How do empirical studies contribute to the development and evolution of theories?

Theories and empirical studies are intertwined and evolve together. Analysing how a specific research paradigm influences empirical research and, vice versa, how theoretical frameworks evolve through empirical studies, appears to be an interesting way to deepen our research methodologies and theoretical frameworks. The relationship can be shown in specific cases, for instance when two methodologies are applied to the same episode or set of data (Gellert, 2007). Workshop activities have also been proposed in this TWG in some years to enable participants to interact with given empirical material or a spontaneous teaching problem (Prediger & Ruthven, 2007). These activities have shown that empirical studies are always designed with respect to the principles of a theoretical framework, and may differ from one approach to another. For instance, the evidence given by one researcher (with a theoretical framework) might appear to be insufficient for another researcher (with another theoretical framework) to address the question. Or, even the types of problems raised by an approach might not coincide with those of another. This can be seen in papers such as Rodríguez, Bosch and Gascón (2007) on metacognition or Bergsten (2007) on the limits of functions. In this last case, the differences shown do not only affect the way the problem is raised (students' difficulties vs. limitations of the knowledge taught) but also the practical implications that can be derived, leading to contradictory recommendations. The way to deal with a diversity of results thus remains an open problem in need of further development.

To summarise, and even if it might appear as obvious, the work done at the TWG constantly shows the dynamic dimension of theories and their different degrees of development, from limited models to more extensive theoretical constructions. However, their existence, need and potential utility in MER are

Theoretical perspectives and approaches **263**

now taken for granted. This enables the discussion to focus on their scope and limitations – especially in the contrast with different perspectives – which is closely related to what is considered as the primary object of study and the specific assumptions made in relation to it.

In the next section we will observe how some interactions between theories in the TWG networking activities and the methodological reflections nourished in the TWG over the years clearly contribute to specific evolutions of some theories. The role of the TWG as a place to confront and develop theories should thus not be neglected.

5 Beyond ERME

5.1 Ideas that originated in one research community, were brought into ERME and as a consequence were taken up more widely

The important role of the theory is well demonstrated in the French school of *didactique des mathématiques*, which plays a particular role in European didactics. It offers two related well-known theories: the Theory of Didactical Situations (TDS) and the Anthropological Theory of the Didactic (ATD). Combining theoretical perspectives is not a new approach in the French school; since the 1980s theoretical approaches have been connected within this didactic community. Issues such as the important role of epistemology and the role of a priori analysis, which are strong theoretical and methodological assumptions in the French school, were brought into the discussions of the TWG and were taken up more widely. We cite, for example, epistemological concerns that were highlighted as a consequence of the networking between TDS and the theoretical framework of Abstraction in Context (AiC). As a consequence of the networking experience between AiC and TDS researchers, AiC researchers decided to further develop the idea of a priori analysis in an explicit way. Strong epistemological concerns in TDS were integrated in AiC in a way that reinforced the underlying assumptions of AiC.

5.2 Ideas that arose within ERME and initiated or influenced development of mathematics education beyond ERME

An initial idea that arose in the TWG on theories was that *towards networking or dialogue between theories, researchers need to start working together on concrete cases*. A call for collaboration was stated at the end of CERME 4:

> [A]s a research community, we need to be aware that discussion between researchers from different research communities is insufficient to achieve networking. Collaboration between teams using different theories with different underlying assumptions is called for.
>
> *(Artigue et al., 2006, p. 1242)*

As a consequence, a 'networking group' was created in which a group of European researchers have collaborated since 2006. This group aimed to advance the networking idea as a research practice. Researchers looked at different aspects of a classroom lesson by means of different theoretical frameworks. Five theories were introduced in the networking process: the theory of Action, Production and Communication (APC), TDS, ATD, AiC, and the theory of Interest-Dense Situations (IDS). The results of eight years of intense collaboration between the members of the networking theories group are described in the book *Networking of Theories as a Research Practice in Mathematics Education* (Bikner-Ahsbahs & Prediger, 2014).

Another idea that arose in the TWG is connected to the fact that theoretical frameworks are culturally situated. The different cultural background of each theory is a source of difficulty in the efforts of networking theories. Nevertheless, at ERME relationships between different theories were established as well as analyses of mutual insights offered by each theory to the other. In spite of the different educational context in which each theory was developed, networking was possible and rewarding. This is an important result that influenced, at least partially, the development of MER beyond ERME.

Since CERME 4 the idea of networking was further developed in different projects. Some of these projects might have been influenced by the work that was done at ERME. For example:

- The networking of theories was a topic of interest in two PME research forums in 2010.
- Methods and methodology in networking projects were further developed and some of them are described in the book *Approaches to Qualitative Research in Mathematics Education: Examples of methodology and methods* (Bikner-Ahsbahs, Knipping & Presmeg, 2015).
- In *Educational Studies in Mathematics* (ESM) special issue *Representing Mathematics with Digital Media: Working across theoretical and contextual boundaries* (Lagrange & Kynigos, 2014) different papers mention the work done in the CERME TWG on theory.
- In some chapters in the book edited by Hodgson, Kuzniak and Lagrange (2016) *The Didactics of Mathematics: Approaches and issues (A homage to Michèle Artigue)* the CERME TWG on theory was mentioned repeatedly.
- Bakker (2016) reframed networking in terms of boundary crossing – a concept that is commonly used in the sociology of science and many other disciplines outside MER.

5.3 Ideas that arose within our TWG and have become relevant for other TWGs

Relationships between different theories were discussed in other TWGs. For example, in Dreyfus' introduction to the plenary session at CERME 6 we read:

Theoretical perspectives and approaches **265**

The development and elaboration of theoretical constructs that allow research in mathematics education to progress has long been a focus of mathematics education researchers in Europe. This focus has found its expression in many CERME working groups: some are focused around a specific theoretical approach and others allow researchers from different theoretical traditions and backgrounds to meet and discuss. For example, the working group on Argumentation and Proof at the present (CERME 6) conference has reported on passionate discussions about different theories and their relationships.

(Dreyfus, 2010, p. 2)

More recently, in a *Research in Mathematics Education* (RME) special issue, in a paper by Nardi, Biza, González-Martín, Gueudet and Winsløw (2014) we read about research into the teaching and learning of mathematics at university level which deploys theoretical frameworks that are seen as increasingly essential in the field. The work is based on the work done at CERME on university mathematics education (see Chapter 5, this volume).

6 Reflection on what has been achieved and a vision for the future

6.1 What has been achieved?

This question can be addressed in several ways. As academics our default approach is to look at the advancement of knowledge, but we should also look toward the development of researchers, both novice and experienced researchers. Many people have participated in the seven (at the time of writing) CERME theoretical perspectives TWGs. Almost all of these participants have presented their work with follow-up discussions. The 'social turn', mentioned in Section 2 of this chapter, which sees thought as the product of social activity, applies to the development of 'theoretical thought' just as much as it applies to the development of mathematical thought. 'Theory' can be a frightening (austere) word to novice researchers and the act of joining a forum where theories are the object of discourse can help make theoretical considerations more accessible. This CERME TWG has been a forum that has enabled this discourse and some of the novice researchers have now become TWG leaders. Some CERME TWGs cease when they feel they have served their purpose but there is an argument that the theoretical perspectives TWG should continue to allow this specialised discourse to continue. We now turn to the advancement of knowledge and focus on two dominant foci in the preceding pages, epistemology and networking.

Addressing how epistemology features in one's research is difficult, even for seasoned theoreticians. The CERME theoretical perspectives TWG benefitted from the presence of representatives from the ATD and TDS at every meeting, to whom questions of epistemology are central: "Pedagogy considers the knowledge

266 Kidron et al.

to be taught as a given, and focuses on the best conditions or practices to teach and learn it: the knowledge is not problematic, the relationships of the students to it are" (Chevallard, 2000, cited in Florensa et al., 2015, p. 2635).

The impact of the focus on epistemology is reported in Section 3 of this chapter in relation to the networking of theories, to which we now turn. Section 3 is the largest section in this chapter and this is fitting as the networking of theories has been the dominant theme of TWGs to date. But we feel we are now at a more mature stage, recent work has focused less on strategies of networking and more on epistemological and methodological reflections. Kidron (2015, p. 2666), for example, notes:

> In the last CERME we discussed cases in which the epistemological dimension permitted the networking . . . by means of the idea of 'reference epistemological model'. In this paper, we notice how, by means of networking, strong epistemological concerns in one theory might be integrated in another theory in a way that reinforces the underlying assumptions of this other theory.

6.2 A vision for the future

We avoid predicting the future but we can look forward to advances in unresolved areas of past work. The networking of theories deserves further attention in itself and its potential in the field of didactics. We look forward to members of future CERME theoretical perspectives TWGs using the experiences of past TWGs to facilitate other kinds of fruitful dialogues. For example, between mathematics education and disciplines close to mathematics education and between mathematics education researchers and mathematics educators who do not see themselves as 'theoreticians' but who nevertheless employ elements of theories in their instructional design. Such networking could enable a better understanding, from the perspectives of different theories, of the source of difficulties students and teachers encounter in their learning and teaching of mathematics. Indeed, the most recent TWG discussed the importance of developing networking with regard to the design dimension. Such networking would, of course, involve 'the epistemological dimension' to which we refer above.

More attention could be given to networking and practice. We could also pay attention to the fact that we, as theoreticians, have a discourse/knowledge that purports to describe and explain practice, but those whom we aspire to work with are likely to be outside of this domain of discourse. Mutual work to improve the teaching and learning of mathematics will likely result in new practices and new ways to describe and explain these new practices. New challenging tasks are offered as well by the dynamic character of theory, especially when we face the changes of our cultural development as well as the changes of the context in which each theory emerged. Our final words express a hope for the future that the CERME theoretical perspectives TWG will continue to be a venue for open discussion with critical friends.

References

Artigue, M. (2007). Digital technologies: A window on theoretical issues in mathematics education. *CERME5* (pp. 68–82).

Artigue, M., Bartolini-Bussi, M., Dreyfus, T., Gray, E., & Prediger, S. (2006). Different theoretical perspectives and approaches in research in mathematics education. *CERME4* (pp. 1239–1244).

Artigue, M., Bosch, M., Gascón, J., & Lenfant, A. (2010). Research problems emerging from a teaching episode: A dialogue between TDS and ATD. *CERME6* (pp. 1535–1544).

Artigue, M., Bosch, M., & Gascón, J. (2011). Research praxeologies and networking theories. *CERME7* (pp. 2381–2390).

Arzarello, F., Bosch, M., Lenfant, A., & Prediger, S. (2007). Different theoretical perspectives in research. *CERME5* (pp. 1618–1627).

Bakker, A. (2016). Networking theories as an example of boundary crossing. *Educational Studies in Mathematics, 93*(2), 265–273.

Bergsten, C. (2007). How do theories influence the research on teaching and learning limits of functions? *CERME5* (pp. 1638–1647).

Bikner-Ahsbahs, A. (2010). Networking of theories: Why and how? *CERME6* (pp. 6–15).

Bikner-Ahsbahs, A., Bakker, A., Haspekian, M., & Maracci, M. (2017). Theoretical perspectives and approaches in mathematics education research. *CERME10* (pp. 2683–2690).

Bikner-Ahsbahs, A., Knipping, C., & Presmeg, N. C. (Eds.) (2015). *Approaches to qualitative research in mathematics education: Examples of methodology and methods*. Dordrecht: Springer.

Bikner-Ahsbahs, A., & Prediger, S. (Eds.) (2014). *Networking of theories as a research practice in mathematics education*. Cham: Springer.

Bingolbali, E., & Bingolbali, F. (2015). Principles of student centred teaching and implications for mathematics teaching. *CERME9* (pp. 2600–2606).

Bosch, M., Chevallard, Y., & Gascón, J. (2006). Science or magic? The use of models and theories in didactics of mathematics. *CERME4* (pp. 1254–1263).

Brousseau, G. (1997). *Theory of Didactical Situations in mathematics*. Dordrecht: Kluwer Academic Publishers Group.

Cerulli, M., Pedemonte, B., & Robotti, E. (2006). An integrated perspective to approach technology in mathematics education. *CERME4* (pp. 1389–1399).

Chevallard, Y. (2000). La recherche en didactique et la formation des professeurs: problématiques, concepts, problèmes. In M. Bailleul (Ed.), *Actes de la Xe École d'été de didactique des mathématiques* (pp. 98–112). Caen: IUFM.

Dreyfus, T. (2010). Ways of working with different theoretical approaches in mathematics education research: An introduction. *CERME6* (pp. 2–5).

Florensa, I., Bosch, M., & Gascón, J. (2015). The epistemological dimension in didactics: Two problematic issues. *CERME9* (pp. 2635–2641).

Font, V., Malaspina, U., Giménez, J., & Wilhelmi, M. R. (2011). Mathematical objects through the lens of three different theoretical perspectives. *CERME7* (pp. 2411–2420).

Gellert, U. (2007). Emergence or structure: A comparison of two sociological perspectives on mathematics classroom practice. *CERME5* (pp. 1668–1677).

Goodchild, S. (2007). An activity theory perspective of didacticians' learning within a mathematics teaching development research project. *CERME5* (pp. 1678–1687).

Hejny, M., Shiu, C., Godino, J. D., & Maier, H. (1999). Research paradigms and methodologies and their relationship to questions in mathematics education. *CERME1* (pp. 211–219).

Hodgson, B. R., Kuzniak, A., & Lagrange, J.-B. (Eds.) (2016). *The didactics of mathematics: Approaches and issues (A homage to Michèle Artigue)*. Switzerland: Springer.

268 Kidron et al.

Jaworski, B. (2007). Theory in developmental research in mathematics teaching and learning: Social practice theory and community of inquiry as analytical tools. *CERME5* (pp. 1688–1697).

Jaworski, B., Serrazina, L., Koop, A. P., & Krainer, K. (2004). Inter-relating theory and practice in mathematics teacher education. *CERME3*: www.mathematik.uni-dortmund. de/~erme/CERME3/Groups/TG11/TG11_introduction_cerme3.pdf.

Kidron, I. (2015). The epistemological dimension revisited. *CERME9* (pp. 2662–2667).

Kidron, I., Bikner-Ahsbahs, A., Monaghan, J., Radford, L., & Sensevy, G. (2011). Different theoretical perspectives and approaches in research in mathematics education. *CERME7* (pp. 2376–2380).

Kidron, I., Bosch, M., Monaghan, J., & Radford, L. (2013). Different theoretical perspectives and approaches in research in mathematics education. *CERME8* (pp. 2785–2789).

Kidron, I., Lenfant, A., Bikner-Ahsbahs, A., Artigue, M., & Dreyfus, T. (2007). Social interaction in learning processes as seen by three theoretical frameworks. *CERME5* (pp. 1708–1724).

Lagrange, J.-B., & Kynigos, C. (2014). Digital technologies to teach and learn mathematics: Context and recontextualization. *Educational Studies in Mathematics, 85*(3), 381–403.

Lagrange, J.-B., & Monaghan, J. (2010). On the adoption of a model to interpret teachers' use of technologies in mathematics lessons. *CERME6* (pp. 1605–1614).

Lerman, S. (2000). The social turn in mathematics education research. In J. Boaler (Ed.), *Multiple perspectives on mathematics teaching and learning* (pp. 19–44). Westport, CT: Ablex Publishing.

Lerman, S. (Ed.) (2014). *Encyclopedia of mathematics education.* Dordrecht: Springer.

Ligozat, F., & Schubauer-Leoni, M. L. (2010). The joint action theory in didactics: Why do we need it in the case of teaching and learning mathematics? *CERME6* (pp. 1615–1624).

Monaghan, J. (2010). People and theories. *CERME6* (pp. 16–23).

Nardi, E., Biza, I., González-Martín, A. S., Gueudet, G., & Winsløw, C. (2014). Institutional, sociocultural and discursive approaches to research in university mathematics education. *Research in Mathematics Education, 16*(2), 91–94.

Prediger, S., Bosch, M., Kidron, I., Monaghan, J., & Sensevy, G. (2010). Different theoretical perspectives and approaches in mathematics education research: Strategies and difficulties when connecting theories. *CERME6* (pp. 1529–1534).

Prediger, S., & Ruthven, K. (2007). From teaching problems to research problems. *CERME5* (pp. 1745–1754).

Radford, L. (2008). Connecting theories in mathematics education: Challenges and possibilities. *ZDM – The International Journal on Mathematics Education, 40*(2), 317–327.

Rodríguez, E., Bosch, M., & Gascón, J. (2007). An anthropological approach to metacognition. *CERME5* (pp. 1798–1807).

Roos, H., & Palmér, H. (2015). Communities of practice: Exploring the diverse use of a theory. *CERME9* (pp. 2702–2708).

Ruiz-Munzón, N., Bosch, M., & Gascón, J. (2013). Comparing approaches through a reference epistemological model: The case of algebra. *CERME8* (pp. 2870–2879).

Sensevy, G. (2010). Outline of a joint action theory in didactics. *CERME6* (pp.1645–1654).

von Glasersfeld, E. (1995). *Radical constructivism: A way of knowing and learning.* London: Falmer Press.

19

ERME AS A GROUP

Questions to mould its identity?

Marcelo C. Borba

Is there anything better than a book to celebrate the 20 years of a continental, scientific organisation? Anthropologists and sociologists might consider our various groups living in this complex society to be a cultural group – groups of men and women, groups of elderly folks, groups that produce different ethnomathematics, and groups of scientists. I believe that we can view ERME, which meets face-to-face during its conferences, as a cultural group, a group of mathematics educators who decided to form a scientific organisation 20 years ago to theorise about mathematics education. Ten conferences throughout these past years have enabled the European community to build the trends summarised in the previous 18 chapters.

But how to define 'trends'? We may think of them as something stylish, and therefore a bit derogatory, perhaps as in something fashionable we simply follow. But in a more productive way, some of us, inspired by discussions raised by Mogens Niss, have characterised a trend as a response to a crisis.

For example, the mathematical trend of modelling responded to the crisis of relevance (Why am I studying mathematics?). Technology and resources were responses to the availability of different artefacts, in particular digital artefacts that seem to pop up every day. In this sense, these trends can be seen as responses to 'crises' faced in (mathematics) education. Thus, the 18 chapters of this book summarise the ways that the various Thematic Working Groups (TWGs) of ERME have dealt with problems during the last 20 years.

Most of the chapters here, written on behalf of ERME's TWGs, would fit the description of having been written in response to a given crisis. One might see the geometrical thinking and number sense groups as a response to topics the community judged important enough for researchers to investigate. In the future, it might be important to author a paper discussing what necessitates or generates a new group in organisations such as ERME. There is no intention in this chapter to develop such a study, but it might be good to identify some motives for one – for

270 Borba

example, the motive behind choosing a specific topic of the curriculum as a means to forming a group on algebraic thinking. Algebra is an important topic at different levels of basic education, and there is a tradition in mathematics education of trying to study specific ways of thinking within the realm of mathematical thinking. A conceptual field, such as number sense, could also be a catalyst for creating a TWG.

Of course, we know that there is a transparent way of creating a new TWG, as was discussed in the preface of this book. One can find the current call on the ERME website. Additionally, different committees of ERME, such as the ERME board and IPC, discuss the proposals and proposer, and then judge according to relevance, actuality, potentiality of attracting people, and distinction to existing groups. Also, some TWGs attract many participants in one or a series of conferences which might end up with the group separating into two sub-groups. But it would be interesting to study the 'objective' and 'subjective' reasons for the creation of a working group within an organisation such as ERME. A sociological or philosophical approach to such an issue could shed some light on factors that influence the creation of a working group other than merely 'responding to a given crisis'.

One might notice that a topic for a TWG may be considered to be one of the 'established topics', as these have comprised part of curriculum for a long time and because the community believes they are critical enough to study. On the other hand, 'new topics' without an established tradition in school curricula may also be a theme. There might be discussion around whether a statistics TWG should be established, and, though I am too unfamiliar with the European curriculum to be certain of this, fractals could have become a group topic, had they become important in the curriculum. In any case, if programming does become a trend in early education, it is very likely that ERME might create a group in response to such an area when it becomes important enough to call for investigation. Many researchers concerned with education in the early years are not familiar with programming, much less how to coordinate this 'new area' with mathematics.

Modelling and technology are, in a different perspective, 'transversal themes' that are not isomorphic to any theme of the curriculum in particular. Although one might think of teaching modelling as a topic, the topic groups feel these two topics are more about the importance of using modelling or digital technology to teach and learn geometry, algebra or fractals. History and mathematics education, argumentation and proof, creativity and language could, equally, be placed in the same class.

There are also social forces and leaders within an organisation who are responsible for the creation of a given group. Response to a given crisis, together with social forces within the community and research financing might be components involved in the creation of a thematic group. And what about those areas that have not formed groups in ERME? Why have they not become a thematic group? Why have fractals and complex numbers not formed working groups? Why have immigration and mathematics education or online mathematics distance education not become themes for a group? These are questions that I as an outsider cannot answer, but my hope is that the reading of this short chapter might result in some discussion among ERME's membership on this issue: What groups should be part of ERME? Why? Thinking about what does not exist is one element of research many of us have developed in order to be creative or to de-naturalise established norms. Much of the creativity in research in mathematics education and elsewhere can be related to the breaking of established rules. This was true in geometry, with the creation of geometries other than the Euclidean one, and it was also true in mathematics education where the idea that mathematics is not culturally bound was brought into discussion. As a matter of fact, the very notion of a mathematics education as an established field or as a discipline is a result of such processes.

A TWG at CERME is important not only as a means of consolidating areas of research, but also as a means of organising the biannual conferences. Group meetings are an important part of the conference and result in the organisation of proceedings, books, book chapters, and special issues of different journals. Papers published at CERME comprise a significant proportion of the references found in the previous 18 chapters of this book.

The high number of citations of CERME papers is normal, due to the nature of the book, which does not try to be an encyclopaedia of ERME, but rather attempts to show the evolving history of the organisation. There is a prevalence of European references in general, not just in papers published at CERME. Only one country outside Europe seems to have a great number of references, but not enough to break the 'pattern' noticed in the chapters of this book. Otherwise, a great number of references seem to be European, for Europeans tend to cite European references. Although I do not elaborate with a table supporting the above sentence, I have read the reference list at the end of each chapter, and my knowledge of scholars from different parts of the world points me in this direction. I admit there could be mistakes in my 'impressionistic' analysis of the drafts on which this commentary is based. Neither good nor bad, citing more European references might be a characteristic of ERME. An opposite and additional characteristic of ERME is to have leaders in publications and in the organisation itself who are not from Europe. In the same manner, the editors of this ERME book had the intentional goal of inviting commentary papers from scholars outside of Europe. This again is neither good nor bad, but it is a characteristic of this 20-year-old organisation.

CERME has attracted participants from outside the European continent, and the internationalisation of these conferences might well be a natural path. There are many examples of conferences with a continental identity or even with an international identity.

But how can we define what is meant by 'international'? This is another tough question, which I do not mean to answer, but instead want to encourage the reader to think about. Does the fact that scholars from outside the European continent attend it make the conference international? Does having some of these scholars in leading conference positions make it international? Would citing and dialoguing with research from different parts of the globe be a criterion for internationalisation? Again, it is not an easy question to answer. But if we have some agreement

272 Borba

on the answers to each of these questions, we might be on the right path to internationalise CERME or to confirm that CERME can already be considered international (of course not considering the diversity of countries in Europe).

But one might ask whether CERME being continental or international is the goal. By the term 'continental', I mean European (as in the case of ERME), and by the term 'international', I understand this as countries outside of the European continent. And again, in this case, there is no correct or incorrect answer. There are options to these definitions, options that a 20-year-old organisation will consciously have to choose, for we need to ponder what identity ERME wants to have in its adulthood.

Another option available, especially in postmodern times, is that CERME be considered as both continental and international. Building a European identity for research in mathematics education can be combined with international dialogue – if we understand a dialogue as a deep relationship of listening and speaking, as suggested by the famous educator Paulo Freire, whose work is quite well known in Europe. Such a dialogue could help to identify which European problems generate trends and are distinct from those that have emerged in South America, Asia, Africa, or elsewhere.

For reasons that might one day be understood, there is a great deal of thematic similarity between the 18 groups of ERME and the 15 groups of the 30-year-old Brazilian Mathematics Education Society (SBEM). Undergraduate mathematics education, history of mathematics, technology, teacher education, linguistic issues, modelling, statistics and probability, and diversity are some of the groups with similar titles. Of course, there are differences that show the perspectives and/or idiosyncrasies of each group. While in Brazil the name of the group on technology stresses digital technology and online education – perhaps due to the fact that the country is one of the pioneers in online mathematics education – the European counterpart emphasises resources, using a perspective in which digital technology is only one resource among many. The group of ERME holds a title similar to one of the discussion groups created to celebrate the centennial of ICMI at the memorable Rome conference in 2008, which shows another possible feature of internationalisation.

In spite of their differences, working groups at both ERME and at SBEM were probably created in response to the crises generated by the advent of personal computers just over 40 years ago in California, USA. Personal computing, computer laboratories, computer software, mobility, and Internet are all terms that have become more or less part of our daily lives. These former novelties – although in 30 years they have already been classified into four phases – became crises that (mathematics) educators have focused on. In doing so, other artefacts, seen as being natural or commonplace, have also become themes of research.

If instead of looking at the commonalities of the two organisations ERME and SBEM, we also focus on what is dissimilar, we note that the Brazilian society has two groups (one on assessment and one on the philosophy of mathematics education) not found in ERME. During the last CERME, a group on assessment was created which reduced this difference, but this still begs the question: what makes one scientific organisation create (or not create) a given working group? While both societies have groups focused on studying the early years of mathematics education, the Brazilian society had, until quite recently, two other groups focused on mathematics education – one in middle school and one in high school. These two groups have now merged, as has a group spotlighting curriculum and mathematics education.

The Brazilian Mathematics Education Society has bylaws for creating and closing down thematic groups. There are no indications of ERME groups disbanding, but one can see that several groups within ERME are not present in the Brazilian mathematics society. Keeping in mind the distinctions between the two organisations, ERME's themes of affect, algebraic thinking and geometrical thinking are themes that are 'solely European'. Naturally the theme 'comparative studies' is international, while the other three themes are present in Brazilian mathematics education, in particular the last two, but they are more likely to be found in the groups that focus on high school and the group that used to exist focusing solely on the second half of elementary education. Is it idiosyncrasy? Are there social forces in action? Did a particular leadership generate a given group? Or is it just a trend, a reaction to a given problem, to a new problem? Or is it a combination of factors? I am now puzzled about some absences on the Brazilian side! Will I get to know Brazilian mathematics education better as I study European Mathematics Education, through the chapters of this book?

By comparing and contrasting the two organisations of ERME and SBEM, I want to show that some themes seem more European. By no means can the comparison of these two organisations raise the claim of one theme being uniquely European or uniquely Brazilian. Like in a trip, going abroad leads to understanding our own country, or continent, in a way that is not possible for someone who has never left their place of birth.

Identifying groups or even the identity of ERME itself might depend on the relationship between pure research and research aiming to have an impact in educational settings. The distinction comes from mathematics, from one of the sources of mathematics education seen as an area of study or a discipline. One may consider pure mathematics, where the idea is to develop mathematics, theory in mathematics, independent of any applications or impact on other disciplines. Is there such a counterpart in mathematics education? As I have asked a few times elsewhere, is there a pure mathematics education? If so, is pure mathematics education considered better by our community than applied mathematics education? In its roots, if we look at the book organised for the centennial of ICMI (and other documents) the concern of mathematics education is with practical problems: how to introduce new mathematical topics into education or how kids can understand some topic. In other words, mathematics education is born to deal with problems that neither mathematics nor education could deal with.

Each one of the previous 18 chapters is concerned with theoretical aspects of a specific theme. I do not intend to discuss the definition of 'theory', and I believe that there are differences in the use of this term within thematic groups. The terms 'model', 'theoretical framework', or 'theory' may be used synonymously, and it is likely that no group is in agreement about a clear distinction between these three terms or others that are used as well.

However, there does seem to be an implicit agreement on the distinct relationship between the words 'practice' and 'theory'. Within the two groups devoted to teachers, one sees a clear concern about linking 'theory' to 'practice'. The report of the group 'history and mathematics education' focuses more on how the history of mathematics has been positioned within mathematics education itself. Others, such as the 'modelling and applications' group, claim that they should develop more theoretical work that links their group's work to other trends in mathematics education.

We can say that mathematics education emerges as an interdisciplinary area of study, an area that deals with practical problems. But like every other field, meta-research develops some perspectives further and these end up being called 'theories', such as Activity Theory used also in knowledge domains other than mathematics education. So it is not harmful to have different perspectives in groups.

Thus it is also important that researchers experience TWGs other than the one they regularly follow. We have already argued that understanding a non-European organisation or going to a non-European conference could help a European better understand ERME and CERME. It would be interesting to have members of various thematic groups understand how other groups define terms such as 'theory'. Yet accomplishing this would not be an easy task. Thematic groups have emerged everywhere to lessen the fragmentation of research in different areas. However, if one attends a conference in mathematics education of only a specific thematic group, research, as well as the interaction between established and developing researchers, could become even more fragmented.

Although we might have no agreement about the definition of 'theory', it is possible to agree that a 'good theory' sheds light on some problems, problems relating to observed phenomena. Good theory in mathematics education might help one see how digital technology is related to human thinking, and, in particular, to mathematical thinking. Some might argue that this perspective is too narrow and that we should broaden our understanding of mathematics education. In this case, norms, communities, goals and artefacts of a given activity should be considered so we can understand the way students and teachers can move from one goal to another as they try, for example, to solve a problem using a piece of software. Activity Theory, now in its third generation, considers these factors and might shed some light on research problems.

For instance, as our research group in Brazil researched the relationship between digital technology and mathematics education, we used different facets of Activity Theory. But we found that this theory was not developed to consider the specificities of digital technology. Moreover, it did not separate technology, human beings and thinking: so the path is open for the development of complements to this theory. As Activity Theory reveals results to our problems, we believe we have also developed Activity Theory itself.

Some could look at this above development and call it a 'new theory', but I suggest we be more careful about creating theories. We developed the theory further, but we did not create a 'new' theory, nor, I would argue, should we create one for each special issue of any journal we organise. I believe some confusion is occurring: a new theory is not necessitated simply because of the use of new terminology, of the creation of new words, or of the use of words utilised by important authors of the 20th century. I believe that members of ERME might be aware of issues regarding theory, but it would be interesting for various groups to discuss what theory is (or is not). In my opinion, the main issue regarding theory, at least in the Brazilian case, is the attempt to develop theories without a commitment to connecting them to fundamental problems that might initiate the development of this theory. This attempt might, further, lead to the creation of wordy theories that do not address the problems they relate to. Thus, pure mathematics education that does not connect to problems might separate our field from problems.

Naturally, not every thesis or paper at ERME needs, in the end, to finalise some kind of application. But papers, books or theses in a given block of time should revisit problems related to issues to justify that mathematics education is a field of study that deals with problems not covered by education or mathematics themselves.

In this short chapter, I wanted to acknowledge the developments of ERME and CERME. By asking questions (for which I have no answers), I want to intensify the dialogue between the members of ERME and the readers of this book and give the mathematics education community an idea of what has taken place in these 10 CERMEs and in the 20 years of ERME. I believe a search for identity is constant – it comprises who we are, whether we are thinking about people or organisations. In the case of scientific associations, I believe that discussing theory and practice is essential to building identity. In the case of ERME, we are being culturally shaped, constantly challenged, and are searching to build a research community that interacts with the broader mathematics education community. Happy birthday, ERME!

20

COMMUNICATION, COOPERATION AND COLLABORATION

ERME's magnificent experiment

Norma Presmeg

1 Context and an anecdote

1.1 Europe at the genesis of mathematics education research

The field of mathematics education research – little more than a century old as a field in its own right – still in its infancy compared to the millennia, and established traditions, of research in mathematics (e.g. Fried, 2014), grew from the insight of several famous mathematicians and other scholars concerned with mathematics education (Klein, Fischbein, Freudenthal, and others) that the teaching and learning of mathematics is indeed a field in its own right, linked with but distinct from the discipline of mathematics (Karp & Schubring, 2014). It was not long ago that mathematics education as a research domain was searching for an identity (Sierpinska & Kilpatrick, 1998), and the field in some respects is still doing so as relevant theories, methodologies and issues continue to evolve.

Europe has been central in the recognition of the necessity for cultural and international cooperation in the search for quality in mathematics education research. Mathematics education research came into its own as a field about half a century ago with the organisation of the first International Congress on Mathematics Education (ICME-1) in Lyon, France, in 1969, and these quadrennial conferences continue to thrive and bring together researchers in venues around the world. But it seems no coincidence that they started in Europe with its cultural diversity, and the first three of these congresses were organised and took place in Europe (ICME-2 was in Exeter, England, in 1972, and ICME-3 was in Karlsruhe, Germany, in 1976). John Conway (1985) characterised the ICME-2 conference in 1972 as having a goal of *thematic unity*. He pointed out that this congress was the venue for the genesis of ideas for the International Group for the Psychology of Mathematics Education (PME), which held its first annual conference (PME-1, with 86 participants) in 1977 in Utrecht, The Netherlands. The communication in

Communication, cooperation, collaboration **277**

PME conferences since then has been truly world-wide, with meetings held each year in a different country or region, and it is appropriate that PME-40 returned to Europe once again in 2016, in Szeged, Hungary. PME has annually between 700 and 800 members from about 60 countries, and the goals remain relevant to those of ERME. The third goal of PME illustrates how the field is still evolving, and the necessity of taking such evolution into account, also as ERME evolves. This goal is "to further a deeper and more correct understanding of the psychology and other aspects of teaching and learning mathematics and the implications thereof". For several years, as the focus of the field changed from psychological theories in mathematics education to sociological ones, it was debated whether the *name* of the organisation (PME) should change to reflect the change in focus. The original name was retained for historical reasons, but for inclusiveness the 'other aspects' were to be taken into account in this mission statement (Balacheff, 1990). This shift in paradigms is addressed specifically with regard to teaching mathematics in Chapter 12 of this book, which describes the change from acquisitionist to participationary notions of learning, and the research focus from *teachers' teaching* to *classroom practice*.

Furinghetti (2014) outlined the history of the significant undertaking of building cooperation, as far as it concerned *mathematics education* – as distinct from mathematics education *research* – as follows. She described

the evolution of international cooperation in mathematics education from its beginning in the nineteenth century, when an international perspective entered the world view, until the post-Second World War origin of the present structured network of activities aimed at mutually helping people. Cooperation is treated by following the path that goes from communication in mathematics to communication in mathematics education and from communication within countries to communication among countries.

(p. 543)

Her characterisation of the journey of mathematics *educators* in this regard, in some ways mirrors the journey taken by ERME *researchers* and those involved in other associations such as PME and the group for the History and Pedagogy of Mathematics (HPM). The spirit of communication, cooperation and collaboration implicit in the activities of these groups is aimed at better mutual understanding among members, with an emphasis on both *inclusion* and *quality* (see the goals of ERME as elaborated in the Introduction). Inclusion extends both to the distinction between established researchers and newcomers to the group in each case, and also to the unseen cultural assumptions that are the concomitant of the *natural attitude* in different counties and societies: we are not aware of our cultural world-views unless we are able to suspend this natural attitude (Husserl, 1970), and view perspectives, events and actions from the outside, as it were. This phenomenon applies emphatically to research in mathematics education. Hence the value of, and also the difficulties associated with, communication, cooperation and collaboration, in

278 Presmeg

furthering knowledge in the field of mathematics education. ERME has embarked on the ambitious initiative of building bridges across cultures both within countries and across their borders. Although European research is the main focus of these endeavours, the initiative is world-wide, as evidenced by the relatively large number of non-European mathematics education researchers that have also presented papers in the CERME conferences since their inception in Osnabrück, Germany, with ideas that started in 1997. The goals of ERME are evidenced in all of the 18 core chapters of this book, but the specific details of the paths that the journeys of the Thematic Working Groups (TWGs) have taken are as varied as their research foci. Some of these paths, and ways that TWGs have met challenges along the way, are explored later in this chapter.

This brief setting of the international context in which ERME started is followed by a personal story, as an introduction to some of the challenging issues involved in communication, cooperation and collaboration across cultures and countries.

1.2 A vignette

In 1982 I started studying for a PhD degree at Cambridge University in England. I attended courses there with the group of MPhil students. (My specific goal was *to understand more about the circumstances that affect the visual pupil's operating in his or her preferred mode in high school, and how the teacher facilitates this or otherwise* (Presmeg, 1985).) In one of the sessions with this group, Alan Bishop gave us an experience that has been engraved on my professional career ever since. Alan distributed cards to each of the students present: on each card was written – in the form of a question – a focus theme or topic for what they were about to witness, and each focus in the group of about eight students was different. Then he proceeded to interview one of the students while we watched and listened. He asked the student to "spell the word 'mathematics' backwards". The student paused, lifted his eyes to the ceiling, hesitated, started "s, c, i, t", paused again, continued "a, m, . . . e, . . . h, t, a, m!" The sense of accomplishment was palpable. Alan's questions revealed that the student was unable to hold the image of the whole word, and broke it into pieces in order to unravel the spelling. After the interview, which included also other visualisation tasks, each person in the group of students spoke in turn, each reporting – according to the focus on his or her card – what had been witnessed. And each student saw a different interview! Some foci included cognitive perspectives, some affective according to the emotions of the student who was interviewed, some involved the goals of the interviewer, and others the goals of the student. I became lucidly aware of the difference between *data sources* in mathematics education research, and *data*, which are constructed by the researcher. In this interview we were privy to the same data source (and the interview could have been audio- or video-recorded and transcribed), but in each case different data evolved according to the foci on the cards. The data were qualitative in nature by virtue of a single unique experience in the interview: mildly numerical data could have been generated, but there was no opportunity for large-scale statistical analysis. I suspect that today a different set

Communication, cooperation, collaboration **279**

of foci could be selected for the foci on the cards, e.g. involving material action, gestures and sociocultural issues that concern the genesis of knowledge (Presmeg, Radford, Roth & Kadunz, 2016). The point is that this small but powerful experiment illustrated clearly the difference between data sources and data, and that the latter require some theoretical framework and a specific question in order to come into being (see Chapter 18).

This experiment illustrates some of the difficult choices that are germane to the efforts of ERME researchers to foster communication, cooperation and collaboration among scholars from different cultures and countries. Some of these choices are the topic of the next section.

2 Relationships in ERME

It seems to me that the essence of ERME is relationships of various kinds, and I shall address some of these in this section. There are of course the human links implied in *the spirit of communication and collaboration* that epitomises ERME (see the Introduction). But then there are also various links that pertain to theory and its relationships in research, e.g. the links between theoretical and empirical aspects of research studies that have been presented at CERME conferences (e.g. Chapter 1[1]). More specifically, as suggested by the vignette in the previous section, I am interested in the links between theories and the data that are thereby enabled from data sources. This type of link is also central to ERME: some of the TWGs (e.g. Chapter 1) aim to *envisage a common research agenda*. Chapter 16, which has comparative studies as a focus, could be taken as in some ways representative of the whole ERME endeavour, because the whole focus of this chapter is the complexity of issues implicated in the bridging of cultures. However, because articulation of theories and concomitant research goals is then vital for this purpose, I take Chapter 18 on 'Theoretical perspectives' as a central case in point, if networking of researchers is to be successful, involving border crossing across local or national orientations, languages and cultures, in some cases using the same data *sources*.

The vision of ERME is *huge*: let me try to untangle whether this vision has been realised in 20 years, or in fact if it is realisable at all. Let me start with theory and its relationships.

2.1 Specification of theories and their compatibility

The authors of Chapter 3 specify that the relationship between theory and design needs to be explicit, and they distinguish between theories *of* and theories *for*, thus highlighting that there are overarching theories (e.g. cognitive or sociocultural frameworks) that serve as umbrellas for specific conceptual designs for research, such as a context-dependent stance on *abstraction* (Freudenthal and Davydov). The authors of Chapter 4 note a maelstrom of repeated research, which does not always build on prior results, without explication, and without taking previous research into account. These concerns echo the need for the nature and use of theories to

280 Presmeg

be spelled out – their research paradigms and methodologies, and their relationship to questions investigated in mathematics education (a need already noted from CERME 1; see Chapter 18).

However, these concerns are complicated by the embeddedness of theories in culture itself. There are two aspects at play here. First, are theories complementary or contradictory? They are, after all, expressive of particular world-views. The authors of Chapter 3 point out that at their TWG at CERME 9 alone, there were 25 different theoretical frameworks represented. They highlight and explore particular theoretical models (TDS, ATD and the RBC model) that have been used in successful collaborative studies, and they stress the need for literature reviews and integration of theories – and also for more cross-national studies on algebraic thinking. These authors believe in the successful integration of theories. Second, as culture is not static but continually evolving, so also theoretical notions accepted as useful are changing over time. This change in our field has been dramatic, as early cognitive and psychological theories gave way to sociocultural ones that are still evolving (even in a topic as specific as semiotics in mathematics education: see Presmeg et al., 2016, 2018).

How, then, is the integration of theories possible? Are umbrella theories from different cultural traditions incommensurable? If so, this conclusion would not bode well for the cooperation and collaboration that are goals of ERME researchers. In exploring the issues involved, I take as paradigmatic three specific parts of a paper from the mid-1990s, by Jere Confrey in the USA. Confrey (1991) had early on embraced radical constructivism as a framework for her research in mathematics education, as reflected in her early papers, e.g. 'Learning to listen: A student's understanding of powers of ten' in a book edited by Ernst von Glasersfeld. By the mid-1990s, Confrey seemed to indicate that radical constructivism, social-cultural approaches, and social constructivism are too fundamentally different to be combined meaningfully. However, in a deeply perceptive paper published in three parts, she continued to unpack the assumptions of these theories, as epitomised in writings of Piaget and Vygotsky:

> In this paper I propose to provide brief summaries of radical constructivism (as one interpretation of Piaget), and the socio-cultural perspective (as one interpretation of Vygotsky). The summaries will include major principles, primary contributions to mathematics education, and potential limitations. In a previous paper [Confrey, 1994b] I warned readers of combining these theories too simplistically. In this paper I introduce a new theoretical perspective which integrates the two theories by means of a feminist perspective.
>
> *(1994a, p. 2)*

Confrey (1994a) pointed out limitations of the constructivist perspective. Vygotskian theory was considered as a foil to these limitations (Confrey, 1995), but it had its own limitations. Confrey (1995) wrote as follows:

Communication, cooperation, collaboration **281**

These two theories, then, express two powerful metaphors for understanding humanity and for modelling and investigating human development. We experience ourselves both as biologically developing beings and as productive members of a collective enterprise . . . This suggests that both are necessary, and that the challenge is to integrate them.

(p. 46)

The point is that any theory is designed in a particular context, with a particular world-view, for a specific purpose, and therefore theories cannot be transported simplistically across cultures. In order to combine a theory with another developed in a different setting, a fundamental new perspective is required that takes into account both lenses. By 1996, several researchers in mathematics education were questioning the problem of intersubjectivity in radical constructivism (e.g. Lerman, 1996). The *emergent perspective* of Yackel and Cobb (1996) was essentially a new theory that took into account the implications of the individualism of the psychological perspective as one lens, and the collective perspective of the socio-cultural tradition as the other. Much good research emerged from the emergent perspective, thus manifesting that the combination of lenses can move the field forward in fruitful ways that build on previous research. It is not that *theories* are combined, but that the complementarity of lenses provides alternative views of the same phenomenon.

This complementarity of research lenses in ERME, epitomised in the idea of *networking* (Bikner-Ahsbahs & Prediger, 2014), is illustrated abundantly in the work reported in Chapter 18, which I take as a paradigm case of the push for connectedness that is emblematic of all the 18 chapters at the core of this book. The authors of Chapter 18 perceive diversity as both an opportunity and a challenge: they express "the necessity of learning how to deal with the diversity, complexity and richness of theoretical perspectives in mathematics education". As one way of coping with this diversity, networking depends on "ways of making theories interact" (Chapter 18), choosing judiciously theories that have the potential to highlight different aspects of a phenomenon in compatible ways, as a first step on the "path toward establishing mathematics education as a scientific discipline" (Chapter 18). Then the specific question arises, at what point do data *sources* provide *data* in this endeavour?

2.2 Theory, data and data sources

In the Cambridge interview (Section 1.2), the words spoken by interviewer and interviewee, their gestures and other body language, their intonations of voice, pauses, etc., are all data *sources*, even if captured in a frame of space and time by video- and audio-recording, and even if a full transcription is performed. At what point in the research process, then, can we start talking about *data* to be analysed? This question is especially important in networking situations in which a dataset is provided by one group of researchers, to be analysed by that group and a different

282 Presmeg

group, using compatible but different theoretical frameworks. The crucial element is that decisions are involved at all stages of the research process, according to some issue that is the focus of the investigation (Bikner-Ahsbahs & Prediger, 2014).

The networking of theories – and related methodologies – was developed by groups of researchers (see Chapter 18) into sets of principles that transform the networking process into a scientific discipline that continued beyond the boundaries of the CERME conferences. Consider one specific example of networking, between collaborating teams in Israel, using the RBC+C theory of Abstraction-in-Context (AiC), and in Germany, using the theory of Interest-Dense Situations (IDS) (Kidron & Bikner-Ahsbahs, 2015). These researchers were aware of potential theoretical and methodological difficulties that could accompany the process:

> One main difficulty relates to the relevance of data and its appropriateness to the different foci of attention that characterise the different theories. In all the cases of networking in which both of us were involved, we observed difficulties already in the first stage of designing common activities or selecting data towards the analysis which will permit the networking.
>
> *(p. 228)*

Certain elements from the data source have to be selected, and others *bracketed* (as Husserl (1970) would characterise the omission of others). These researchers are aware of a spectrum of networking strategies in a 'landscape' of degrees of integration of theories, which extends from ignorance of other theories at one extreme, to full global integration at the other (p. 224). They recognise that optimal networking occurs in the middle ranges of this spectrum, which they epitomise in their decision-making from the early stages of the process. They ascertained that, "The two approaches [AiC and IDS] share a certain view on knowing and therefore the networking of the two theories could fruitfully be done" (Bikner-Ahsbahs & Kidron, 2015, p. 236). Every step in their collaboration was conducted by a series of *five cross-over stages*, which guided the research toward common results:

- *deciding cooperatively* about what will be done, for example deciding about the part of the transcript to be analysed;
- *separate processing . . .*;
- *exchanging the results and working with alien results . . .*;
- *reworking home results*, for example re-analysing the part of the transcript; and
- finally a collaborative meeting aimed at *building consensus* about the work done. (p. 237)

The careful details of this process are provided by these authors (pp. 237–248). Through *understanding* the other theory and *making one's own understandable* through offering analysis of the same transcript; implicit *comparing and contrasting*; more explicitly *coordinating* the process through the five cross-over stages; and finally through *local integration*, the authors concluded that "the two teams learned from

Communication, cooperation, collaboration **283**

each other, and in this way deepened their analyses" (p. 248), at the same time improving also their theoretical approaches. This example epitomises two words that are stressed in Chapter 18 as being of importance: *dialogue* describes "that which enables mutual understanding in the way we communicate our theories but also . . . differences in the use of language" (Chapter 18), not necessarily implying changes in the theories involved. The word *transformation* describes that the researchers involved are thereby enabled "to see things [that] they could not see before" (Chapter 18). Chapter 18 thus highlights ways in which such networking might turn the plurality of cultures into a resource for scientific advancement of research in mathematics education in all specific areas of focus.

2.3 Relationships among researchers from different cultures

The previous two sections have focused on relationships concerning different theoretical frameworks and methodologies. But just as important are the human relationships involved as researchers come together with a vision such as ERME's. There are two challenges at play: first, researchers come to CERME with different cultural backgrounds and research traditions (addressed to some extent in the discussion above). The diversity in multicultural classrooms, as well as diversities of researchers in this context, are at the core of Chapter 15: 'Diversity in mathematics education'. Second, each meeting of CERME welcomes both experienced and beginning researchers. Chapter 17 ('History and mathematics education') spells out the ERME procedures to turn this challenge of diversity into a resource. Beginning with the call for papers in CERME 6, this group spells out precisely the nine points to be included in the proposals. Further, rules about 'how to behave' in the TWG are specified: 1. Submit a paper; 2. Review a paper or papers; and 3. Read the papers before the conference. This sense of structure is helpful, but even more important is the sense of ERME *inclusiveness* that is apparent in the reports of all the chapters.

The Introduction summarises *inclusiveness* and *quality* as two key aspects that are vital to the success of the ERME enterprise, operationalised in four cells of a table: A, starting to communicate; B, developing cooperation in engaging with debate; C, Developing cooperation in recognising ideas; and D, enabling collaboration in the development of key ideas and new shared constructs. How have the TWGs coped with this challenging but rewarding progression, which has human relationships at its core?

Chapter 1 authors lament the volatility inherent in the inclusiveness relation – members leave, other members join – in their mission to establish a common research agenda for research on geometrical thinking. However, they are able to report 'sustainable development' among researchers from six countries, three in Europe and three in the Americas. The authors of the TWG on 'Number Sense' (Chapter 2) report the difficulties of terminology associated with two distinct communities with regard to definitions in their field, highlighting the necessity for the *willingness* of members to communicate deeply. A shared language is needed (Chapter 9,

284 Presmeg

'Mathematical potential, creativity and talent')! Such constructive willingness is also stressed in Chapters 3 ('Algebraic thinking'), 4 (on probability and statistics education) and 5 (on university mathematics education), and is implicit in other reports. It is apparent that a certain *commitment* to the goals of ERME is necessary for the TWGs to succeed in the inclusion-quality progression, thus eliminating the "maelstrom of repeated research, which does not always build on prior results" (Chapter 4).

Some TWGs (Chapters 5, 8, 10 and 15) report the prevalence of small-scale studies among their members, and would like to see more of the larger international studies that bridge cultures (e.g. those reported in Chapter 16). Other TWGs (e.g. 'Argumentation and Proof', Chapter 6) have attained the stability of establishing a community of researchers, with consequent 'good practices' manifesting both inclusiveness and quality. Chapter 7 (on applications and modelling) reports an increase in co-authored papers to more than 80%, from their inception at CERME 4, to CERME 9.

Language, both as a window into cultures and practices, and as a tool for the cultural-historical analysis of classrooms (Chapter 14) has been present as a focus in all CERMEs from the first – and is indispensable in building cooperative research relationships. Europe, with its diverse languages, cultures and educational systems, is a "wonderful testbed" for the goals of ERME (Chapter 10). This TWG on 'Affect' highlights the evolution of concepts and theories in their field, suggesting again both the need for unified language and theory, and the issue of compatibility of theories. These authors report on a useful linguistic structure for research on beliefs, attitudes and emotions, which has been developed over the course of CERME 3 to CERME 9, in their journey to find common ground. Such common ground is also an ongoing aim of the TWG on 'Technology and Resources' (Chapter 11), in order to eliminate the *compartmentalisation* that could hinder communication and collaboration in TWGs. *Paradigms* might change, but the beliefs of practitioners in the field could take longer to evolve (Chapters 12 and 13), highlighting the need also for constructive and open relationships between researchers and teachers of mathematics.

This section has been able only to hint at the importance of open and stable relationships in the TWGs in their journey to accomplish the ERME goals of communication, cooperation and collaboration.

3 Conclusion

Finally, let me return to the question that lies behind all the preceding sections: to what extent has ERME research over the last 20 years been successful in fostering a spirit of communication, cooperation and collaboration among researchers from a diversity of language, cultural and national backgrounds? Europe is a prime example of a continent built on such diversity. The evolution of methods of coping with differences and turning them into strengths is described in all 18 core chapters in this book. Despite difficulties – or perhaps because it is not easy – ERME has manifested that unity in diversity is a stable and attainable goal, for the benefit of research in mathematics education. This accomplishment has repercussions far beyond the borders of Europe.

Note

1 For brevity I am referring to the 18 chapters in this book only by each chapter number. However, I respect the careful and painstaking work done by the authors, and their insights into what has been accomplished in 20 years of ERME research. My omission of their names in no way diminishes my appreciation of their work.

References

Balacheff, N. (1990). Beyond a psychological approach: The psychology of mathematics education. *For the Learning of Mathematics, 10*(3), 2–8.

Bikner-Ahsbahs, A., & Kidron, I. (2015). A cross-methodology for the networking of theories: The General Epistemic Need (GEN) as a new concept at the boundary of two theories. In A. Bikner-Ahsbahs, C. Knipping, & N. Presmeg (Eds.), *Approaches to qualitative research in mathematics education: Examples of methodology and methods* (pp. 233–250). Dordrecht: Springer.

Bikner-Ahsbahs, A., Prediger, S., & Networking Theories Group (Eds.) (2014). *Networking of theories as a research practice.* Heidelberg/New York: Springer.

Confrey, J. (1991). Learning to listen: A student's understanding of powers of ten. In E. von Glasersfeld (Ed.), *Radical constructivism in mathematics education* (pp. 111–138). Dordrecht: Kluwer.

Confrey, J. (1994a). A theory of intellectual development (Part 1). *Mathematical Thinking and Learning, 14*(3), 2–8.

Confrey, J. (1994b). Voice and perspective: Hearing epistemological innovation in students' words. In N. Bednarz, M. Larochelle, & J. Désautels (Eds.), *Revue de sciences de l'education. Special Issue: Constructivism in education, 20*(1), 115–133.

Confrey, J. (1995). A theory of intellectual development (Part 2). *Mathematical Thinking and Learning, 15*(1), 38–48.

Conway, J. (1985). Four views of ICME-5: ICME past, ICME still to come. *For the Learning of Mathematics, 5*(1), 29–36.

Fried, M. (2014). Mathematics and mathematics education: Searching for common ground (Introduction). In M. Fried & T. Dreyfus (Eds.), *Mathematics and mathematics education: Searching for common ground* (pp. 3–22). Dordrecht: Springer.

Furinghetti, F. (2014). History of international cooperation in mathematics education. In A. Karp & G. Schubring (Eds.), *Handbook on the history of mathematics education* (pp. 543–564). Dordrecht: Springer.

Husserl, E. (1970). *The crisis of European sciences and transcendental phenomenology.* (Translated from German: original 1954.) Chicago, IL: Northwestern University Press.

Karp, A., & Schubring, G. (Eds.) (2014). *Handbook on the history of mathematics education.* Dordrecht: Springer.

Kidron, I., & Bikner-Ahsbahs, A. (2015). Advancing research by means of the networking of theories. In A. Bikner-Ahsbahs, C. Knipping, & N. Presmeg (Eds.), *Approaches to qualitative research in mathematics education: Examples of methodology and methods* (pp. 221–232). Dordrecht: Springer.

Lerman, S. (1996). Intersubjectivity in mathematics learning: A challenge to the radical constructivist paradigm? *Journal for Research in Mathematics Education, 27*(2), 133–150.

Presmeg, N. (1985). *The role of visually mediated processes in high school mathematics: A classroom investigation.* Unpublished PhD dissertation, University of Cambridge.

Presmeg, N., Radford, L., Roth, W.-M., & Kadunz, G. (2016). *Semiotics in mathematics education.* (ICME-13 Topical Surveys, Ed. G. Kaiser.) Hamburg: Springer Open.

Presmeg, N., Radford, L., Roth, W.-M., & Kadunz, G. (2018). *Signs of signification: Semiotics in mathematics education research.* (ICME-13 Topical Surveys, Ed. G. Kaiser.) Hamburg: Springer Open.

Sierpinska, A., & Kilpatrick, J. (1998). *Mathematics education as a research domain: A search for identity.* Dordrecht: Kluwer.

Yackel, E., & Cobb, P. (1996). Sociomathematical norms, argumentation, and autonomy in mathematics. *Journal for Research in Mathematics Education, 27*(4), 458–477.

INDEX

abstraction 34, 53, 65, 68, 107, 153, 232, 255, 263, 279, 282

academic emotion 131, 133

advanced mathematical thinking 60, 86, 115

affect 64–65, 121, 123, 128–138, 163, 170, 204, 225, 230, 242, 273, 278, 284

affective factors of giftedness 121

affective state 134, 136, 138

affective trait 134–136

aggregate thinking 53–54

algebra 10, 11, 20, 32–43, 61, 68, 100, 169, 172, 174, 183, 228, 249, 270; early algebra 36–37, 184

algebraic thinking 32–43, 109, 270, 273, 280, 284

analysis 65–66, 115

Anthropological Theory of the Didactic (ATD) 35, 63, 66, 68–69, 91, 95, 100–101, 103, 231, 255, 259, 263–265, 280

applications 60, 90–95, 98–103, 274, 284

Approximate Number System (ANS) 25

argumentation 10, 12, 16, 19, 23, 75–87, 97, 172, 198, 203, 265, 270, 284

arithmetic 20, 23–26, 28, 30, 35–40, 46, 107, 109, 149, 241, 245

artefact 9, 14–15, 32, 34, 40–41, 110, 123, 151, 154, 205, 207–209, 269, 272, 274

attitude 47–48, 54, 64, 67, 85, 121, 128–134, 136, 156, 182, 184, 204, 244, 277, 284

autonomy 66, 181–183

belief 29, 51, 54, 62–63, 67, 69, 97–98, 108, 111, 123, 128–133, 136, 162–164, 166, 168–171, 173–176, 181, 184, 188, 212, 227, 229, 230, 284

calculus 61, 65, 80, 231, 241

children 23, 25, 27, 47, 50, 55, 106–112, 116, 120, 121, 123, 135, 151, 165, 202, 204, 206, 208, 216, 218, 249

classroom interaction 48, 99, 167–168, 172, 199, 203–206, 228

classroom practice 97, 147, 156, 162–164, 167–169, 171, 173–175, 225, 277

classroom-based research 83, 197, 201

Cognitive Unity 77–79, 84

collaboration 1, 4, 8, 12, 16–18, 53, 70, 71, 90, 95, 138, 148, 171, 173, 182, 184, 186, 190, 219, 239, 240, 246, 250–252, 254, 263, 264, 276–80, 282–284

collaborative environments 182, 185

collective argumentation 203

commognition 62, 65, 176

communication 1, 4, 8, 9, 11, 16–18, 27, 30, 41, 43, 70, 71, 77, 110, 123, 142, 151, 167, 169, 172, 186, 191, 197–199, 201, 202, 205–207, 209, 219, 223, 239, 240, 250, 251, 254, 255–257, 264, 276–279, 284

community of practice 157, 186

comparative analysis 233

comparative approach 226, 233

comparative research/studies 117, 138, 223–224, 227, 232–234, 273, 279

288 Index

compatibility of theories 284
computer algebra system (CAS) 40, 62, 144
connecting theories 185, 257, 258
construction 12, 14–15, 17, 19, 47, 53, 65,
    78, 84, 108, 112, 120, 144, 150, 154,
    164, 187, 199, 202, 205, 206, 215–219,
    224–225, 234, 250, 256, 262
context 12, 15, 27–28, 34, 38, 47–48, 50,
    51, 55–56, 61–62, 65–66, 71, 80, 82,
    93–101, 108, 111, 116–117, 122, 129,
    132, 135, 138, 143, 149, 153, 167–168,
    170, 172, 173, 181, 183–191, 196–200,
    203, 205–206, 208–209, 212, 214–219,
    223–228, 233–234, 243, 247, 255–256,
    263–264, 266, 276, 278–279, 281, 283
continuum of complexity 197, 200
cooperation 1, 4, 107, 184, 186, 235,
    239–241, 250–252, 254, 276–280,
    283–284
creativity 115–121, 123–125
cross-cultural research 234
cross-system studies 225
cultural affiliation 234
cultural authorship 233
cultural distance 226, 234
cultural heritage 224, 235
cultural proximity 226, 234
culturally sensitive research 223, 224, 234
culture 17, 27, 47, 108, 111–112, 118, 123,
    134, 196, 199–201, 203, 208, 211–212,
    214, 216, 218, 224–227, 233–235, 247,
    251–252, 256, 259, 260, 278–281,
    283–284
curriculum 9–11, 18, 36, 39, 47, 52,
    56, 82, 99, 106–108, 111, 123, 142,
    145–147, 149, 156, 165, 168, 187, 189,
    198, 224–226, 229, 232–233, 243–244,
    246, 251, 270

design-based research 39–40
dialogical and narrative approaches 217
dialogue between theories 263
didactic engineering 68
didactics of geometry 8
digital artefact 269
digital environment 11, 149, 151
digital technology 40, 47, 52, 144–145,
    154, 158, 270, 262, 272, 274
digital tools 11, 14, 48, 52, 142, 144, 155, 262
discourse approaches 216
discursive and sociological approaches 215
diversity i, xxii, 2, 9, 196–197, 200, 202,
    204–205, 211–218, 252, 256–258,
    260–262, 272, 276, 281, 283–284

documentational approach 146, 155
dropping out 135
dynamic geometry software (DGS) 19,
    149, 151

e-book 147, 151
e-learning 151, 158
ecological fallacy 234
educational systems 16, 33, 90, 99, 109,
    138, 224, 227, 232, 250, 284
emotion 64, 128, *129–130*, 131–134, 136,
    225, 278, 284
empirical studies 115, 117–118, 120,
    125, 145, 147, 170, 224–225,
    241–242, 262
enrichment 123–124
epistemology 35, 67, 76, 82, 256, 259–260,
    263, 265–266
equations 33–38, 42, 65, 117, 230–231
ERME spirit xx, xxi, 1–2, 4, 90, 138,
    186, 191
estimation 25–26, 30
ethics 209, 218
ethnographic approaches 216
everyday language 110, 198, 207
experiment 12, 19, 36, 38–39, 49–51,
    67–68, 83, 96, 111, 214, 242–244,
    250–251, 276, 279
exploratory data analysis (EDA) 47,
    53–54
expressions 33, 35–36, 38, 40–41, 76, 87,
    109, 132, 153

flexibility 23–24, 26, 117–120, 181
fluency 117–119, 122
fraction 28–29, 75, 110, 117, 208–209,
    230–231
frequentist 49–50
function 33–35, 37–39, 63–64, 95–96, 100,
    144, 147, 154, 166, 173, 174, 208, 244,
    249, 262

generalisation 11, 33–34, 36, 50, 107,
    122, 226
geometric competencies 9
geometric work 14
geometrical paradigm 9, 11–13, 16–17
geometrical thinking 8, 10–13, 16, 110,
    269, 273, 283
geometrical working space 9, 11, 14
geometry i, 8–20, 41, 69, 79, 83–84,
    109–112, 117, 119, 144, 146–147, 149,
    151–152, 270–271
giftedness 116–117, 120–125

Index

Habermas' model 78–79, 84
hiccup (in technology-mediated lessons) 155
historical text 244
history in mathematics education 239–244, 247, 251–252
history of mathematics education 239–242, 245, 247–252
HPM Study Group 240

identity xx, xxii, 131–132, 137–138, 162, 175, 191, 215–217, 232, 234, 243, 269, 271–273, 275–276
inclusion 1, 3–5, 124, 203, 212–213, 216, 219, 277, 284
inferential reasoning 50
innovation 53–54, 62, 68, 71, 124, 157, 167, 173, 184, 186, 188–189
inquiry 4, 50–51, 62, 102, 116, 156, 167
instrument 12, 14–15, 19, 35, 40, 42, 97–98, 124, 136–137, 145, 154–155, 230, 233–234
instrumental approach 145, 154–157
instrumental genesis *14*
instrumental orchestration 155
instrumentation 10–11, 14, 154
interest 51, 186, 198, 200–201, 206, 211, 215, 264, 282

knowledge quartet 166

language, conceptualisations of 196, 200, 208
language, forms of 206, 208
language, functioning of 198, 203, 206–207
language as culture 196, 200, 209
language as discourse 196, 200, 209
language as system 196, 200, 208
language of mathematics 196–200, 205–209
language of the learner 196–200, 200–203
language of the teacher/classroom 196–205
lectures 61, 62, 66
linear algebra 61, 63–65, 69
linguistic paradigm 196, 203
logic 75, 77, 79–82, 86
logic and language 80–82

manipulatives 110–111
mathematical abilities 116
mathematical communication 198–199, 205, 207
mathematical competencies 47, 243, 251
mathematical creativity 115–125
mathematical giftedness 120–123

mathematical knowledge for teaching 165, 230
mathematical language 198, 205, 207–209
mathematical talent 115–117, 120–123
mathematisation 97, 101, 122
media 68–69, 143–145
mental calculation 23, 26–28
metacognition 129, 262
methodology 135–137, 227–232, 259
misconception 38–39
modelling 47–49, 67–68, 90–103, 270
modelling project 97–99
motivation 51, 129–138
multicultural 211–212
multilingual 202–203, 211
multimodal 199, 201, 205

natural numbers 28–29, 35
negative numbers 28, 35
networking theories 256–260
neuroscience 29–30
number sense 23–30
numbers 23–30, 35, 39, 40, 202, 208

online course 146–147, 151
outsider perspectives 223–224, 234–235

paradigm 10–13, 16–19, 68–69, 255–256, 262, 277
play, guided 109
playful learning 108–109
political 212–213, 217–218
preschool 106–112
primary school 107–109
problem posing 115, 118, 120
problem solving 20, 41, 66–67, 98, 116–122, 133–135, 144–145, 163–164, 202
professional competence 181
professional development 70–71, 98–102, 108, 156, 181–191, 209, 262
proof 10–12, 19–20, 64–65, 75–87, 149
proving 10, 14–15, 64–65, 75–87, 153, 201

randomness 47–55
rational numbers 23–24, 110
Realistic Mathematics Education (RME) 93, 103
reasoning 13–20, 34–42, 48–56, 64, 77–87, 107, 120–125, 135, 175, 198, 203–209, 245, 255
reform 10, 11, 19, 102, 163–174
representation 11–20, 23–30, 37–41, 65, 110, 129, 145–149, 180, 190, 198, 206–208, 215–218, 225–233

290 Index

resilience 131–138, 228
resource 4, 19, 21, 52, 61–72, 102–103, 110, 142–161, 171, 197–209, 213–219, 269, 272, 283–284
risk 47–55

sampling (statistics) 48–55
scaling, scaling-up 43, 147–148
semantics 68, 80
semiotics 15, 33–34, 68, 199, 205, 233, 280
service mathematics courses 66–67
simulation 49–55
social 12, 64–72, 83, 108, 117, 123, 129–138, 153–157, 163, 171, 175–176, 191, 196–209, 211–219, 223, 232, 247, 252, 255–266, 270, 280
social context 129, 132
socio-cultural 38, 171, 174, 213, 218, 280, 281
statistical inference 50, 55
story 248–249
strategy 16–17, 24–30, 38, 66, 118–125, 133, 144, 188, 206–207, 217, 224–227, 243, 258–260, 266, 282
symbol 25, 33–43, 87, 144–145, 153, 175, 199, 201, 206, 208
syntax 68, 80

talented students 115–127
task design 12, 39–40, 111–112, 143–158

teacher education 10, 18, 34, 39, 41, 47–48, 55, 70–71, 82, 86, 91, 99, 102, 116, 155, 209, 162–195, 228–230, 242, 247–252, 256, 272; in-service 16, 54–55, 66–67, 70, 97–98, 119–120, 125, 182–191, 213, 225, 229–230, 242–243, 252; pre-service 16, 54–55, 70, 97, 119–120, 182–191, 225, 229–230, 242
teacher knowledge 46, 48, 54, 56, 164–168, 185
teaching of proof 82–84
technology 10, 14, 18, 40–42, 46–49, 52–56, 62–63, 102, 118–119, 135, 142–161, 199, 262, 269–274, 284
textbook 38, 66–67, 86, 98–99, 122–123, 142, 146–147, 155, 158, 231, 235, 241–244
theorem 15, 55, 67–68, 77, 84, 144, 153
theoretical assumptions 257–258
theoretical framework 26, 69, 78, 83, 99, 100, 118, 145, 154, 182, 243, 254–268, 274, 279
theoretical perspective 5, 63, 66, 254–268, 280
theory and practice 91, 94, 124, 157, 182–186, 190, 192, 254, 256, 275
tool: mathematical, digital, technological 8–20, 33–40, 43, 47–58, 65, 77, 78–85, 98, 104
transition 17, 36, 40, 53, 63–71, 87, 107, 134, 188, 202, 212, 215